SODIC SOILS

Topics in Sustainable Agronomy

SERIES EDITORS

Rattan Lal
Pedro Sanchez
Malcolm Sumner
M. E. Swisher
P. B. Tinker
Robert E. White

VOLUMES

T.R. Yu
Chemistry of Variable Charge Soils

M.E. Sumner and R. Naidu
*Sodic Soils: Distribution Properties, Management,
and Environmental Consequences*

P.B. Tinker and P. Nye
Solute Movement in the Rhizosphere

SODIC SOILS

Distribution, Properties, Management, and Environmental Consequences

Edited by
Malcolm E. Sumner
Ravendra Naidu

New York   Oxford

Oxford University Press

1998

Oxford University Press

Oxford   New York
Athens   Auckland   Bangkok   Bogotá   Buenos Aires   Calcutta
Cape Town   Chennai   Dar es Salaam   Delhi   Florence   Hong Kong   Istanbul
Karachi   Kuala Lumpur   Madrid   Melbourne   Mexico City   Mumbai
Nairobi   Paris   São Paulo   Singapore   Taipei   Tokyo   Toronto   Warsaw

and associated companies in
Berlin   Ibadan

Copyright © 1998 by Oxford University Press, Inc.

Published by Oxford University Press, Inc.
198 Madison Avenue, New York, New York 10016

Oxford is a registered trademark of Oxford University Press

Library of Congress Cataloging-in-Publication Data
Sodic soils : distribution, properties, management, and
environmental consequences / edited by Malcolm E. Sumner, Ravendra Naidu.
p.   cm. — (Topics in sustainable agronomy)
Includes bibliographical references and index.
ISBN 0-19-509655-X
1. Sodic soils.   I. Sumner, M. E. (Malcolm E.), 1933–
II. Naidu, R.   III. Series.
S595.S63   1997
631.4'16—dc20   96-15906

9 8 7 6 5 4 3 2 1
Printed in the United States of America
on acid-free paper

To

our wives, Priscilla and Shamilla,

without whose sacrifices, understanding, support, and encourage-
ment this project would not have
come to fruition

Foreword

The literature on sodic soils is fragmented to say the least and to the knowledge of the writer, there is currently no book on the subject. Added to this, there recently has been a mini revolution in the way sodic soils and sodicity are viewed, with the two editors of this book being at the center of this reappraisal.

This book aims to fill the void in current literature and to review, in a scholarly way, much of the new thinking on sodic soils and sodicity in terms of the mechanisms involved in the degradation of soils resulting from accessions of sodium (Na). More importantly, it devotes much attention to soils which were previously thought to be nonsodic but which, in fact, show sodic behavior. These are the soils which have lost much of their organic matter with cultivation and which disperse, often quite readily, when mechanically disturbed, for example by raindrop action. Recent work indicates that provided the electrolyte concentration is sufficiently low, a soil with an exchangeable sodium percentage (ESP) of as low as five can show sodic behavior. It is under these conditions that crusting and hardsetting become serious problems, causing reduced germination and seedling emergence, and increased surface runoff and erosion.

One other novel aspect of the book is that the environmental consequences of sodicity are considered. Increasing concern with the effects of colloid-assisted transport of contaminants, such as certain nutrients, heavy metals, radionuclides, and pesticides, in both surface runoff and subsurface flow, has focused attention on the need for a better understanding of the mechanisms and processes involved in interaction and transport. Sodicity and dispersion are key contributors to this pervasive form of contaminant transport. The emphasis placed on this and other environmental aspects is thus timely.

The writer has one concern with regard to the impact of research on sodic soils on current practices. Despite much excellent work done on Na-affected soils, in terms of the causes and effects of salinity and sodicity, including the mechanisms involved and the chemical and physical properties of such soils, remarkably little of this work has been translated into successful management practices at the farm level. A major exception is the use of gypsum, the reactions and effects of which in soils are well understood. Perhaps there is a need to rethink the research approach to the management of saline and sodic soils, and in particular how this could be improved by taking on board the social, economic, and environmental dimensions of the farming situation. The present book will not provide solutions to this perceived problem, but it will provide a useful basis for the planning and conduct of the necessary adaptive research.

It is a pleasure to provide a foreword for a book which both fills a major gap in the literature and which addresses new issues of agricultural and environmental importance. It is even more of a pleasure when the two editors are close professional colleagues who have made a substantial contribution to our understanding of the subject matter of this timely and important book.

J. Keith Syers
Newcastle-upon-Tyne, UK
October 1996

Preface

Although the consequences of the accumulation of sodium in soils have been known for more than a century, no single text has appeared which has been entirely devoted to this topic. In the past, sodic (sodium-affected) soils and the phenomenon of sodicity (accumulation of sodium) have usually been dealt with as a subsection of books dealing with salt-affected soils. The appearance of a comprehensive treatise on the subject of sodic soils is therefore timely. Traditionally sodic soils have been classified as having an exchangeable sodium percentage (ESP) which exceeds 15, but over the years, this threshold level has been called into question by workers in Australia, Israel, and Africa who proposed limits which were much lower. Using the former value, the worldwide extent of sodic soils has been estimated to exceed 3.5 million hectares, but if a lower value were to be used, vast tracts of land would fall into this category. In addition in recent times, soils with only slightly elevated levels of exchangeable sodium have been shown to exhibit considerable sodic character in terms of physical properties such as crust and seal formation and reduced internal drainage. Consequently, many soils comprising the Alfisol, Aridisol, Inceptisol, Vertisol and even Ultisol Orders exhibit sodic behavior making an understanding of this phenomenon of great importance to workers in areas which have traditionally not been included in the sodic soils category.

Under intensive cultivation, soils lose a large proportion of their organic matter content which is responsible for water-stable aggregation, and is therefore the primary structure-building component in many soils. With progressively increasing time under cultivation, many soils have been shown to become more sensitive to the adverse effects of sodium whether introduced in irrigation water or originally present in the soil. This increased sodium sensitivity results in soils with poor physical condition which are prone to seal formation and erosion rendering them more difficult to manage in a sustainable manner. Indeed, large areas have been abandoned as a result of severe soil loss. These degradative processes continue today in many parts of the world; regions with deep fertile soils and clear waters have been transformed into eroded and less productive areas with exposed subsoils and polluted surface waters which have choked streams and lakes. In addition, clay and organic matter dispersion is promoted by elevated levels of sodium which has important consequences for the environment. Not only does this lead to increased sediment loads in surface waters resulting in degraded aquatic environments, but the sorption of pollutants such as pesticides and toxic organic and inorganic compounds on these mobile colloids further contaminates both surface and groundwaters by means of colloid-assisted transport.

The purpose of this book is to bring together and critically examine present knowledge of the subject of sodic soils in the traditional vein and to extend this knowledge to the soils which have recently been identified as exhibiting sodic behavior. Despite the fact that much is already known, many aspects are still obscure, and hopefully this discussion will lead to the identification of facets which need further examination and research. All contributions to this work, which were invited based on the expertise and experience of the authors in the field of sodic soils, have been peer reviewed prior to publication.

The opening chapter deals with a critical review of the development of the changing concepts of sodic soils culminating in a proposal for a new scheme for the categorization of sodic soils based on the sodium adsorption ratio (SAR) and electrical conductivity (EC) of a 1:5 soil: water extract. This has been chosen based on ease of measurement. This new scheme, being based mainly on topsoil characteristics, should be viewed as an attempt to classify soils showing sodic behavior for management rather than for genetic or taxonomic classification purposes. The subsequent chapter deals with the global distribution and areal extent of sodic soils. In the third chapter, the processes involved in sodic behavior are examined starting with the slaking of aggregates on wetting through clay swelling and dispersion. Particular attention has been paid to the roles played by hydration, ionic bonding, attractive and repulsive forces in relation to the level of sodium in the system. Chapter four examines the role played by organic matter in sodic soils and the effects of sodicity on the behavior of organic matter. Special attention is paid to the mechanisms involved which are used in the development of strategies for the biological amelioration of sodic soils. Chapter five explores the effects of sodicity on physical properties such as crusting and seal formation, particularly in soils only slightly affected by sodium. Amelioration is readily effected by the surface application of by-product gypsum which reduces clay dispersion responsible for clogging pore space. In chapter six, the question of subsoil sodicity is addressed in conjunction with salinity which often are covariant. The major focus is the hydraulic behavior of the root zone determined by soil wetting, available water storage, salt accumulation, and deep drainage. A unified soil property model for predicting the behavior of sodic subsoils with respect to changing ESP levels is developed. The fertility constraints to crop production on sodic soils are described in chapter seven focussing mainly on the effects of sodicity on nutrient availability. In chapter eight, the management considerations required to crop sodic soils culminating in a few examples of successful crop production systems developed in various parts of the world on sodic soils, are presented. Finally, chapter nine deals with the environmental impacts of sodic soils on surface and groundwaters in terms of sediment load delivered and associated colloid-assisted pollutant transport. In addition, engineering considerations required for the successful use of sodic soils for construction purposed are discussed.

Finally, we wish to acknowledge support in the form of study leave and expenses provided by the College of Agricultural and Environmental Sciences, the University of Georgia and the Cooperative Research Centre for Soil and Land Management and the Commonwealth Scientific and Industrial Research Organization, Australia which greatly facilitated the production of this book. This book was typeset by Mrs. Pam Wilson whose services are gratefully acknowledged. And last but certainly not least, we are grateful to our colleagues who contributed the material contained in this volume.

Malcolm E. Sumner
Athens, GA
June 1997

Ravi Naidu
Adelaide, Australia
June 1997

Editors

M.E. Sumner, B.Sc., M.Sc., D.Phil., was born in Pinetown, South Africa. He received a bachelor's and a master's degree in Agriculture from the University of Natal, Pietermaritzburg, South Africa and a doctorate of Philosophy from the University of Oxford, England. He spent the first 23 years of his career at the University of Natal in the Department of Soil Science and Agrometeorology, where he eventually became Professor and Head of the department. Since 1977, he has been Professor of Agronomy at the University of Georgia, and in 1991, he was named Regents' Professor of Environmental Soil Science. He is also an Honorary Professor in the Department of Agricultural and Environmental Science at the University of Newcastle-upon-Tyne in England. He has spent sabbatical leaves at the Universities of Missouri, Wisconsin, Newcastle-upon-Tyne, the Agricultural University of Wageningen, and the Cooperative Research Centre for Soil and Land Management, Adelaide, Australia.

Dr. Sumner has been an invited speaker at many international conferences, universities, and research institutes throughout the world. He has authored or edited 6 books, 29 book chapters, 200 refereed journal articles, and 120 other scientific papers. He has trained 56 M.S. and Ph.D. candidates. He is a Fellow in the American Society of Agronomy and the Soil Science Society of America. He was awarded the Werner L. Nelson Award for Diagnosis of Yield Limiting Factors, the Agronomic Research Award, and the Soil Science Research Award by these two societies.

R. Naidu, B.Sc., M.Sc. Ph.D., is a native of Fiji. He received his bachelor's degree from the University of the South Pacific, Suva, Fiji, his master's degree from the University of Aberdeen, Aberdeen, Scotland and his doctorate from Massey University, Palmerston North, New Zealand. From 1976 to 1989 he was Senior Lecturer in Chemistry and then Head of the School of Pure and Applied Sciences at the University of the South Pacific. In 1989, he moved to the Division of Soils of the Commonwealth Scientific and Industrial Research Organization in Adelaide, Australia where he currently holds the position of Principal Research Scientist. Dr. Naidu has been an invited speaker at international conferences and is at present the Foundation Chair of the International Network on Soil Contamination Research in the Australia-Pacific Region. He has received a number of international travel awards and has cooperative research projects with scientists in the US and New Zealand. He has edited 5 books, authored 14 book chapters, 59 refereed journal and 50 other technical papers.

Contributors

L. C. Bell, Professor of Soil Science, Department of Agriculture, University of Queensland, St. Lucia, Q 4067, Australia

E. N. Bui, Senior Research Scientist, Division of Soils, CSIRO, Canberra, Q 4814, Australia

K. J. Coughlan, Senior Research Fellow, Faculty of Environmental Sciences, Griffith University, Nathan, Q 4111, Australia

D. Curtin, Soil Scientist, Agriculture and Agri-Food Canada, Swift Current, Saskatchewan, S9H 3X2, Canada

R. W. Fitzpatrick, Senior Principal Research Scientist, Division of Soils, CSIRO, Glen Osmond, SA 5064, Australia

P. Hazelton, Senior Lecturer, School of Civil Engineering, University of Technology, Sydney, NSW 2007, Australia

N. S. Jayawardane, Principal Research Scientist, Division of Water Resources, CSIRO, Griffith, NSW 2680, Australia

R. S. Kookana, Senior Research Scientist, Cooperative Research Centre for Soil and Land Management, Glen Osmond, SA 5064, Australia

L. Krogh, Assistant Research Professor, Institute of Geography, University of Copenhagen, DK 1350 Copenhagen K, Denmark

R. S. Lavado, Professor, Facultad de Agronomía, Universidad de Buenos Aires, 1417 Buenos Aires, Argentina

G. J. Levy, Soil Scientist, Institute of Soils and Water, Agricultural Research Organization, The Volcani Center, Bet Dagan, Israel

W. P. Miller, Professor, Department of Crop and Soil Sciences, University of Georgia, Athens, GA 30602, USA

F. O. Nachtergaele, Technical Officer, Soil Resources, Management, and Conservation Service, Land and Water Development Division, Food and Agriculture Organization, 0010 Rome, Italy

R. Naidu, Principal Research Scientist, Division of Soils, CSIRO, Glen Osmond, SA 5064, Australia

P. N. Nelson, Bureau of Sugar Experiment Stations, P. O. Box 117, Ayr, Qld. 4670, Australia

J. M. Oades, Department of Soil Science, University of Adelaide, Glen Osmond, SA 5064, Australia

J. D. Oster, Specialist and Adjunct Professor, Department of Soil and Environmental Sciences, Riverside, CA 92521, USA

P. Rengasamy, Senior Research Fellow, Department of Soil Science, University of Adelaide, Glen Osmond, SA 5064, Australia

I. Shainberg, Soil Scientist, Institute of Soils and Water, Agricultural Research Organization, The Volcani Center, Bet Dagan, Israel

R. J. Shaw, Principal Scientist, Salinity and Contaminant Hydrology, Queensland Department of Primary Industries, Indooroopilly, Q 4068, Australia

M. E. Sumner, Regents' Professor, Department of Crop and Soil Sciences, University of Georgia, Athens, GA 30602, USA

J. K. Syers, Professor, Department of Agricultural & Environmental Science, University of Newcastle, Newcastle-upon-Tyne, UK

T. Tóth, Research Fellow, Research Institute for Soil Science and Agricultural Chemistry, Hungarian Academy of Sciences, Budapest H-1022, Hungary

Contents

List of Abbreviations

μg	microgram
ΔG^o	standard free energy change
ϵ	dielectric constant
ς	zeta potential
ρ_b	bulk density
σ	surface charge
ψ_0	surface potential
ψ_d	potential midway between plates
$1/\kappa$	effective thickness of double layer
A	upland erosion potential
A	Hamaker constant; activity
a	activity
ADMC	air dry moisture content
ANC	acid neutralizing capacity
APS	anionic polysaccharide
BGR	Bundesantalt für Geowissenschaften und Rohstoffe
BUNASOL	Bureau National des Sols
C	concentration of ion
C/C_0	fraction of initial concentration
CBD	citrate bicarbonate dithionite
CCR	CEC-to-clay ratio
CEC	cation exchange capacity
CFC	critical flocculation concentration
C_L	salt concentration of input water
cm	centimeter
C_O	salt concentration of water below root zone
C_o	concentration of output water
CPCS	Commission de P dologie et de Cartographie des Sols
C_r	runoff sedminet concentration
C_s	sediment content in original soil
d	distance between particles; also half-distance
DDT	dichlorodiphenyltrichlorolethane
DLVO	Derjaquin-Landau-Verwey-Overbeek
DOC	dissolved organic carbon
DOM	dissolved organic matter
DR	delivery ratio
dS	decisiemens
D_s	depth of soil
D_w	depth of infiltrated water
DW	distilled water
e	electronic charge
EAED	electron-acceptor electron-donor
EC	electrical conductivity
$EC_{1:5}$	electrical conductivity of 1:5 soil:water extract
EC_e	electrical conductivity of saturation extract
ECEC	effective cation exchange capacity
Ec_{iw}	electrical conductivity of irrigation water
EDTA	ethylene diamine tetraacetic acid
E_h	redox potential

EMBRAPA	Empresa Brasileira de Pesquisa Agropecuária
EmgP	exchangeable magnesium percentage
EPA	Environmental Protection Agency
EPP	exchangeable potassium percentage
ESP	exchangeable sodium percentage
ESR	exchangeable sodium ratio
ESSS	equivalent salt solution series
ET	evapotranspiration
F	Ca-Na exchange efficiency factor
FA	fulvic acid
FAO	Food and Agriculture Organization of the United Nations
FGDG	flue gas desulfurization gypsum
FIR	final infiltration rate
f_{oc}	fraction of organic carbon in soil
g	gram
G^O	Gibbs standard free energy
GR	gypsum requirement
h	hour
ha	hectare
HA	humic acid
HC	hydraulic conductivity
HPAN	hydrolyzed polyecrylonitrile
HSAB	hand-soft acid-base
INTA	Instituto de Nutrición y technología de los Alimentos
IR	infiltration rate
I, I_z	ionization potential
J	joule
k	Boltzmann's constant
K	degrees Kelvin
K_g	Gapon constant (exchangeable sodium ratio divided by SAR of equilibrium solution)
KJ	kilojoule
K_{oc}	octanol-water partition coefficient
K_{ow}	dissociation constant
L	liter
LC_{50}	lethal concentration
LF	leaching fraction
m	meter
M	molar
mg	milligram
MgL	megaliter
min	minute
mL	milliliter
MOR	modulus of rupture
MW	molecular weight
M^{z+}	metal ion of valency $^{z+}$
n	electrolyte concentration
NLWR	nonlimiting water range
nm	nanometer
NOM	natural organic matter
NSW	New South Wales
NMR	nuclear magnetic resonance
NTU	nephelometric turbidity unit
ORSTOM	Office de la Recherche Scientifique et Technique OutreMer
Pa	Pascal
PAH	Polyaromatic hydrocarbon
PAM	Polyacrylamide
PAR	potassium adsorption ratio
PAWC	plant available water capacity

PCB	polychlorinated biphenyl
P_{CO2}	partial pressure of carbon dioxide
P_{dis}	dispersive potential
PG	phosphogypsum
pK_a	negative log of dissociation constant of an acid to a proton and its conjugate base
P_{osm}	osmotic pressure
PSD	polysaccharide
P_{sol}	osmotic pressure of soil solution
P_{tcc}	osmotic pressure at TEC
PZNC	point of zero net change
Q_i	quantity of input water
Q_o	quantity of output water
R	gas constant
r^2	(15)
RFC	relative flocculation concentration
r_i	ionic radius
RSC	residual sodium carbonate ($[HCO_3^- + CO_3^{2-}] - [Ca^{2+} + Mg^{2+}]$ concentration in $mmol_c\ L^{-1}$)
s	second
Sal	Saline
Sal_H	Very Saline
Sal_L	Nonsaline
SAR	sodium adsorption ratio ($Na^+]/[Ca^{2+} + Mg^{2+}]^{0.5}$)
SAR_{adj}	adjusted sodium adsorption ratio
SAR_c	corrected sodium adsorption ratio
SAR_e	sodium adsorption ratio of saturation extract
SAR_{iw}	sodium adsorption ratio of irrigation water
SAR_p	practical sodium adsorption ratio
SCD	surface charge density
SI	langelier saturation index
Sod	Sodic
Sod_H	Very sodic
Sod_L	Nonsodic
SOTER	soils and terrain database
SP	saturation percentage
t	metric ton
T	temperature (K)
TCC	total cation concentration
TEA	triethanolamine
TEC	threshold electrolyte concentration
TOC	toxic organic compound
UNESCO	United Nations Educational, Scientific, and Cultural Organization
V_a	attractive energy
VAMA	vinyl-acetate maleic acid
V_r	repulsive energy
W_{max}	maximum field gravimetric water content measured 2–3 days after soil profile wetting
Y	Misono factor
Y	basin sediment yield
yr	year
z	charge; valence of concentration

SODIC SOILS

1

Sodic Soils: A Reappraisal

M.E. Sumner, P. Rengasamy and R. Naidu

1.1 SALT-AFFECTED SOILS AND AGRICULTURAL PRODUCTIVITY

Excessive levels of salts occur in large areas of soil around the world and profoundly affect land use. Usually two major types of soil have been defined within this group, namely, saline and sodic (alkali) soils. The former contain excess salts, comprised largely of the chlorides (Cl) and sulfates (SO_4) of sodium (Na), magnesium (Mg), and calcium (Ca) in the soil solution, whereas the latter have elevated levels of Na on the exchange complex and may contain free sodium carbonate (Na_2CO_3). Estimates of the areas of the globe covered by salt-affected soils are presented in Table 1.1. In terms of the total area of salt-affected soils, sodic soils occupy the largest proportion. However, the data in Table 1.1 are a little misleading because dissimilar criteria

Table 1.1 Global distribution of salt-effected soils [Szabolcs, *Review of Research on Salt Affected Soils*, UNESCO, Paris, 1979; *Salt-affected Soils*, CRC Press, Boca Raton, FL, (1989)]

Continent	Area in millions of ha		
	Saline	Sodic (Alkali)	Total
North America	6.2	9.6	15.8
Central America	2.0	--	2.0
South America	69.4	59.6	129.0
Africa	53.5	27.0	80.5
South Asia	83.3	1.8	85.1
North & Central Asia	91.6	120.1	211.7
Southeast Asia	20.0	--	20.0
Australasia	17.4	340.0	357.4
Europe	7.8	22.9	30.7
TOTAL	351.5	581.0	932.2

for sodicity were used in arriving at the areas involved in different geographic regions. For example, in the United States, and subsequently throughout much of the rest of the world, a value of ESP (exchangeable sodium percentage) greater than 15 was the criterion for separating sodic soils whereas in Australasia, a value of ESP above 6 has been used (Section 1.2.2). This explains the disproportionately large areas of sodic soils in Australasia. Based on these definitions, the most important regions in which sodic soils occur, are Australasia, North and Central Asia, South America, and Africa. However, if the Australian definition were used on a worldwide basis, an enormous additional area, in which most of the world's cereal crops are grown, would be involved. A more detailed discussion of the occurrence and distribution of sodic soils is presented in Chapter 2.

To facilitate the initial part of this discussion, a sodic soil will simply be considered to be one that has been adversely affected by Na, particularly as far as its physical condition is concerned. Later, a more precise definition will be attempted. While such soils have been found largely in arid and semiarid regions, there has been an increasing realization, recently, that soils exhibiting similar properties, but not the required levels of Na, also occur in more humid climates (Shainberg and Letey, 1984; Shainberg, *et al.*, 1989; Sumner, 1993a). This proposition will be developed later. In any event, because Na, at almost any level in a soil, tends to promote the dispersion of particles in the clay-sized fraction, Na-affected soils with low levels of salt exhibit poor physical condition, which is manifest in very weak structural stability, and low hydraulic conductivities (HC) and infiltration rates (IR), all of which translate into reduced crop productivity due to poor aeration and water supply. In addition, the very poor acceptance of water at the soil surface leads to increased runoff and, consequently, to erosion, which can often be very severe. However, the effects of Na on soil structural and hydraulic properties vary widely with soil type (McNeal and Coleman, 1966). For example, smectite clays show marked changes while soils that are kaolinitic, sandy, or soils rich in sesquioxides and organic matter exhibit greater stability (Rhoades and Loveday, 1990). Thus a better understanding of the role that Na plays in soils will lead to the development of appropriate management strategies for such soils, which, it is hoped, will result in improved crop growth and the preservation of our soil resource in an improved condition, as well as reduction in sediment and pollutant transport to water bodies. These objectives will form the focus of the remainder of this treatise.

1.2 WHAT IS A SODIC SOIL?

1.2.1 Sodic Soils and Sodicity

During the past 40 years or more, the different terms, such as alkali, black alkali, nonsaline sodic, saline sodic, saline alkali, which are being used to describe what today are generally accepted as sodic soils, have undergone various permutations, with a considerable degree of latitude in the values of the parameters used for their definition. At the outset, it is important to review these variations and lack of precision in the defining parameters, which have been accepted with time, as well as the underlying reasons that made these changes necessary. Furthermore, many agriculturalists, agronomists, and even some soil scientists have experienced much confusion in differentiating between saline soils on the one hand and their sodic counterparts on the other. Part of this problem stems from the fact that, in many cases, soils are simultaneously saline and sodic, with the latter sometimes evolving from the former as a result of the removal of salts by leaching. Therefore it is first necessary to draw a clear distinction between these two groups of soils, which, fortunately, is relatively easy, before attempting to rationalize and clarify the definition of what a sodic soil might be. However, it is impossible to separate the effects of salinity totally from those of sodicity as they often go hand in hand, while the level of salt that might be present in a sodic soil is of the utmost importance in determining how the soil will behave from a physical point of view.

1.2.2 Saline Soils

As this treatise is to be devoted to sodic soils, only a brief mention of their saline counterparts is necessary. Saline soils contain elevated levels of the Cl and SO_4 salts of Na, Ca, Mg, and potassium (K). Little controversy exists in the literature over the definition of a saline soil. A value of 4 dS m^{-1} for the electrical conductivity (EC_e) of the saturated paste extract of a soil (USSL Staff, 1954) which corresponds to a total cation concentration (TCC) of roughly 40 mmol$_c$ L^{-1} (an osmotic potential of -145 kPa), is generally accepted as the level that differentiates between saline and nonsaline soils, although even below this level, plants, particularly sensitive types, can be adversely affected. In some parts of the world (Szabolcs, 1979; Rengasamy *et al.*, 1984b), the $EC_{1:5}$ in an extract with a wider soil-to-solution ratio (1:5) is often used. Because of the different dilutions involved, both the concentration and the nature of the solution will differ from that obtained by the saturated paste extract procedure. Consequently, a different set of norms is required for the interpretation of 1:5 soil-to-solution extracts, and care should be taken in selecting the appropriate norms, depending on the method used. The $EC_{1:5}$ values corresponding to 4 dS m^{-1} in the saturated paste extract can be calculated from the model that Shaw (1994) has developed and validated. It is sound with realistic boundary conditions and specifically accounts for the influence of partially soluble salts and soil properties. This conversion is illustrated in Table 1.2, in which average root zone $EC_{1:5}$ values, corresponding to the EC_e values for different plant salt tolerance groupings have been calculated. When the value of EC_e = 4 dS m^{-1}, selected as the division between saline and nonsaline soils, is compared with the plant salt tolerance groupings of Maas and Hoffman (1977) in Table 1.2, it is clear that this value is well founded in terms of plant response. Salt in the soil solution increases the osmotic pressure (decreases osmotic potential), making it more difficult for plants to extract water. Consequently, crop production decreases due to moisture stress as salt concentration increases.

For the remainder of the discussion, the subject of saline soils and salinity will not be discussed further, except insofar as salt concentration has a modifying effect on sodic behavior. Those readers interested in a more comprehensive discussion of saline soils and salinity are referred to Szabolcs (1979, 1989), Bresler *et al.* (1982), Shainberg and Shalhevet (1984), and Gupta and Abrol (1990a).

1.2.3 Parameters Necessary for Definition

In order to understand the problems involved in the definition of sodic soils and sodicity, it is necessary to trace historically the evolution of the concepts and the nature of the evidence on which the differentiation from nonsodic soils was made. However, before doing so, the parameters used to describe the soil conditions associated with sodic soils must be defined.

Solid Phase
The level of Na in a soil system has usually been described by one of two parameters, either the ESP, which reflects

TABLE 1.2 Electrical conductivity measured in a saturated paste extract (EC$_e$) with calculated values for 1:5 soil–water extracts (EC$_{1:5}$) of soils of varying clay contents for various plant salt tolerance groupings [Shaw, *Understanding Soils and Soil Data*, Queensland Branch, Australian Society of Soil Science, Brisbane, (1988)

			Soil salinity measurement			
				EC$_{1:5}$		
Plant salt tolerance sensitivity grouping[a]	Soil salinity rating	EC$_e$	10–20% clay	20–40% clay	40-60% clay	60-80% clay
				dS m^{-1}		
Sensitive	Very low	<0.95	<0.05	<0.08	<0.12	<0.18
Moderately Sensitive	Low	0.95–1.9	0.10	0.165	0.25	0.37
Moderately tolerant	Medium	1.9–4.5	0.25	0.40	0.58	0.85
Tolerant	High	4.5–7.7	0.45	0.67	1.00	1.50
Very tolerant	Very High	7.7–12.2	0.70	1.05	1.58	2.40
Generally too saline	Extreme	>12.2	>0.70	>1.05	>1.58	>2.40

[a]Plant salt tolerance groupings for a 10% yield reduction (Maas and Hoffman, 1977)

the saturation of the exchange complex with Na relative to the other cations present, or the sodium adsorption ratio (SAR) of the soil solution (or irrigation water), which reflects the relative balance between Na and Ca plus Mg in the solution phase. These parameters are defined as follows:

$$ESP = (100Na_{ex})/CEC \qquad [1.1]$$

or

$$ESP = (100Na_{ex})/ECEC \qquad [1.2]$$

where

CEC = cation exchange capacity in cmol$_c$ kg^{-1}
ECEC = effective cation exchange capacity equal to exchangeable Ca + Mg + K + Na + Al in cmol$_c$ kg^{-1}
Na$_{ex}$ = exchangeable Na in cmol$_c$ kg^{-1}

Even the formulation in Equation [1.1] is not without ambiguity, arising from the method of CEC measurement, the nature of the charges on the soil, and the propensity of certain extractants to remove Na from sources not truly exchangeable (zeolites, Na fixation) (Schulz *et al.*, 1964; Sposito and Le Vesque, 1985; Gupta and Abrol, 1990b). For example, in variable charge soils, the value of the CEC obtained is crucially dependent on the pH, concentration, and buffer capacity of the extractant used (Grove *et al.*, 1982), which becomes particularly important in view of the increasing incidence of acid sodic soils being reported in the literature (Ford *et al.*, 1993; McKenzie *et al.*, 1993;

Naidu *et al.*, 1993b). Thus, if the traditional ammonium acetate (NH$_4$OOCCH$_3$, pH 7) or barium chloride-triethanolamine (BaCl$_2$-TEA, pH 8.2) methods (Sumner and Miller, 1997) are used to measure the CEC of an acid sodic soil, grossly inflated CEC values will be obtained, resulting in much lower calculated ESP values. Although this is a serious problem, which requires immediate attention, the solution is relatively easy. Replacement of the buffered by unbuffered methods of CEC determination, such as the Gillman and Sumpter (1986) procedure, and the use of techniques to overcome the problems associated with excessive extraction of Na from nonexchangeable sites have solved this problem. This is discussed in detail in Rayment and Higginson (1992). In calcareous soils, the use of Equation [1.2] presents problems because most extractants for exchangeable cations also dissolve CaCO$_3$ and MgCO$_3$, which erroneously inflates the values for exchangeable Ca and Mg in the estimation of ECEC. As a result ESP values measured on such soils in this way would be too low. Because of these problems, the use of ESP should be avoided unless the conditions of measurement are clearly defined.

Because Mg^{2+} and K$^+$ have been shown to affect soil behavior in a manner similar to Na$^+$, albeit to a lesser extent, similar formulations have been proposed, namely, EMgP and EPP, as follows:

$$EMgP = (100Mg_{ex})/CEC \qquad [1.3]$$

$$EPP = (100K_{ex})/CEC \qquad [1.4]$$

where

Mg$_{ex}$ and K$_{ex}$ = exchangeable Mg and K,
                         respectively, in cmol$_c$ kg^{-1}
EMgP   = exchangeable Mg percentage
EPP   = exchangeable K percentage

Solution Phase

In order to avoid some of the problems associated with the calculation of ESP arising from problems in the estimation of CEC, the sodium adsorption ratio (SAR) of the soil solution, which arose from a modification of the Gapon equation (see Equation [1.13]), was proposed as an alternative,

$$SAR = Na^+/[(Ca^{2+} + Mg^{2+})/2]^{1/2} \qquad [1.5]$$

where

Na$^+$, Ca^{2+}, and Mg^{2+} refer to soluble ionic concentrations in mmol$_c$ L^{-1}

Even though SAR has been used widely and, generally, successfully as an indirect measure of ESP, there are even problems associated with the use of SAR related to the manner in which it is calculated. The summation of Ca^{2+} and Mg^{2+} makes the inherent assumption that they both contribute equally in overcoming the negative effects of Na$^+$ on soil physical properties, which has been shown in many instances to be untrue (Alperovitch *et al.*, 1981, 1986; Rengasamy *et al.*, 1986; Levy *et al.*, 1988; Chapter 3). On the contrary, Mg^{2+} in substantial proportions sometimes promotes dispersive effects on soil colloids affected by sodicity (Emerson and Chi, 1977; Rengasamy, 1983). However, this is not always the case, as has been suggested by van der Merwe and Burger (1969), Rahman and Rowell (1979), Ali *et al.* (1987), Rengasamy (1987), Yousaf *et al.* (1987), and Palaveyev and Penkov (1990). Clearly, consideration should be given, in certain circumstances, to calculating SAR by not giving equal weight to Mg^{2+} and Ca^{2+}. A further complication arises from the method used to obtain the solution in equilibrium with the soil. Two approaches have been used widely, namely, the saturated paste extract (USSL Staff, 1954) and the 1:5 soil-to-solution extract (Rengasamy *et al.*, 1984b), resulting in different values for the SAR. Rengasamy *et al.* (1984b) developed a relationship between the two SAR values for Australian Red-Brown Earth topsoils (Alfisols), which is approximately

$$SAR1:5 \approx SAR_e \qquad [1.6]$$

where

SAR$_{1:5}$ and SAR$_e$ are the SAR values in the 1:5 and saturated paste extracts, respectively

However, this relationship is unlikely to hold for all soils, particularly those containing partially soluble salts such as gypsum where, on dilution, exchange reactions take place which change the composition of the soil solution. Therefore, this approximation is likely to be limited to a narrow range of soils. In any event, confusion can be avoided by stating the method used for obtaining the equilibrium solution from which SAR values are calculated. Fortunately, this problem does not apply to the calculation of the SAR of an irrigation water.

A further problem arises in that the "practical" SAR, as most commonly used, does not take into account corrections for activity coefficients, ion pairs, and complexes. Although Sposito and Mattigod (1979) showed that the "practical" SAR bears no exact chemical relation to the theory of cation exchange, the "true" SAR could be estimated empirically from the "practical" SAR (SAR$_p$) at values below 20 as follows:

$$SAR = 0.08 + (1.115 SAR_p) \qquad [1.7]$$

However, this is seldom done in field applications, and for most practical purposes, the SAR used is that calculated from the concentrations of the ions in solution, without any corrections for complexation and ion pair formation. On a routine basis, it is impractical to calculate the SAR using activity coefficients.

Where CaCO$_3$ is likely to precipitate in a soil due to changes in the partial pressure of CO$_2$ (P$_{CO2}$), an adjusted SAR (SAR$_{adj}$) value of the irrigation water is sometimes calculated (Frenkel, 1984) using the Langelier saturation index (SI) as follows:

$$SI = pH_a - [(pK_2' - pK_s') + p(Ca) + p(Alk)] \qquad [1.8]$$

where

pH$_a$ = actual pH of the irrigation water
pK$_2'$ = negative logarithm of second dissociation
            constant of H$_2$CO$_3$ corrected for ionic strength
pK$_s'$ = negative logarithm of solubility product
            of CaCO$_3$ corrected for ionic strength
p(Ca)   = negative logarithm of molar concentration
            of Ca
p(Alk)= negative logarithm of equivalent concentration of titratable base (CO$_3^{2-}$ + HCO$_3^-$)

When Equation [1.8] is applied to soils, soil pH in equilibrium with $CaCO_3$ (8.4) is substituted for pH_a, and both Ca and Mg can be combined in the p(Ca) term (Bower, 1961) on the assumption that they behave similarly which is not true (Emerson and Chi, 1977), giving

$$SI = 8.4 - [(pK_2' - pK_s') + p(Ca + Mg) + p(Alk)] \quad [1.9]$$

Equation [1.9] has been criticized by Suarez (1981) on the grounds that the value of 8.4 varies with solution composition, leaching fraction (LF), and P_{CO2} and the expressed doubt that Ca and Mg behave similarly. Shaw (1994) has experienced great difficulty in using this concept on soils and irrigation waters with high levels of Mg because $MgCO_3$ does not precipitate with $CaCO_3$ and the precipitated Mg compounds are much more soluble than $MgCO_3$. Despite these limitations, these equations might have some use in practice as semiquantitative indications of what is likely to take place. Positive and negative values of SI reflect conditions under which $CaCO_3$ will precipitate and dissolve, respectively. Rhoades (1972) calculated the SAR_{adj} from

$$SAR_{adj} = SAR_{iw} (1 + SI) \quad [1.10]$$

where

SAR_{iw} = SAR of the irrigation water

Because of the problems involved in obtaining meaningful values of SAR_{adj}, Ayers and Westcot (1976) have proposed that its use be discontinued.

Oster and Rhoades (1975) calculated a corrected SAR (SAR_c) where the effective solubility of $CaCO_3$ and its effect on irrigation water quality were computed from ion activities as corrected for ion pair formation. However, this approach is rather impractical.

Because electrolyte concentration plays a vital role in determining the physical condition of Na-affected soils, as will be seen in Chapter 5, it is also necessary to define the parameters used. The major difference in approach between workers in different parts of the world lies in the method used to make an extract of the soil in which the concentration of the salts is measured. Two basic extracts have been used, namely, the saturated paste extract (USSL Staff, 1954) and the 1:5 soil-water extract (Rengasamy et al., 1984b). In both cases, the EC measured is roughly related to the TCC by the following equation:

$$EC (dS\ m^{-1}) \approx 10\ TCC\ (mmol_c\ L^{-1}) \quad [1.11]$$

However, the value of the measured EC varies depending on the method used to obtain the extract, with the relationship between the two values being given by the equation developed by Shaw (1994):

$$EC_e = EC_{1:5} [(500 + 6ADMC)/SP]^b \quad [1.12]$$

where

EC_e, $EC_{1:5}$	= electrical conductivities in saturated paste and 1:5 extracts, respectively, in $dS\ m^{-1}$
ADMC	= air dry moisture content
SP	= saturation percentage
b	= coefficient, $1 > b > 0$

Relationship between ESP and SAR

On the basis of the Gapon equation, the following theoretical relationship between SAR and ESP can be derived:

$$ESP = (100SAR)/(SAR + 1/K_g) \quad [1.13]$$

where

K_g = Gapon constant, equal to 0.015 (Sumner and Miller, 1997)

Recently, Naidu et al. (1995b) have demonstrated for Australian duplex soils that the value of K_g varies with the EC of the soil solution. Various approximate relationships derived by regression analysis exist between ESP and SAR:

$$ESP = \frac{[-1.26 + (1.475\ SAR_e)]}{[0.9874 + (0.0147\ SAR_e)]}$$
(USSL Staff, 1954) \quad [1.14]

$$ESP = (0.52SAR_{adj}) + 5.16$$
(Oster and Schroer, 1979) \quad [1.15]

$$ESP = (1.25SAR_c) + 1.76$$
(Oster and Schroer, 1979) \quad [1.16]

and

$$ESP = (1.95\ SAR_{1:5}) + 1.8$$
(Rengasamy et al., 1984b) \quad [1.17]

where

SAR_{adj} and SAR_c are the adjusted and corrected SAR values for the irrigation water with which the soil comes into equilibrium

Despite the fact that these relationships between ESP and SAR have a sound theoretical basis (Bolt, 1967), they should be used with some caution because the slopes and intercepts depend on clay mineralogy (Levy and Hillel, 1968) and vary with salinity and saturation percentage (Frenkel and Alperovitch, 1983). Furthermore, Equation [1.14] was based only on a group of about 60 soils from the western United States, whereas Equation [1.17] was derived from 138 Australian Red-Brown Earth (Alfisol) surface samples. In addition, the coefficients of determination r^2 were of the order of 0.64–0.89, indicating that these relationships, while good, are not particularly tight. It is, therefore, quite possible that extrapolations outside the range of soils used in each respective study may result in substantial errors. Unfortunately, both SAR and ESP were developed on the basis of exchange chemistry and bear no *ab initio* relationship to clay dispersion.

1.3 PROBLEMS ASSOCIATED WITH THE DEFINITION OF SODIC SOILS

1.3.1 Experience in the United States and Great Britain

Having developed the required parameters for definition, we can now proceed to explore the problems in delineating sodic soils by tracing the historical development of the topic. The first report of alkali soil in the United States was made by Hilgard (1877), at which time the perception was one of accumulation of salts. He believed that an alkali soil would be restored to its normal state by leaching. Subsequently, although Hilgard (1889) modified this view, realizing that soil physical properties were also impacted, he stated:

> the time is not far distant when in California the laying of underdrains will be considered an excellent investment on any land as valuable as all irrigated land is likely to be; and when that day comes, alkali will be at an end on irrigated lands in this state.

This statement implies that only the leaching of salts was required for the reclamation of alkali soils. Although he recognized that the physical properties of some alkali soils were extremely poor, Hilgard failed to understand the exchange chemistry involved when such soils are leached. Subsequently, Kelley (1937) stated:

> Despite the importance of Hilgard's work and the brilliance of his ideas, investigation in recent years has established the fact that an additional factor, not recognized by him, also plays an important part in many alkali soils. The reference here is to the chemical reactions and physical effects that are produced in the soil by soluble salts. As a

result of reactions between soluble sodium salts and the soil, sodium becomes adsorbed. These chemical (reactions) render the soil highly toxic to ordinary crops and ... produce extremely bad physical conditions in the soil. These adverse conditions persist after all the soluble salts have been removed by leaching. Under such conditions some special treatment may be necessary in addition to the removal of salts.

This was the first realization that, to maintain acceptable physical conditions, a balance must be struck between the level of exchangeable Na and the equilibrium electrolyte concentration. Bodman (1937), who made a comprehensive study of the effect of long-continued percolation, stated:

> Explanation of the great decreases in saturated water-permeability of all of the soils examined seems to lie in the early removal of electrolytes and subsequent gradual dispersion and rearrangement of the clay particles so that the conducting pores are reduced in size more or less permanently.

This was confirmed a little later by Fireman and Bodman (1939), who concluded:

> Velocity of flow depended to a much greater extent upon the concentration of soluble salts in the water than upon base status of the soil. High sodium water (Ca/Na = 0.17) increased exchangeable sodium in the soil and also very greatly reduced its permeability, but the latter occurred only when irrigation with high sodium water was followed by irrigation with water of much lower content of total soluble salts.

Fireman and Bodman (1939) even demonstrated that a Yolo clay loam from California with an ESP of 2.7 exhibited a marked decline in HC when leached with distilled water. The importance of electrolyte in maintaining IR and HC in California soils was further demonstrated by Huberty and Pillsbury (1941) and the existing state of knowledge reviewed by Magistad and Christiansen (1944), who stated:

> If the accumulated salts are principally sodium salts, the finer soil particles or soil colloids will contain larger quantities of sodium than normal, and this will change their physical properties as well, particularly if most of the salts are removed.

Further work by Fireman (1944) clearly demonstrated the effect of electrolyte on the permeability of the Hesperia sandy loam (ESP 20) (Fig. 1.1). Note that permeability is plotted on a logarithmic axis. Permeability was readily maintained at a high level when the soil was leached with a solution containing 800 μg g^{-1} CaCl$_2$ or Riverside tap

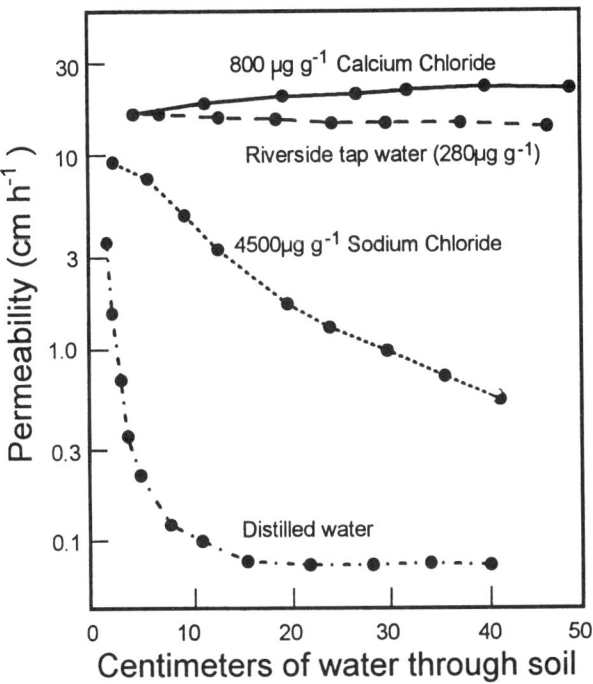

Figure 1.1 Effect of salt concentration on permeability of a Hesperia sandy loam to distilled water and various solutions [Fireman, *Soil Sci.* 58:337-353, (1944)].

water (280 μg g^{-1} mixed salts) but rapidly declined with distilled water. He concluded:

> These and other data indicate that the chemical composition of the percolating water and the chemical changes brought about by it may be of the utmost importance in permeability tests on colloid-containing materials. The rate of movement of water through soils is the resultant of a large number of complex factors, both chemical and physical.

The significance of these results, showing the difference in permeability between tap and distilled water, will become apparent a little later in the discussion. Although, by this time, soil permeability was well known to be a function of the ESP of the soil and the TCC of the percolating solution, the first attempt at a rigorous definition was made by the staff of the U.S. Regional Salinity Laboratory, Riverside, California (Kelley, 1951):

> Based on the ideas of de'Sigmond and Gedroiz ..soils which contain only a low concentration of soluble salts (unspecified) with a pH more than 8.5 and whose exchange material is more than 15 percent Na$^+$-saturated, are called alkali soils.

In retrospect, this definition does not do justice to the knowledge of sodic soil behavior extant at the time, particularly in view of the known effect of electrolyte con-

centration on physical behavior. It is a pity that no attempt was made in this definition to specify the concentration of salts because later definitions omitted mention of it altogether. It would appear that this definition was the integrated experience of the workers at the U.S. Salinity Laboratory at the time and the reasons why the effects of electrolyte concentration on soil stability did not feature more prominently are a little difficult to ascertain from the literature.

Little direct evidence in terms of the relationship between HC decline and ESP and EC could be found in the literature published at that time to support the critical value of ESP = 15 established in the preceding definition (de'Sigmond, 1927b; de'Sigmond *et al.*, 1927; Kelley, 1948; Reeve *et al.*, 1954; Brooks *et al.*, 1956). This is supported by a quotation from Kelley (1951):

> Another physical peculiarity of alkali soils is that the effect of exchangeable Na$^+$ varies from soil to soil. Certain soils seem to be much more highly dispersed when high in exchangeable Na$^+$ than others. For example, upon bringing about an equal degree of Na$^+$ saturation, soils of similar texture but of different series have shown rather wide differences in permeability.

In hindsight, it would appear that one of the factors responsible for this differential behavior might be the difference in electrolyte concentration among the soils, although differences in clay mineralogy could also play a part. Calculations based on the results of Doneen (1948) have also shown that reductions in permeability greater than an order of magnitude were obtained on a soil from the San Joaquin Valley of California when the ESP of the soil was increased from 9 to 13 with water having a concentration of 2.7 mmol$_c$ L^{-1}. This reflected the well-established practice of the local farmers of applying gypsum to promote infiltration, the usage of which reached 300 000 tons in the valley by 1945. However, in terms of the views prevailing at the time, this gypsum was probably added with the aim of reducing the ESP of the soil and not to increase the electrolyte concentration *per se*. Subsequently, in the then definitive text entitled *Diagnosis and Improvement of Saline and Alkali Soils* (USSL Staff, 1954), a nonsaline-alkali soil was defined as one in which ESP exceeded 15 and EC$_e$ was less than 4 dS m^{-1} with a pH of between 8.5 and 10. A saline-alkali soil has ESP greater than 15 and EC$_e$ above 4 dS m^{-1}. The full classification of saline and sodic soils as proposed by the USSL Staff (1954) is presented in Table 1.3.

The category of nonsaline nonsodic soils that could be construed as "normal" soils has a very wide range of per-

TABLE 1.3 Classification of saline and sodic soils [USSL Staff, *Diagnosis and Improvement of Saline and Alkali Soils*, USDA, U. S. Government Printing Office, Washington, D.C, 1954)]

Saline Nonsodic	Sodic Nonsaline	Saline Sodic	Nonsaline Nonsodic
ESP<15	ESP>15	ESP>15	ESP<15
EC_e^a>4	EC_e<4	EC_e>4	EC_e<4

^{a}in dS m^{-1}

missible limits for ESP and EC_e with correspondingly wide variations in possible physical behavior patterns. Although these definitions contain an upper limit for salts in a nonsaline sodic soil, the fact that variations in electrolyte concentration below this level can have profound effects on soil behavior appears to have been totally ignored. However, this effect was well known by farmers on the eastern side of the Central Valley of California, where gypsum dissolution in the dilute irrigation waters used to irrigate the nonsodic soils was a common practice in the 1940s and 1950s (Fireman, 1944; Oster, 1982). Fireman (1944) states "A factor which has received much attention in soil reclamation, but which is often overlooked in permeability studies, is the chemical composition of the water used for the permeability test." Furthermore, it would appear that, at the time, soil physicists at Riverside, California measured HC using tap water, which varied in concentration between 4 and 10 mmol$_c$ L^{-1} (Shainberg *et al.*, 1989). This would have reduced hydraulic decay even in soils with moderately high ESP values, as was fully realized at the time by Fireman (1944) who goes on to say: "Some alkali soils tested at this laboratory are hundreds of times more permeable to tap water . . . than they are to distilled water or low-salt waters containing a high percentage of sodium." This is probably the reason why such a high ESP value was selected to differentiate between nonsodic and sodic soils. Even the data for air-to-water permeability ratios presented by the USSL Staff (1954) in support of their definition contained examples of soils where a sharp rise in this ratio occurred between ESP 0 and 10, indicating sensitivity to hydraulic failure for a number of soils (Reeve, 1953; Reeve *et al.*, 1954). Brooks *et al.* (1956) presented data for the ratio of air to water permeability which showed this trend clearly (Fig. 1.2). The Chino, Huntley, and Sebree soils are particularly sensitive to hydraulic failure at ESP values below 10. At the time, the importance of water quality in determining permeability was certainly well known, as illustrated by Reeve (1953), who stated:

Inasmuch as deflocculation and dispersion, which are processes that affect structural stability, are functions of both

the solution concentration and the cation exchange status of the soil, it would be expected that differences in the quality of the water used in the determination (of permeability) will produce differences in the permeability ratio.

Reeve *et al.* (1957) also concluded: "The quality of the water that percolates through the soil has a marked effect upon the permeability of soils. Both the electrolyte concentration and composition of the water influence the permeability or hydraulic conductivity of soils."

Meanwhile in Great Britain, Quirk and Schofield (1955), whom McNeal (1974) credits with being the first to conduct definitive quantitative work on the effects of salinity and Na on soil permeability, developed the threshold electrolyte concentration (TEC) concept. This arose from an exhaustive study of the effect of various cation saturations and electrolyte concentrations on the permeability of soil. The TEC is described by a line on an ESP-EC plot which separates soils exhibiting stable hydraulic behavior from those exhibiting unstable behavior (Fig. 1.3). They stated

that there is no particular basis for the division of soils into alkali and non-alkali classes at 15 per cent exchangeable sodium. As the exchangeable sodium percentages increase, the permeability to solutions considerably less concentrated than the threshold concentration decreases continuously.

At the U.S. Salinity Laboratory, Reeve (1958) reworked some of Quirk and Schofield's (1955) data (Fig. 1.4):

It is apparent that hydraulic conductivity drops off markedly for all values of ESP and tends toward some maxi-

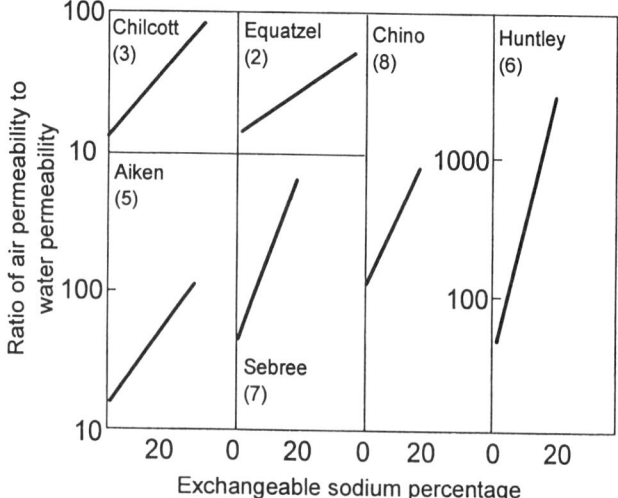

Figure 1.2 Ratio of air to water permeability as influenced by exchangeable sodium percentage (ESP). ESP values for each soil are in parentheses [Brooks, *et al., Soil Sci. Soc. Am. Proc.* 20:325-327, (1956)].

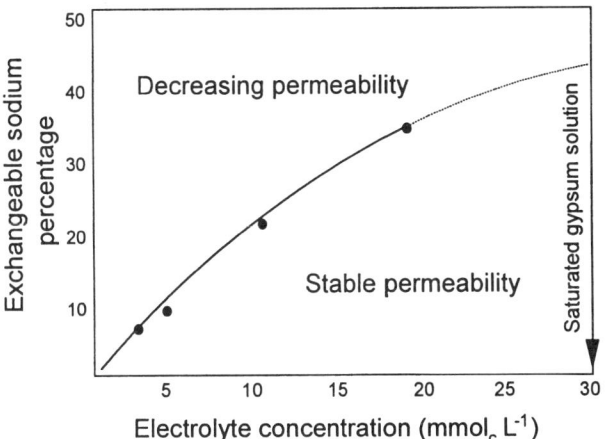

Figure 1.3 Threshold electrolyte concentration (TEC) curve illustrating zones of stable and unstable permeability [Quirk and Schofield, *J. Soil Sci.* 6:163-178, (1955)].

mum value as the electrolyte concentration of the water is increased. There are other factors and processes that are operative, but the effects of exchangeable sodium and electrolyte concentration are by far the greater effects involved.

This statement clearly indicates that the workers at the U.S. Salinity Laboratory were aware of the implications of Quirk and Schofield's (1955) work. For example, the HC of a soil with an ESP value as low as 5.8 (Fig. 1.4) is particularly sensitive to low electrolyte concentrations. Nevertheless in 1958, a recommendation was made in the United States to replace the term alkali by sodic, originally coined by Overstreet in 1951 (McNeal and Coleman, 1966) as a particular category of alkali soils characterized by a high amount of Na, with the following definition (Bower *et al.*, 1958): "Soil that contains sufficient exchangeable sodium to interfere with the growth of most crop plants. For purposes of definition, soil for which ESP is 15% or more." Note the complete absence of any mention of electrolyte concentration in this definition. The apparent neglect of the substantial body of evidence that had been accumulated both in the United States and overseas, clearly demonstrating the importance of incorporating electrolyte concentration into the definition of a sodic soil, is hard to understand. Had the work of Quirk and Schofield (1955) been adopted in the United States, a much clearer understanding of sodic soils and sodicity would have resulted.

At about this time, in California Martin and Richards (1959) demonstrated on a Hanford sandy loam, presumably using southern California tap water (4–10 mmol$_c$ L^{-1}), that increasing the ESP of a base saturated soil from 0 to 5 reduced HC from 40 to 11 mm h^{-1}, and at ESP 10, HC was

only 5 mm h^{-1}. These reductions were much greater as the soil became more acid. These results clearly conflict with the earlier definition, but the HC values are high in comparison to the value of 1 mm h^{-1} considered to be adequate by the USSL Staff (1954). (Care should be taken in setting arbitrary threshold values for HC to define sodic conditions because low HC could also arise as a result of compaction, no-tillage, and other factors. What is more important is the relative effect of sodicity on the initial HC value.) Nevertheless, the Hanford soil unquestionably suffered severe physical degradation, even at a relatively high EC$_e$. Gardner *et al.* (1959) also showed that the diffusivity of soil water under unsaturated conditions was a function of ESP and EC$_e$. The question now arises: Are these observations of practical importance? There is no doubt that the answer today would be a resounding yes, and in view of the knowledge of sodicity in the late 1950s, it should also have been in the affirmative at that time. One of the curious aspects of this apparent conflict between the data and the definition of a sodic soil is that most of the authors who produced evidence that did not support the critical ESP value of 15 were also authors of the definition.

Later, McNeal and Coleman (1966) pointed out that the ESP > 15 requirement had only been tentatively established by the USSL Staff (1954) and that they set out to obtain supporting evidence for this limit.

For soils with clay mineralogy which can be most readily construed as "typical" for the major agricultural soils of

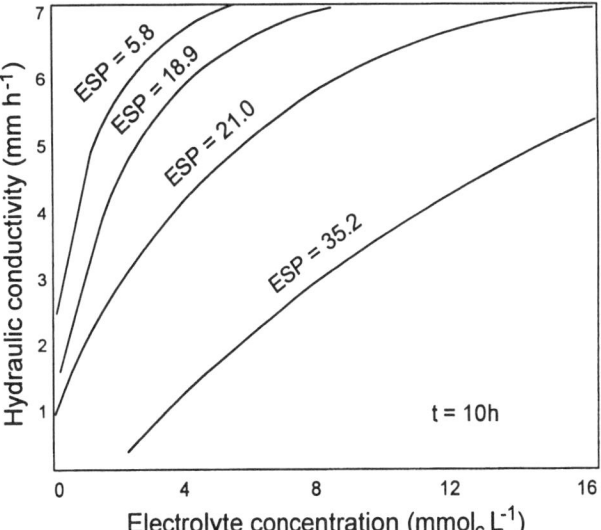

Figure 1.4 Hydraulic conductivity of silty loam soil (Sawyers Field 1, Rothamsted Farm, England) as a function of electrolyte concentration for several levels of exchangeable sodium percentage (ESP) [Reeve, *Trans. 5th Int. Congr. Agric. Eng.* pp. 21-32, (1958)].

arid lands (i.e. having clay mineralogy dominated by the presence of 2:1 layer-silicates, but containing only moderate amounts of montmorillonite), ESP values of 15 or greater can generally be tolerated before serious reductions in hydraulic conductivity (>25%) will occur, providing the salt concentration of the percolating solution exceeds 3 mmol$_c$ L^{-1}.

At concentrations above this, many soils exhibit little decline in HC, whereas below it rapid decreases result (Shainberg and Letey, 1984). However, under rainfed conditions in the field on many topsoils, such an electrolyte concentration level will seldom be sustained. Subsequently, Felhendler *et al.* (1974) demonstrated, for two Israeli Alfisols, that leaching with distilled water (equivalent to rain) caused very marked decreases in HC (> 25%) after the soils had been equilibrated with solutions of SAR between 5 and 10. Despite all this accumulated evidence, suggesting that there might be a problem with the current definition of a sodic soil, no significant action was taken. Bernstein (1974) at the U.S. Salinity Laboratory subsequently introduced a change in the critical ESP value to 10 for fine and to 20 for coarse textured soils. By 1979, the term alkali was considered to be obsolete in the United States, with the following modified definition of a sodic soil being accepted by the Soil Science Society of America (Anon., 1979):

> A non-saline soil (undefined) containing sufficient exchangeable Na to adversely affect crop production and soil structure under most conditions of soil and plant type. The lower limit of SAR$_e$ of such soils is conventionally set at 13.

Using Equation [1.14], this would correspond to an ESP of 16. The reference to nonsaline in this definition implies that the soil has an EC less than 4 dS m^{-1} (~ 40 mmol$_c$ L^{-1}). Again this definition ignores the fact that many soils below an ESP of 15 respond very favorably to increases in electrolyte concentration of the order of 3–6 mmol$_c$ L^{-1}, very much lower than the implied limit. The change to using SAR instead of ESP as the diagnostic criterion was introduced in an attempt to eliminate the numerous potential errors in traditional CEC and ESP determinations.

As a result of recent contact with Bower (1994, personal communication), who was a director of the U.S. Salinity Laboratory at the time the ESP = 15 threshold was established for a sodic soil, more light has been cast on the situation:

> Prior to and during preparation of the Handbook, Milt Freeman, Ron Reeve and I, in cooperation with State Experiment Station and Bureau of Reclamation workers, were

involved in field work on salt affected soils in 8 western States. My field experience indicated that exchangeable Na became a significant problem, mainly from the standpoint of irrigation water infiltration, when the ESP was in the range 10–20, depending on soil texture. ...In my opinion, two things mitigate against the use of data such as that of Quirk and Schofield for classifying U.S. salt affected soils. First, such data are obtained in the laboratory on disturbed, ESP adjusted soil samples packed in small containers and saturated flow of water prevails whereas in the field there are usually some natural structure benefits and unsaturated flow prevails; moreover, dispersion seems to be greater in the disturbed sample. In any case, my experience indicates that laboratory studies markedly exaggerate the effects of exchangeable Na on soil physical properties relative to the field situation...Second, while Handbook 60 (USSL Staff, 1954) recognizes saline and nonsaline alkali soils for discussion purposes, owing to evapotranspiration, variable leaching during irrigation, upward movement of saline groundwater and other reasons, the salt concentration of the soil solution is too transient to use as a factor in a comprehensive classification of sodic soils under irrigation.

The problems associated with the use of disturbed samples to estimate the HC of entire profiles are real and are discussed in detail in Chapter 6. However, as far as the IR at the soil surface is concerned, such problems are likely to be minimal (Chapter 5). In addition, consideration of only the exchanger phase in sodic soils (ESP) without a knowledge of the electrolyte concentration has led to much confusion in the prediction of sodic behavior in soils, as will be discussed later.

Based on their vast experience with soils in India, Gupta and Abrol (1990a) reintroduced the term alkali soil to refer to those sodic soils which have ESP greater than 15 and contain free Na$_2$CO$_3$ indicated by pH values above 8.4 and a Na/(Cl + SO$_4$) ratio greater than 1. Strongly alkaline soils (pH > 8.4) usually have high ESP values, but not all soils with high ESP values necessarily have high pH. High pH is usually associated with high levels of Na in the presence of correspondingly high concentrations of HCO$_3^-$ and CO$_3^{2-}$.

1.3.2 Experience in Australia and Africa

The earliest work cited in Africa, in the Gezira of the Sudan, was by Greene and Snow (1939), who demonstrated the marked sensitivity of the soils (ESP 8–13) to physical degradation in the field when irrigated with Nile water (SAR = 32, TCC = 2 mmol$_c$ L^{-1}). They found that the application of modest rates of gypsum had a profound initial effect in improving soil physical conditions as well as cotton yields, but the effect was not long-lasting. In Australia,

Davidson and Quirk (1961), following the pioneering work of the latter, demonstrated that when soils with moderate to high ESP (9–23) were irrigated with water of low EC, severe crusting resulted in exceedingly poor emergence of pasture seedlings. However, the addition of as little as 2.5 $mmol_c L^{-1}$ of electrolyte in the form of gypsum to the irrigation water resulted in substantial improvement in seedling emergence as a result of maintaining the clay in a flocculated condition. This concentration of gypsum is far less than that which would be required to reclaim a sodic soil by removing the Na in a reasonable period of time. The foregoing results highlight the fact that dispersive behavior can take place in soils with ESP values below 15 as well as the importance of the electrolyte effect in dealing with Na-affected soils.

Later on, Cass (1972) and Northcote and Skene (1972) encountered problems in both South Africa and Australia with the use of the American definition to describe the behavior of Na-affected soils in that the critical value separating sodic from nonsodic soils appeared to be far too high. In Australia, the benefits of increasing electrolyte concentration supplied by gypsum on infiltration into soils with ESP values below 10 had been demonstrated between 1921 and 1933 under field conditions (Simms and Rooney, 1965). Consequently, Northcote and Skene (1972), on the basis of morphological evidence and supported by the data of Loveday and Pyle (1973) adopted an ESP value above 6 in the top meter of soil as the Australian definition of a sodic soil with a strongly sodic soil having an ESP greater than 15 (Loveday and Bridge, 1983), whereas McIntyre (1979) proposed a value of greater than 5 based on a detailed study of 71 Australian soils. McIntyre (1979) demonstrated that, at low values of SAR (ESP), TCC is particularly important in determining the dispersive behavior of soils. At about the same time, Cass (1972), in southern Africa, observed very similar behavior on soils from northern Namibia, indicating that the threshold ESP value of 15 set by the USSL Staff (1954) was too high. A comparison of these two data sets is presented in Fig. 1.5. In both cases, the sensitivity of HC to increases in ESP from 0 to 5 can be readily seen. Cass (1972) argued that soil would have to have an ESP less than 2 to retain an acceptable level of HC (<10% reduction on leaching with irrigation water having an EC of 0.04 dS m^{-1}), whereas McIntyre (1979) demonstrated that HC = 1 mm h^{-1} considered to be acceptable by the USSL Staff (1954) was reached at an ESP value of 3–5. The level of ESP used to differentiate between sodic and nonsodic soils is of critical importance. First of all, soils with an ESP greater than 15 in the topsoil are relatively rare in areas of intensive agricultural production in semiarid and arid regions and almost nonexistant in subhumid

and humid areas of the world. Second, this apparently contradictory situation arose from the fact that different-quality tap waters (TCC = 3–14 $mmol_c L^{-1}$ in California, 0.4–0.7 $mmol_c L^{-1}$ in South Africa and Australia) had been used by all the investigators to study the decay of soil hydraulic properties under the influence of increasing ESP (Fireman, 1944; Shainberg *et al.*, 1989). The much higher TCC used in California is apparently one of the reasons why a greater threshold value (ESP > 15) was established to separate sodic from nonsodic soils. In addition, the Australian and South African work involved mainly clay soils, whereas in the United States the soils studied were lighter in texture, which could also clearly have affected the outcome to some extent. Because Na-affected soils are likely to exhibit their worst behavior physically under rainfed conditions (very low EC), there can be little justification for the retention of the ESP >15 criterion established at an elevated EC and supported by little confirmatory data as a universal threshold for sodic behavior. Quirk (1986) pointed out the arbitrary nature of such definitions particularly in view of the fact that there is no abrupt change in soil properties as the ESP increases. Supporting this view, Sumner (1993a), based on the work of many others, has demonstrated that sodic behavior (clay dispersion and swelling) and its effects on HC and IR form a continuum, increasing in severity from ESP = 0, depending on the TCC level. Although both ESP (SAR) and EC values usually increase in a very qualitative parallel fashion in many soils (Chapter 6), the problems of physical condition only

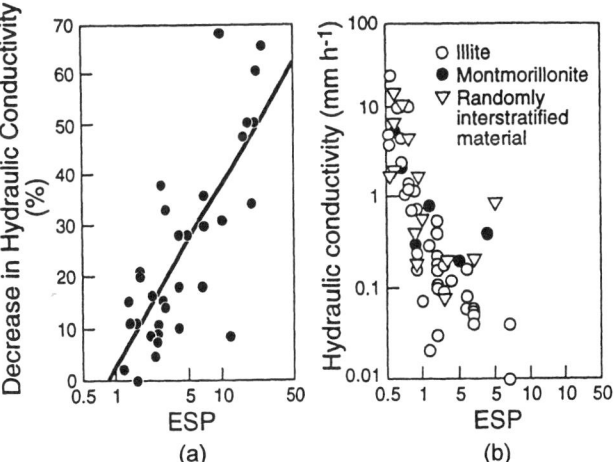

Figure 1.5 (a) Reduction in hydraulic conductivity (HC) caused by leaching a group of South African soils of varying exchangeable sodium percentage (ESP) with tap water (0.7 $mmol_c L^{-1}$) [Cass, M.Sc. Agric. Thesis University of Natal, South Africa, (1972)]; (b) HC of Australian soils having a range of ESP values, leached with tap water (0.7 $mmol_c L^{-1}$) [McIntyre, *Aust. J. Soil Res.* 17:115-120, (1979)].

arise when the EC value is too low to maintain the soil clay in a flocculated condition at a given ESP (SAR) value. Such situations usually occur when soils are subjected to rainfall during the wet season, which results in alternating high and low electrolyte concentrations. The deleterious effects of these conditions on the hydraulic properties of soils were clearly demonstrated by Oster and Schroer (1979). This led Shanmuganathan and Oades (1983b) to suggest the use of the ESP/EC ratio as a guide to evaluating soil behavior. Thus any attempt to set critical ESP values would be entirely arbitrary and, unless TCC or EC is taken into account simultaneously, also meaningless from a management point of view. In an attempt to account for the effect of water quality on soil hydraulic behavior, Cass and Sumner (1982a,b,c) demonstrated successfully the use of Na stability values derived from a model relating SAR and TCC with the measured decline in HC as a means of separating stable soils from unstable ones.

1.3.3 Need for Change

It would, therefore, appear that some new system of nomenclature to catagorize Na-affected soils for management purposes must be developed based on soil behavior rather than on arbitrary threshold ESP criteria ignoring electrolyte concentration. It is really somewhat contradictory to call a soil with only a low percentage saturation with Na (ESP > 6) a sodic soil when, in fact, the exchange sites are almost completely saturated with Ca and Mg. This points to a need to take account of soil behavior. Furthermore, for management purposes, topsoil properties and behavior are often of equal importance with those in the subsurface. On the other hand, from a soil classification standpoint, pedologists favor the use of subsoil horizon criteria and are reluctant to incorporate TCC or EC into their definitions because of their ephemeral nature (Isbell, 1995a). Thus the terminology needs of classification and management seriously conflict when dealing with Na-affected soils, and the solution probably lies in the development of two separate systems with a minimum of overlapping terms. From a management point of view, it would probably be better to base the differentiating criteria on readily measured soil behavior parameters, such as spontaneous or mechanical dispersibility rather than chemical composition of the soil. Rengasamy *et al.* (1991, 1992) have proposed the use of the dispersive potential, defined as the difference in osmotic pressures in the diffuse double layer between the critical flocculation concentration (CFC) of electrolyte and the existing soil solution concentration, to describe sodic behavior without setting arbitrary limits of ESP and EC. This concept has merit and is being devel-

oped further in Chapter 3. Additional work is required to integrate the effects of mechanical energy inputs such as from raindrops so that dispersive potential would cover the full range of conditions encountered in the field. Until such time as this concept is fully developed, an interim arbitrary classification, taking into account both soil composition and behavior, will have to suffice.

Furthermore, the soils of greatest interest as far as agricultural production is concerned are those having lower levels of ESP (SAR) because of the greater ease of management that such values confer. Consequently, the remainder of this treatise will emphasize the distribution, properties, and management of the soils falling into this category.

1.4 PROPOSAL FOR CATEGORIZATION OF SODIUM-AFFECTED SOILS

As a result of an extensive review of the literature, Sumner (1993a) came to the conclusion that there was a definite need for a review of the nomenclature used to describe and categorize Na-affected soils. This view was supported by the outcome of First Australian National Conference on Sodic Soils (Naidu, et al. 1995b). This need had previously been recognized by Rengasamy *et al.,* (1984b), who proposed a scheme based on the original TEC concept of Quirk and Schofield (1955), involving $SAR_{1:5}$, TCC, and the dispersivity of the soil to place soils in a number of categories (Fig. 1.6). Details of each class are paraphrased below.

Class 1: Dispersive Soils. Soils that disperse spontaneously without shaking will have severe problems associated with crusting, reduced porosity, etc., even when subjected to minimum mechanical stress, e.g. under zero tillage.

Class 2: Potentially Dispersive Soils. Soils that require inputs of mechanical energy (raindrop impact, tillage) to bring about dispersion will experience problems of crusting, reduced porosity, etc., when mechanically disturbed by impacting raindrops or intensive cultivation.

Class 3: Flocculated Soils. Soils that contain more than the minimum electrolyte concentration required for flocculation of the clay fraction will present few physical problems, but the level of salts could be excessive and limit productivity by drought stress.

The current proposal is a further modification of this scheme to describe soil structural stability. Sumner (1993a)

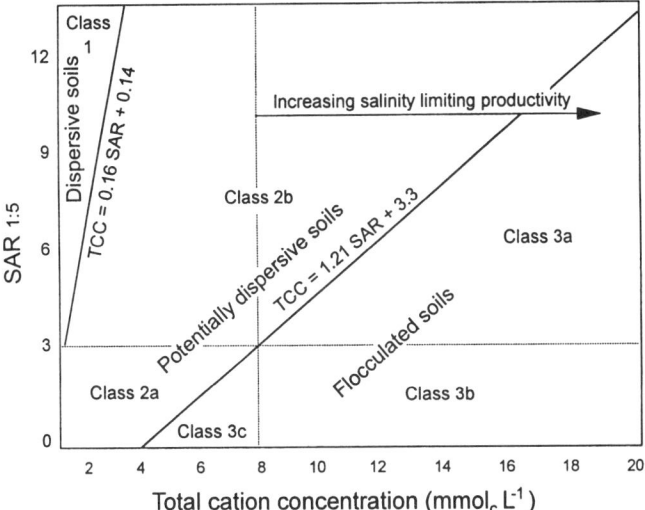

Figure 1.6 Classification scheme for prediction of dispersive behavior of A horizons of Red-Brown Earths. Lines reflect different threshold electrolyte concentrations (TEC) conditions for soils of differing stabilities. [Rengasamy *et al.*, *Aust. J. Soil Res.* 22:413-431, (1984)].

matter, positive charge, sesquioxides, kaolinite, and salt from weathering of soil minerals. Thus the position of the TEC lines in Fig. 1.6 must be considered to be flexible and not fixed. Rengasamy *et al.* (1984a) found that when their scheme was tested, there were a number of aberrant soils, particularly in Class 2a, namely, soils of low $SAR_{1:5}$ (ESP) and very low TCC. The aberrant soils had higher organic matter contents and lower pH values than those that fitted the scheme. Thus the state of soil degradation reflected in reduced organic matter contents plays an important role in determining the dispersive behavior of soils.

The proposed new scheme is presented in Fig. 1.7 for nonswelling 2:1 clay (illitic type) soils commonly found in the United States, Australia, Africa, and elsewhere. The applicability of this classification to other soils will have to be tested in the future, and modifications may have to be made. Calibrations are in terms of $SAR_{1:5}$ and $EC_{1:5}$, with the various categories being named in accordance with levels of Na and salt present. The selection of $SAR_{1:5}$ and $EC_{1:5}$, has been based on their much greater ease of measurement compared to SAR_e and EC_e which involve the tedious saturated paste extraction procedure. In addition, ESP has not been used as a measure of sodicity in the classification because of the difficulties associated with its measurement discussed in Section 1.2.1. The use of *Non* to describe situations where the levels of salt or Na are very low has been retained, and the adjective *Very* (capitalized) has been added to extend the range. In addi-

showed that the position of the threshold concentration line separating dispersed from flocculated soils depended on a number of factors and moved in such a way as to produce an increased dispersive field with increases in mechanical energy, negative charge, smectite or illite content, exchangeable K and Mg, and specifically adsorbed anions (phosphate) as well as decreases in P_{CO2}, organic

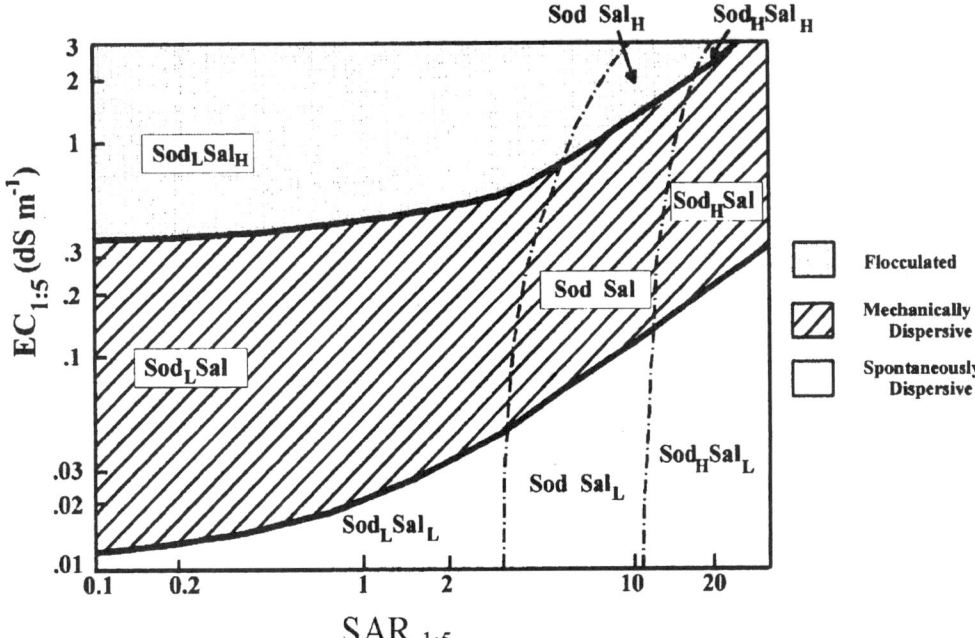

Figure 1.7 Proposed scheme for description of Na-affected soils in terms of physical behavior (dispersibility) and sodium (Sod) and salinity (Sal) classes.

tion, because the new classification involves categories where changes in EC and SAR occur concurrently, consideration was given to coining a new term to describe the electrolyte regime in the soil. The word *salic* was considered to be a possibility but unfortunately it has already been used in soil classification (U.S. Soil Survey Staff, 1975). Because there are no other words in the English language which are suitable for the purpose, the term *saline* has been retained, but given a new definition. In the past, saline has referred to soils having an EC_e greater than 4 dS m^{-1}, but in terms of the requirements of the new classification, *Saline*, now with a Capital S, is defined in terms of $EC_{1:5}$ and has continuously variable limits, more or less in proportion to the variable salt level required to maintain a soil in a particular behavioral category (flocculated, mechanically dispersible, or spontaneously dispersible). Thus the term *Saline* with the descriptors *Non* and *Very* now indicates the range of salt contents required to maintain a soil with a certain range of Na saturations in a given dispersibility class. The salt level required to maintain the soil in a given behavioral category increases with $SAR_{1:5}$ and consequently, the fields for the different Na and salt classes do not have the same shape. The term *Sodic* is also redefined in terms of $SAR_{1:5}$ instead of ESP, also now with a Capital S. Thus on the $SAR_{1:5}$ axis, the subdivisions are *Nonsodic* ($SAR_{1:5} \lesssim 3$), *Sodic* ($SAR_{1:5}$ ~3–10), and *Very Sodic* ($SAR_{1:5} \gtrsim 10$). These divisions are arbitrary and correspond roughly to the ESP values of 5, 10, and 15 selected by others in the past to differentiate levels of sodicity (Bernstein, 1974; McIntyre, 1979; Loveday and Bridge, 1983). On the $EC_{1:5}$ axis, *Nonsaline* corresponds to very low salt contents, which vary with increasing levels of $SAR_{1:5}$ to develop a field in which most of the soils would be *spontaneously dispersive* in water. *Saline* corresponds to higher salt contents, developing a field expanding with increasing $SAR_{1:5}$ in which the soils would be *mechanically dispersive* with energy inputs. Finally, *Very Saline* corresponds to high salt contents, also with increasing with $SAR_{1:5}$, creating a field in which the soils would be *floccu-*

lated. In these descriptions the abbreviations Sod and Sal refer to Sodic and Saline, and the subscripts H and L to high and low, respectively. Thus the descriptors in the classification now become:

Spontaneously Dispersive Soils
  Nonsodic Nonsaline [$Sod_L Sal_L$](low $SAR_{1:5}$, low $EC_{1:5}$)
  Sodic Nonsaline [Sod Sal_L](medium $SAR_{1:5}$, low $EC_{1:5}$)
  Very Sodic Nonsaline [$Sod_H Sal_L$](high $SAR_{1:5}$, low $EC_{1:5}$)

Mechanically Dispersive Soils
  Nonsodic Saline [Sod_L Sal](low $SAR_{1:5}$, medium $EC_{1:5}$)
  Sodic Saline [Sod Sal](medium $SAR_{1:5}$, medium $EC_{1:5}$)
  Very Sodic Saline [Sod_H Sal](high $SAR_{1:5}$, medium $EC_{1:5}$)

Flocculated
  Nonsodic Very Saline [$Sod_L Sal_H$](low $SAR_{1:5}$, high $EC_{1:5}$)
  Sodic Very Saline [Sod Sal_H](medium $SAR_{1:5}$, high $EC_{1:5}$)
  Very Sodic Very Saline [$Sod_H Sal_H$](high $SAR_{1:5}$, high $EC_{1:5}$)

The positions of the dotted lines will shift markedly according to soil type. For smectitic clays, the lines will shift toward the upper left corner of Fig. 1.7, whereas for the more stable soils, such as those dominated by kaolinite and sesquioxides, the lines will be displaced toward the bottom right corner. It is obvious from this scheme that most arable topsoils of the world would be situated toward the bottom left-hand corner of Figure 1.7. The categories of sodic behavior would then be described as illustrated in Table 1.4. A brief description of the soils in each category is presented hereafter:

Nonsodic Nonsaline [$Sod_L Sal_L$]
These soils would have extremely low $SAR_{1:5}$ values (≤ 3), but with exceedingly low $EC_{1:5}$ values (≤ 0.05 dS m^{-1}) and would disperse spontaneously. They will be highly degraded soils, having lost much of their organic content as a result of cultivation, and are most likely to be found in

TABLE 1.4 Relationship between sodic behavior and sodium and salt classes

Behavior class	Sodium [Sod] and salt [Sal] class		
Spontaneously dispersive	*Nonsodic Nonsaline* [$Sod_L Sal_L$]	*Sodic Nonsaline* [Sod Sal_L]	*Very Sodic Nonsaline* [$Sod_H Sal_L$]
Mechanically dispersive	*Nonsodic Saline* [Sod_L Sal]	*Sodic Saline* [Sod Sal]	*Very Sodic Saline* [Sod_H Sal]
Flocculated	*Nonsodic Very Saline* [$Sod_L Sal_H$]	*Sodic Very Saline* [Sod Sal_H]	*Very Sodic Very Saline* [$Sod_H Sal_H$]

areas with excess water available for leaching to maintain the EC at very low levels.

Nonsodic Saline [Sod$_L$ Sal]
These soils, which would cover a large proportion of the arable area of the world, have low SAR$_{1:5}$ values (< 6) and a slightly higher EC$_{1:5}$ than the above category, but insufficient to maintain the soil in a flocculated state when subjected to mechanical energy inputs. Many of these soils will exhibit crusting and reduced porosity, particularly once cultivation has resulted in a reduction in the organic matter content, rendering the aggregates less stable.

Nonsodic Very Saline [Sod$_L$ Sal$_H$]
These soils contain sufficient salt to maintain the clay in a flocculated state under all conditions and present few physical problems save moisture stress when very high EC$_{1:5}$ values are found.

Sodic Nonsaline [Sod Sal$_L$]
Such soils have an SAR$_{1:5}$ of between about 3 and 10 but insufficient salt to maintain the clay in a flocculated condition, and consequently, they will disperse spontaneously. These soils are likely to occur in humid and subhumid areas where the rainfall is sufficient to maintain a low EC. They will be difficult to cultivate and will present problems even under permanent pasture.

Sodic Saline [Sod Sal]
These soils have SAR$_{1:5}$ values between about 3 and 15 but contain a higher level of salt than the preceding category, which is sufficient to prevent spontaneous dispersion. However, with mechanical energy inputs they could disperse and crust.

Sodic Very Saline [Sod Sal$_H$]
This category contains sufficient salt to prevent dispersion under all conditions despite the relatively high Na levels (SAR ~ 6 -> 18). However, drought stress due to the high salt content could be a limitation to production. These soils are likely to occur in subhumid to arid regions where rainfall is insufficient to remove the salts.

Very Sodic Nonsaline [Sod$_H$ Sal$_L$]
Soils in this group, having high SAR$_{1:5}$ (> 10) and very low EC$_{1:5}$ values, would be highly spontaneously dispersive and have exceedingly poor physical properties.

Very Sodic Saline [Sod$_H$ Sal]
With higher EC$_{1:5}$ values than the preceding category but with high SAR$_{1:5}$ values (> 12), many of these soils are likely to still be dispersive but would require varying degrees of mechanical energy inputs to achieve dispersion and would exhibit poor physical properties such as crusting.

Very Sodic Very Saline [Sod$_H$Sal$_H$]
Soils in this category would contain sufficient salt to offset the negative effects of high SAR$_{1:5}$ (> 15) on physical properties and would be flocculated under all conditions. The high salt contents could induce severe drought stress in certain crops.

A similar scheme could be proposed to describe the more gradual changes in permeability of a soil due to sodic and saline effects, using the Equivalent Salt Solution Series (ESSS) concept of Jayawardane (1979, 1983). He defined an ESSS as combinations of SAR and EC which produce the same extent of swelling and dispersion and therefore result in the parallel changes in soil permeability and flow rates. Plotting ESSS values in Fig. 1.7 would produce a series of lines approximately parallel to the two solid lines shown. The lines at the top left corner would represent combinations of SAR and EC, which permit high soil permeabilities, and those towards the bottom right would represent low permeabilities, with a series of intermediate situations between the extremes.

Because soil reaction also affects the dispersive character of a soil, the terms *Acidic* [pH$_L$], *Neutral* [pH$_N$], and *Alkaline* [pH$_H$] will be used to denote soils with pH values below 5.5, between 5.5 and 8.5, and above 8.5, respectively. Thus examples of typical descriptors for soils using this system would be: *Sodic Nonsaline Acidic* [Sod Sal$_L$ pH$_L$], *Very Sodic Saline Alkaline* [Sod$_H$ Sal pH$_H$], or *Sodic Saline Neutral* [Sod Sal pH$_N$].

2

DISTRIBUTION OF SODIC SOILS:
THE WORLD SCENE

E. N. Bui, L. Krogh, R. S. Lavado,
F. O. Nachtergaele, T. Tóth, and
R. W. Fitzpatrick

2.1 INTRODUCTION

The world distribution of sodic soils, which are usually considered a subset of salt-affected soils (Shainberg and Letey, 1984; FAO, 1988; Szabolcs, 1989; Gupta and Abrol, 1990a), has been discussed by Szabolcs (1989) in combination with salinesoils. However, much confusion exists in the literature concerning the use of such terms as saline sodic, alkaline or not, saline nonsodic, and sodic nonsaline soils (Kelley, 1951). In most cases, sodic soils have been defined on the basis of morphological (columnar structure) and chemical properties [exchangeable sodium (Na) percentage (ESP) or sodium adsorption ratio (SAR)] of the B horizon.

Sodic soils are characterized by dense impermeable sodic B horizons. These soils are frequently truncated as a result of the removal of the overlaying surface horizons by water erosion due to the impermeable sodic horizon. The sodic B horizon is very erodible on account of the highly dispersed and easily mobilized nature of the clay fraction (Levy and van der Watt, 1988; van der Watt and Valentin, 1992; Chapter 3). Hence, both sheet and gully erosion are common on these soils (Chapter 5).

Many sodic soils are unsuitable for agriculture, but heavier textured profiles, in which the sodic horizon is as deep as 0.8 m, can be utilized for production, usually under irrigation, provided that great care is taken in their management. For the most part, sodic soils support native grasses for grazing and small grain production. In the dry state, sodic soil materials are difficult to wet with applied water, which is not readily absorbed due to crust formation. This will result in runoff (Levy and van der Watt, 1988), first removing dispersed clay and subsequently, the other particle-size fractions of the material. These properties are responsible for the highly erodible nature of the sodic materials (Chapter 5). Their resistance to wetting

further renders sodic material unsuitable for the construction of dams and roads and for most other civil engineering purposes, since such material cannot be compacted by mechanical means. Dams constructed from sodic materials are very prone to piping, and in the case of roads, uneven subsidence leads to a breakup of the surface mat (Chapter 9).

An attempt will be made to estimate the extent of soils showing sodic characteristics and behavior using three levels of ESP – < 6, 6–15, and > 15 – as the criteria to group soils. These threshold values have been selected on the basis of much published work, indicating that moderately strong sodic characteristics are exhibited when soils reach ESP greater than 6 anywhere in the upper 1 m of soil (Northcote and Skene, 1972) (Chapter 1). These subdivisions correspond roughly to the Nonsodic ($SAR_{1:5} \leq 3$, ESP ≤ 6), Sodic ($SAR_{1:5} \sim 3$–10, ESP $\sim$ 6–15) and Very Sodic ($SAR_{1:5} \geq 10$, ESP ≥ 15) categories proposed in Chapter 1. The reason why these categories are approximate is because the intensity of sodic behavior is modified by the presence of salts (see Fig. 1.7). The latest map and attribute data available from the FAO (1988, 1991) will be used to estimate the extent of soils exhibiting sodic behavior.

2.2 A WORLD SODIC SOILS MAP

2.2.1 Problems Associated with Development

The production of such a map using new definitions as suggested in Chapter 1, is an undertaking fraught with many difficulties. First of all, problems arise because the criteria used to define map units are different depending on factors such as the classification system and scale used, the purpose of the map, the availability of chemical and physical data, and inadequate or incomplete profile descriptions. For example, 600 soil maps of different scales and legends, involving the collation of systematic and reconnaissance surveys, were used to compile the FAO/ UNESCO *Soil Map of the World* (FAO, 1974; 1988). One of the most crucial aspects of the project was the correlation of soil map units to produce a legend involving the use of physiography, vegetation, climate, geology, and land use information as surrogates for limited soil data. Criteria used to define map units were pedogenesis, characteristics and distribution of major soil groups as identified in the major soil classification systems used, significance of soil resources for production, and feasibility of representation on a small scale (1:5 000 000) map. The map units consisted of soil associations, each of which was composed of dominant (occupying the largest area) soil, subdominant soils (occupying more than 20% of the map unit), and inclusions (occupying less than 20% of the map unit).

Another source of confusion is the terminology used to describe pedogenic processes leading to the development of sodic soils, e.g., salinization, solonization, alkalization, solodization, sodification, and the terminology used to refer to the soils themselves. Because these terms have been used somewhat loosely, the separation of soils exhibiting sodic behavior becomes difficult. For example, Isbell (1958) discusses changes in the usage of the term Solonetz and the introduction of the term Solodized Solonetz as the emphasis of the classification shifted from chemical properties to morphology. The evolution of the terms used to describe sodic soils has been discussed by Kelley (1951) and in Chapter 1, to which interested readers are referred.

2.2.2 Differences in Soil Classification Systems

Szabolcs (1991) has reviewed the properties of salt-affected soils as they impact on some of the major soil classifications. Although most classifications consider soil profile morphology (presence of a diagnostic horizon, soil structure) and chemical properties [exchangeable sodium percentage (ESP), sodium adsorption ratio (SAR), and pH] in the definition of sodicity, there are marked differences between classification systems, all of which have been influenced to some extent by the 1938 Great Soil Group system (Baldwin *et al.*, 1938). However, several soil classification systems have changed significantly since 1990.

In the current FAO/UNESCO *Soil Map of the World* legend (FAO, 1988), Solonetz soil units are those that have a natric B horizon and lack an albic E horizon, and which exhibit hydromorphic characteristics in at least part

of the horizon. A natric horizon which is a special type of argillic (now redefined and called argic) horizon has the following additional properties:

1. a columnar or prismatic structure in some part of the B horizon, or a blocky structure with tongues of an eluvial horizon in which there are uncoated silt or sand grains extending more than 25 mm into the horizon;

2. an ESP above 15 within the upper 0.4 m of the horizon, or more exchangeable magnesium (Mg) plus Na than calcium (Ca) plus exchange acidity (at pH 8.2) within the upper 0.4 m of the horizon provided that ESP is greater than 15 in some subhorizon within 2 m of the surface.

While soils with a natric horizon could be classified as Orthic/Mollic/Gleyic Solonetz and/or Solodic Planosols according to the 1974 FAO legend (FAO, 1974), sodic soils have been accommodated exclusively as Solonetz (Gleyic/Stagnic/Mollic/Gypsic/Calcic/Haplic) in the revised legend (FAO, 1988). The diagnostic natric horizon is no longer permitted in the Planosol unit (i.e., Solodic Planosols no longer exist) but the occurrence of diagnostic sodic soil properties is recognized. *Sodic properties*, defined as ESP greater than 15 or ESP + EMgP (exchangeable Mg percentage) about 50, are used to identify Sodic Solonchaks (FAO, 1988). The *sodic phase* describes soils in which ESP exceeds 6 in some horizons within 1 m of the surface. These subdivisions are fortuitous because they are essentially the same as those proposed in Chapter 1.

Solonchak soils, which are the other major component of salt-affected soils, are seldom if ever cultivated because of their very high salt contents and will not be discussed at length here. They have salic but not fluvic properties and contain no diagnostic horizons other than an A, a histic H, a cambic B, a calcic or a gypsic horizon. To exhibit *salic properties*, the electrical conductivity of the saturation extract (EC_e) within 0.3 m of the surface must exceed 15, dS m^{-1} at some time during the year, or if the pH is above 8.5, 4 dS m^{-1}. Sodic Solonchak soils must have sodic properties at least between 0.2 and 0.5 m of the surface, and lack gleyic properties and permafrost within 1 and 2 m of the surface, respectively. In the *salic phase*, the EC_e must exceed 4 dS m^{-1} in some horizon within 1 m of the surface.

Recently the World Reference Base for Soil Resources was introduced as a new basis for correlation between national systems of soil classification, eventually to supersede the revised FAO/UNESCO legend (FAO, 1988).

Under this system, the following eight units are proposed for Solonetz, in key order: Gleyic, Stagnic, Salic, Albic, Mollic, Gypsic, Calcic, and Orthic (Spaargaren, 1994).

In the latest Soil Taxonomy (US Soil Survey Staff, 1992), the definition of a natric horizon is essentially identical to that of the FAO/UNESCO legend with sodic (natric) characteristics impacting at the Great Group rather than at the highest (Order) level. Szabolcs (1991) points out that the 2 m depth requirement in the definition of natric horizons is too deep because the diagnostic horizon must be higher in the profile to determine the soil type. Moreover, Soil Taxonomy does not exploit the relationship between sodicity and the soil water regime (Seelig et al., 1990).

In Australia, Isbell (1995b) has reviewed the development and use of sodicity in classification systems. In the new Australian Soil Classification System, Isbell (1994) recognizes an order (highest categorical level) called Sodosols which has "a clear or abrupt textural B horizon which is sodic (ESP of 6 or more) in the major part of the upper 0.2 m of the B2 horizon (or the whole B2 horizon if it is less than 0.2 m thick) and the pH (1:5 H$_2$O) is 5.5 or greater." Further use of sodicity is made at the Great Group level of Sodosols: subnatric (ESP 6–14), natric (ESP 15–25), and hypernatric (ESP > 25). In seven other orders, B horizon sodicity is used at the Great Group or Subgroup level. The new Sodosols include most of the soils previously called Solodic, Solodized Solonetz, Sodic Red-Brown Earths, and Desert Loams as well as some of the Soloths in the Great Soil Group classification (Stace et al., 1968). Many of the duplex soils in the Factual Key (Northcote, 1979) also fall into this Order.

The French *Référentiel Pédologique* (Baize, 1990), which is an open-ended pedological reference base rather than a hierarchical classification system, defines *Salsodic* soils by the presence of salic or sodic horizons. A sodic horizon is defined as

a horizon at least 0.1 m thick that is present within 0.8 m of the surface and is characterized by either a massive structure or coarse polyhedric, prismatic or columnar structure (which is the case in certain solums that have evolved in more humid pedoclimates) but always with very low intra-pedal porosity in both rainy and dry seasons. This structural degradation is provoked by a more or less elevated level of exchangeable and hydrolyzable Na of the order of 15% of the CEC. This limit can be lower when the 'missing' Na is compensated by a high level of Mg on the exchange complex, especially if the level of Mg is disproportional to the level of Ca. Depending on the nature of clay minerals present, an ESP less than 15% can

also cause structural degradation. The level of soluble salts present in this horizon is nil or very low.

Two terms are used to distinguish saline soils (Salisols chloruro-sulfatés and Salisols carbonatés), three for sodic soils (Sodisols indifferenciés, Sodisols solonetziques, Sodisols solodisés), and two reflect the presence of both salic and sodic horizons (Sodisalisols and Salisodisols), depending on the sequential order of the two diagnostic horizons.

In the Canadian System of Soil Classification (Agriculture Canada Expert Committee on Soil Survey, 1987), the definition of a solonetzic B horizon is based on morphological and chemical criteria:

These horizons have prismatic or columnar primary structure that breaks down to blocky secondary structure; both structural units have hard to extremely hard consistence when dry. The ratio of exchangeable Ca to Na is 10 or less.

The Canadian System of Soil Classification is unique in using the Ca:Na ratio rather than ESP as a classification criterion.

The Orders used to classify sodic soils in different classification systems are presented in Table 2.1.

2.3 GENESIS AND CATENA

Naturally sodic soils as represented by the Solonetz are thought to develop as a result of a sequence of pedogenic processes according to the model developed by the early Russian pedologist K. K. Gedroiz [summarized by Miller and Pawluk (1994)]:

1. Salinization due to saline parent material or capillary rise of saline groundwater tables is the source of soluble Na necessary for the formation of solonetzic soils. Key to the existence of saline conditions is a net evaporative hydrological regime as found in arid or semiarid environments. Subsequent desalinization marks the onset of solonetzic soil formation.

2. Solonization starts with the leaching of salt by dilute percolating rainwater, which leads to the dispersion of clays if ESP exceeds 10–15 and total soluble salt content is 0.1–0.15% or less. The illuviation of dispersed clays leads to an abrupt textural change between A and B horizons.

3. Solodization is driven by Na-induced hydrolysis and eluviation at the top of the (increasingly) slowly permeable B horizon.

TABLE 2.1 Orders in which sodic soils occur in various classification systems

FAO Revised Legend[a]		Australian[b]	French[c]		
Solonchak	Solonetz	Sodosol	Salisol	Sodisol	Sodisalisol/Salisodisol
Haplic	Haplic	Red	Chlorosulfatés	indifferenciés	
Molic	Molic	Brown	Carbonatés	Solonetzques	
Calcic	Calcic	Yellow		Solodisés	
Gypsic	Gypsic	Grey			
Glayic	Stagnic	Black			
Gelic	Gleyic				

Soil Taxonomy[d]				Russian System[e]	
Alfisol	Aridsol	Mollisol	Vertisol	Solonetz	
Natraqualfs	Natrargids	Natralbolls	Natraquerts		Chernozemic
Natriboralfs	Natrigypsids	Natraquolls	Sodic Endoaquerts		Chernozemic Solonchak
Natrudalfs		Natrustolls	Sodic Calciusterts		Solonchakic Chernozemic
Natrustalfs		Natrixerolls	Sodic Gypsiusterts		Deep-solonchakic Chernozemic
Natrixeralfs		Natrudolls	Sodic Haplusterts		Deep-Salinized Chernozemic
			Sodic Salusterts		Chestnut Solonchak
			Sodic Haplloxererts		Solonchakic Chestnut
			Sodic Durixererts	Automorphic Solonetz	Semidesert
					Meadow Chernozemic
					Meadow Chernozemic Solonchak
					Solonchak Meadow Chernozemic
					Deep Solonchakic Meadow
					Meadow Chestnut
				Semi-hydromorphic Solonetz	Meadow Semidesert
					Semhydromorphic Cryogenic
				Hydromorphic Solonetz	

aFAO (1988)
bIsbell (1994)
cAFES (1990)
dUS Soil Survey Staff (1992)
eEgorov (1987)

The catenary relationship of solonetzic soils that fits this genetic model, from hillcrest to footslope, Solod → Solodized Solonetz → Solonetz has been observed in Canada (Miller and Pawluk, 1994), Australia (Oertel and Blackburn, 1970), and North America (Munn and Boehm, 1983). The reverse sequence, with Solod at the lowest and Solonetz at the highest position has also been observed in Canada (Cairns, 1961; Anderson, 1987). According to Szabolcs (1989), the most characteristic Solods usually develop in microdepressions.

Sodium carbonate (Na_2CO_3), which is generally the dominant salt involved in the genesis of alkaline sodic soils, can form in two ways: (a) by evaporation of water having an excess of Na bicarbonate ($NaHCO_3$), or (b) biochemically by microbial reduction of sodium sulfate (Na_2SO_4) (FAO, 1991). A major source of $NaHCO_3$ is the weathering of albite to kaolinite,

$$4NaAlSi_3O_8 + 4H_2CO_3 + 18H_2O \Rightarrow 4Na^+ + 4HCO_3^- + 8H_4SiO_4 + Al_4Si_4O_{10}(OH)_8 \qquad [2.1]$$

and montmorillonite,

$$2NaAlSi_3O_8 + 2H_2CO_3 + 4H_2O \Rightarrow 2Na^+ + 2HCO_3^- + 2H_4SiO_4 + Al_2Si_4O_{10}(OH)_2 \qquad [2.2]$$

Both these reactions generate equivalent amounts of Na^+ and HCO_3^- and substantial amounts of silicic acid (H_4SiO_4).

On the other hand, Na-induced hydrolysis would only release H_4SiO_4 if clay minerals broke down. Because water-soluble silicon (Si) levels and quantitative mineralogy are rarely determined during soil analyses, it is impossible to determine which one of these mechanisms might be the more dominant.

Although the presence of "silica (or quartz) powdering" along cracks and ped faces in the upper soil profile has been interpreted as evidence of solodization (Rozanov, 1961; Szabolcs, 1989), Hallsworth and Waring (1964) have postulated that the silt-sized siliceous material accumulating at the top of the B horizon in Solodized Solonetz consists of finely divided opaline silica phytoliths that wash through the coarser A horizon and accumulate at the top of the finer textured B horizon. Many authors (see Munn and Boehm, 1983) have reported on the degradation of smectite in the upper horizons of sodic soils, but recent work by Kohut and Dudas (1994) suggests that some of the X-ray diffraction (XRD) evidence of smectite degradation may be an artifact, due to interaction of clay with organic matter. Actually, some evidence for the enrichment of sodic soils with smectite due to mineral transformations under alkaline conditions has been presented (Cheverry, 1974; Tardy et al., 1974; Droubi et al., 1976; Szabolcs, 1989) .

2.4 DISTRIBUTION OF SODIC SOILS

2.4.1 Approach Used

In order to plot the distribution of sodic soils on a world and continent basis, the CD-ROM version of the FAO/UNESCO Soil Map of the World was used to construct the maps and estimate the areas presented in Table 2.2 using the terminology corresponding to the definitions in the 1974 and revised FAO/UNESCO Soil Map of the World legends. The total worldwide extent of salt-affected soils (Solonchak, Solonetz, and salic and sodic phases combined) and of those with sodic characteristics (Solonetz plus sodic phases) is illustrated in Figs. 2.1 and 2.2, respectively, while total areas affected are presented in Tables 2.2 and 2.3. These revised estimates, based on the full composition of each map unit, lead to a significant increase in the areal estimates for Solonchak soils and to a near doubling in the estimated extent of Solonetz (Table 2.2) as compared to previous estimates of 187 million ha for Solonchak and 135 million ha for Solonetz (FAO, 1991). The reason for these differences is that because large areas of Solonetz are rare, they were mapped as associated and included soils and, consequently, overlooked in earlier estimates.

The definition of sodic soils used in this treatise (Chapter 1) is based on land management considerations rather than the pedological approach used to produce most soil maps. Fortunately, the sodic soils classes (Nonsodic, Sodic and Very Sodic) in Chapter 1 can be approximated by combining the sodic phases of the FAO/UNESCO Soil Map of the World with the Solonetz map units and separating soils into groups with ESP < 6, 6–15 and > 15 in some horizon within 1 m of the surface for the world (Fig. 2.2) and the continents on which sodic soils occur frequently (see Figs. 2.3–2.6). It has not been possible to identify areas of spontaneous or mechanically dispersible soils (Chapter 1) because such data are seldom measured during routine soil characterizations. However, some new initiatives toward the development of international digital soil data bases, such as SOTER (FAO, 1993), and national data bases (Lytle, 1993) will make it possible, in the future, to produce interpretative soil maps that are custom-made to answer specific problems independent of the initial mapping criteria. However, differences in data collection and methods of analysis (Chapter 1) will continue to cause difficulties in integrating quantitative data to generate new maps.

TABLE 2.2 Worldwide areal extent of Solonchak and Solonetz

Great Group	Total area of Solonchak ('000 ha) Texture				Total Area of Solonetz ('000 ha) Texture			
	Coarse	Medium	Fine	Total	Coarse	Medium	Fine	Total
Orthic	1 765	150 681	28 460	180 906	14 149	130 649	16 647	161 445
Mollic	0	8 032	1 863	9 895	0	19 280	17 973	37 253
Takyric	0	95	13 172	15 267				
Gleyic	252	32 319	12 688	45 259	0	11 163	355	11 518
TOTAL	2 017	191 127	56 183	249 927	14 149	161 092	34 975	210 216

TABLE 2.3 Total area of saline and sodic phases by FAO Soil Group (1994 legend) in '000 ha

FAO Soil Group	Saline Phase (EC >4 but no Solonchak)	Sodic Phase (ESP >6 but no Solonetz)
Fluvisols	16 800	2 940
Gleysols	26 520	1 640
Regosols	845	165
Lithosols	10 805	0
Andosols	40	0
Vertisols	555	5 580
Solonchaks	--	4 095
Solonetz	9 295	--
Yermosols	19 260	2 760
Xerosols	29 080	17 560
Kastanozems	16 825	77 115
Chernozems	1 015	7 220
Phaeosems	620	1 045
Cambisols	4 995	350
Luvisols	80	13 740
Planosols	5 310	80
Histosols	2 385	0
TOTAL	144 430	134 290

Neither saline nor sodic phases have been mapped in Arenosols, Podzols, Acrisols, Ferralsols, Nitisols, Podzoluvisols, Greyzems, Rankers, and Rendzinas.

The distribution and nature of sodic soils will now be discussed on a continental basis. An attempt will be made to keep the discussion proportional to the relative importance of sodic soils on that continent. The focus of the discussion will be on environmental conditions and pedogenic processes involved in the formation of sodic soils.

The areas of sodic soils on each continent and in some countries where they form an important proportion of the soils are presented in Table 2.4. Australia accounts for the greatest area of sodic soils, followed by the former USSR.

2.4.2 Sodic Soils in Australia

Australia, which is often said to be the driest continent on Earth, has the largest areal extent of sodic soils of any continent (340 million ha) (Table 2.4) (FAO, 1988), of which an estimated 38 million ha are Solonetz (FAO, 1991). Generally, Australian landscapes show little relief, with many sodic soils being associated with late Cenozoic landscapes (Beckmann, 1983). The distribution of sodic soils (Fig. 2.3) is approximately related to the average annual rainfall according to Northcote (1988); however, this is not a universally accepted view (Isbell *et al.*, 1983).

Soils that are sodic throughout the profile are generally found where average annual rainfall is less than 500 mm; those with only sodic subsoils generally occur where annual rainfall is < 900 mm (Chartres, 1993). Generally, sodicity increases with depth, often abruptly, and many soils with sodic subsoils are not sodic in their A horizons (Isbell, 1995b). However, many sandy A horizons can have ESP greater than 6 because their cation exchange capacity (CEC) is very low.

The most comprehensive estimate of the extent of sodic soils in Australia was published by Northcote and Skene (1972). They defined sodicity on the basis of ESP values anywhere within the upper 1 m depth; nonsodic soils were those with ESP < 6, sodic with ESP 6–15, and strongly sodic with ESP > 15. They presented the distribution of six mapping units (saline soils, alkaline strongly sodic to sodic clay soils, alkaline strongly sodic to sodic coarse- to medium-textured soils, alkaline strongly sodic to sodic duplex soils, nonalkaline sodic to strongly sodic neutral duplex soils, nonalkaline sodic to strongly sodic acid duplex soils) using the *Atlas of Australian Soils* (1:2 000 000) as a base (Northcote *et al.*, 1960–1968). Although Solonchak soils are not widespread, many of the areas with a high proportion (> 50%) of Solonetz (Fig. 2.3) have a salic phase (Fig. 2.1).

Although no improved version of Northcote and Skene's (1972) map, which overcomes the constraints to map production they encountered, is available, the knowledge on the distribution of sodic soils in each state in Australia has been updated (McKenzie *et al.*, 1993; Ford *et al.*, 1993; Naidu *et al.*, 1993b; Doyle and Habraken, 1993; Shaw et al., 1994; Cochrane *et al.*, 1994). Profile descriptions presented by Northcote and Skene (1972) show that sodic soils in Australia occur on a wide range of parent materials. In Queensland, Shaw *et al.* (1994) were unable to find a relationship between sodic soils and parent material composition beyond a reduced incidence of strongly sodic soils on calcic lithologies. In Tasmania, sodic soils which occur on lowland plains, river terraces, and valley floors, have formed mainly from Triassic and Permian mudstones and sandstones, Tertiary clays, and Quaternary deposits, but are also found on granite and basalt (Doyle and Habraken, 1993). In New South Wales, sodic soils are concentrated in cracking clays west of the Great Dividing Range (McKenzie *et al.*, 1993). Large areas of the Murray-Darling river basin, which covers much of southern Queensland, New South Wales, Victoria, and South Australia, are sodic. Little published information is available for Western Australia and the Northern Territory, but an estimated 3.5 million ha in the southwest of Western Australia, or 34% of the agricultural land, are

TABLE 2.4 World distribution of sodic soils [Massoud, *Proc. Int. Conf. on Management of Saline Water for Irrigation*, Texas Technical University, Lubbock, (1977)]

Continent	Country	Area of sodic soils ('000 ha)
North America	Canada	6 974
	United	2 590
South America	Argentina	53 139
	Bolivia	716
	Brazil	362
	Chile	3 642
Africa	Algeria	129
	Angola	86
	Botswana	670
	Cameroon	671
	Chad	5 950
	Ethiopia	425
	Ghana	118
	Kenya	448
	Liberia	44
	Madagascar	1 287
	Namibia	1 751
	Niger	1 389
	Nigeria	5 837
	Somalia	4 033
	Sudan	2 736
	Tanzania	583
	Zambia	863
	Zimbabwe	26
South Asia	Bangladesh	538
	India	574
	Iran	686
North and Central	China	437
	USSR	119 628
Australasia	Australia	339 971

sodic (Tennant *et al.*, 1992). Most of the texturally contrasted soils of southwest Western Australia have sodic subsoils, small areas have sodic topsoils and, in the wheatbelt, hardsetting (Mullins *et al.*, 1990) is a widespread surface soil problem in which sodicity can play a part (Cochrane *et al.*, 1994).

Sodicity has often been implicated in crusting and hardsetting of surface soils. However, this behavior is also observed in Nonsodic Nonsaline (Sod$_L$ Sal$_L$) soils that are mechanically dispersive (Rengasamy *et al.*, 1984b; Cochrane *et al.*, 1994; Chapter 1). The A horizons of many texturally contrasted soils with strongly sodic B horizons

often have low ESP, and Isbell (1995b) suggests that where relatively thin A horizons overlie sodic B horizons, long-term cultivation may have led to mixing of the horizons. In northern Victoria, Queensland and New South Wales, land-leveling for irrigation may expose sodic subsoils that are more prone to crusting than the original A horizons (McKenzie *et al.*, 1993; Isbell, 1995a).

Rengasamy and Olsson (1991) suggest that two types of sodic soils occur in Australia: those in which the sodicity is a natural phenomenon related to the nature of the parent material and those in which sodification is secondary, arising from human activity. Irrigation without proper drainage can increase the SAR of the soil solution, exacerbate sodicity, and even render soils sodic (Rengasamy and Olsson, 1991, 1993). Large areas of irrigated soils in the Murray-Darling river basin are sodic. Therefore, management of irrigation is closely related with management of sodicity.

Tree clearing and other land management practices that alter the water balance in soils can lead to waterlogging and sodicity problems. For example, Fitzpatrick *et al.* (1994) demonstrated that changes in land management could induce sodicity-related environmental degradation in a toposequence of red, yellow, and grey duplex soils (Palexeralfs and Natraqualfs) in South Australia. Changes in catchment hydrology resulting from forest clearing in upslope positions created salt seepage zones, which resulted in sheet, rill, tunnel, and gully erosion in downslope positions. Fitzpatrick *et al.* (1994) highlight the sequential train of events resulting in salinization and sodification, which arise from the disturbance of natural ecosystems.

In Australia, the spatial distribution of sodic soils often coincides with that of alkaline duplex (texturally contrasted) soils (Northcote and Skene, 1972; Chittleborough, 1992; Chartres, 1993). However, is the duplex nature due to sodicity or *vice versa*? This proposition will now be discussed.

Oertel (1961) concluded that the clay profiles of six sodic Red-Brown Earths [according to the classification of Stace *et al.* (1968)] from South Australia were not the result of eluviation-illuviation processes, but rather that most of the clay in the B horizons had formed *in situ*. Similarly, Sleeman (1964) and Oertel and Blackburn (1970) demonstrated that the abrupt textural contrast in other sodic soils was due to lithologic discontinuities between the sedimentary layers of the parent material. Thus the textural contrast was not due to sodic soil genetic processes.

In a central Queensland catena, Gunn (1967) described Red and Yellow Earths [according to the classification of Stace *et al.* (1968)] on uplands, sodic and related soils with

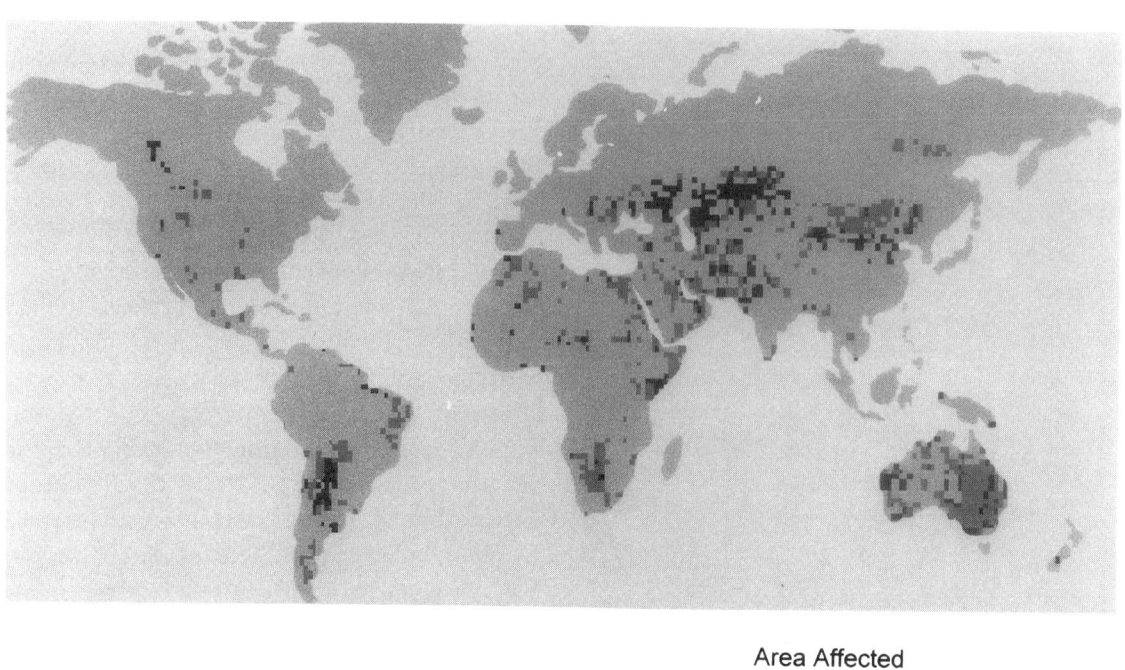

Figure 2.1 World distribution of saline and sodic soils

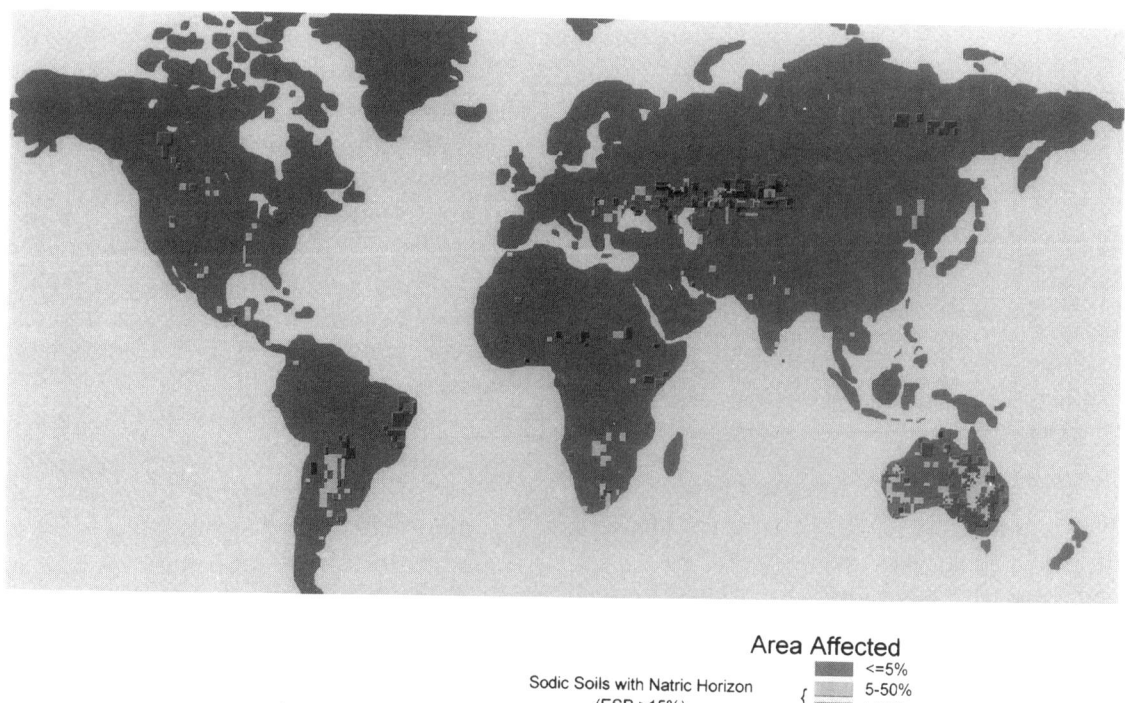

Figure 2.2 World distribution of sodic soils

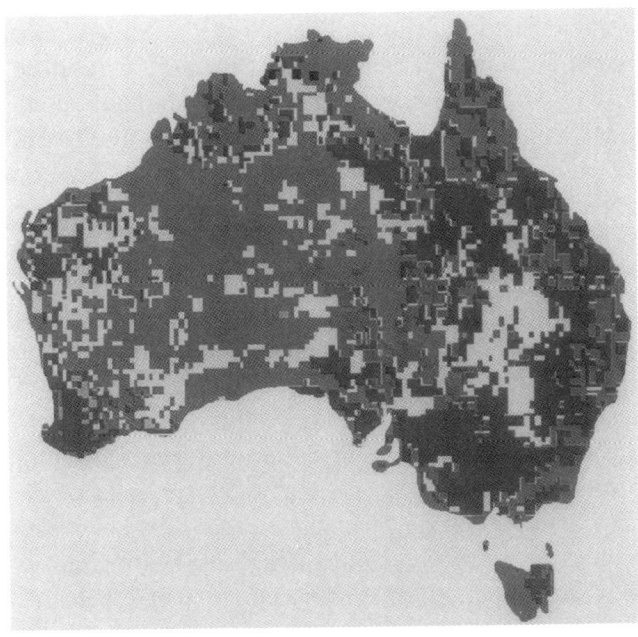

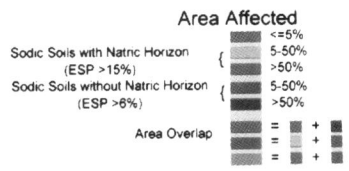

Figure 2.3 Distribution of sodic soils in Australia

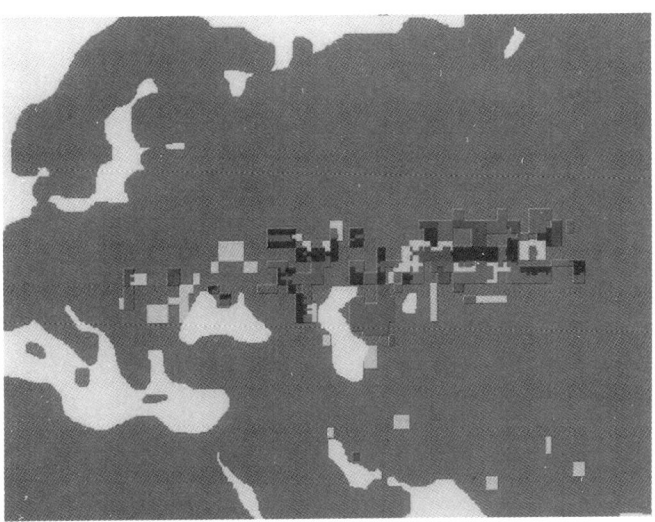

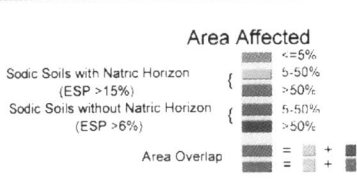

Figure 2.5 Distribution of sodic soils in Eurasia

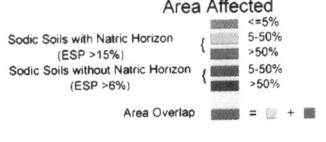

Figure 2.4 Distribution of sodic soils in South America

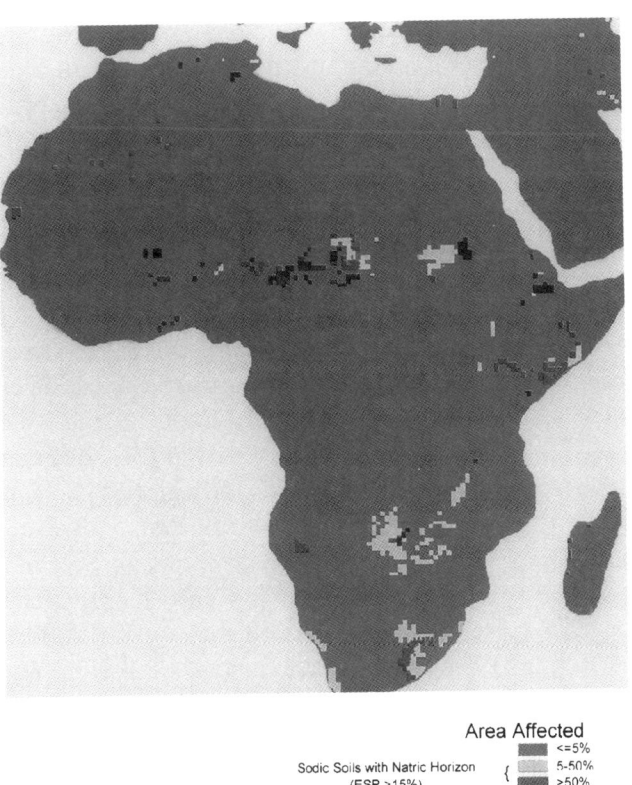

Figure 2.6 Distribution of sodic soils in Africa

abrupt textural contrast on intermediate slopes in gently undulating land or below scarps on moderately sloping land (< 10%), and cracking clay soils, often sodic, with gilgai microrelief in the lower positions. Surface horizons of the duplex sodic soils ranged from sand to sandy clay loam in texture, with a thickness of a few centimeters to 0.6 m. Bleached A horizons were sometimes present. In B horizons with high clay content, Mg dominated the exchange complex with ESP between 15 and 30. The soluble salt content was low throughout the profiles. Gunn (1967) postulated that the textural contrast in these sodic soils developed in different cases as a result of soil genesis from weathering *in situ* or as a consequence of polycyclic deposition. This catena is very similar to that described by Bocquier (1964, 1968, 1971) in Chad in Section 2.4.6.

Hallsworth and Waring (1964) believed that, although the textural contrast in sodic soils developed on Jurassic sandstone alluvium in New South Wales formed during soil genesis, these soils did not reflect the classical Solodized Solonetz evolution and, in fact, were the normal soils found on coarse-textured parent materials in semiarid to subhumid, temperate to subtropical climates. Isbell *et al.* (1983) point to the widespread occurrence of sodic soils in South Australia which lack pronounced textural B horizons and are permeable on the one hand, and to soils with strong textural contrast that do not have impermeable saline and sodic B horizons on the other. These authors conclude that "[it is] unlikely that many of the present diverse and widespread varieties of sodic soils ever went through the complete classical sequence."

2.4.3 Sodic Soils in North America

In North America, most sodic soils occur in the Great Plains of western Canada and the northern United States under cold, semiarid to subhumid climates comprising about 11 million ha of Solonetz (FAO, 1991). Because many are developed on tills and other glacial deposits derived from saline shales (Munn and Boehm, 1983; Heck and Mermut, 1992a,b), the classical theory of Solonetz formation still holds much sway in Canada (Agriculture Canada Expert Committee on Soil Survey, 1987; Heck and Mermut, 1992a,b; Miller and Pawluk, 1994). However, Munn and Boehm (1983) showed that a classical catenary sequence Natrargid → Haplargid in Montana can be interpreted in two ways. Interpretation along the lines of the classical theory necessitates the presence of a shallow, saline water table, implying that the Natrargid soils formed under earlier, wetter climatic conditions and that the sequence is presently in disequilibrium with the climate and

undergoing solodization. The alternative hypothesis suggests ongoing soil genesis under the existing climate involving lateral subsurface movement of salty, silica-rich water into the subsoils of the depressions. The lateral water movement is driven by matric suction and osmotic potential gradients from the moist subsoils of the Haplargid into the drier subsoil of the incipient Natrargid. As the proportion of Na in solution increases, Na replaces Ca and Mg on the exchange complex and dispersion of the subsoil proceeds. The dispersed zone expands, and *in situ* smectitic clay formation is enhanced due to high concentrations of silica and cations.

In the United States in general, a greater variety of environmental conditions has been shown to lead to sodic soil formation. For instance, some sodic soils have been shown to have formed on loess rich in Na-bearing minerals (Wilding *et al.*, 1963). The occurrence of ephemeral, perched water tables has been implicated in the current genesis of sodic soils on loess in Nebraska (Lewis and Drew, 1973). Differences in water movement with landscape position, and topography and permeability of the underlying parent material have been shown to be the driving forces in sodic soil genesis on glacial deposits in all cases (Wilding *et al.*, 1963; Lewis and Drew, 1973; Munn and Boehm, 1983; Seelig *et al.*, 1990; Richardson *et al.*, 1992). In addition, Whittig and Janitsky (1963, 1964) have demonstrated the role of sulfate (SO_4)-reducing microbes in biologically driving sodic soil genesis on the edge of wetland environments.

In the southeastern United States, large areas of mechanically dispersible soils, which would fall in the Nonsodic Nonsaline (Sod_L Sal_L) category (Chapter 1), have been identified (Miller and Baharuddin, 1986; Sumner and Miller, 1992). Although these soils would not fall in any classical sodic soil category, they exhibit many properties in common with soils containing higher levels of Na, such as crusting, hardsetting, and erodibility. This is particularly true in the topsoil where structural degradation due to loss of organic matter allows clay to become dispersed even at very low ESP (SAR) levels because the EC of the soil solution is exceedingly low (Miller, 1987; Miller and Radcliffe, 1992; Chapter 5). These soils are sensitive to even the slightest increases in Na levels (Chiang *et al.*, 1987; Miller and Scifres, 1988).

2.4.4 Sodic Soils in South America

South America has an estimated 34 million ha of Solonetz (FAO, 1991) and 25 million ha of sodic phases of other soil groups (FAO, 1988), mostly in the Argentinian and Paraguayan pampas and in northeastern Brazil (Fig. 2.4).

Argentina

In Argentina, sodic soils are dominant in the Pampa Deprimida and the Bajos Submeridionales regions with salt-affected soils also occurring in the Oeste Bonaerense. All three regions are on the periphery of the Pampean prairie and are used mainly for cattle grazing on native grasslands (Soriano, 1991).

In the Pampa Deprimida, an area of about 80 000 km² east of Buenos Aires Province, most soils have developed on loesslike sediments derived from the Andes Mountains under a subtropical to temperate, subhumid (west) to humid (coast) climate with unevenly distributed rainfall exhibiting no seasonality but cyclic variations. Droughts are common in the summer when evapotranspiration is high (Alconada *et al.*, 1993). The landform, which is extremely flat and at low elevation, has resulted in a lack of equilibrium with the present wet climate. As a result, a significant portion of the region is covered by permanent and temporary ponds generally connected by groundwater, resulting in frequent deep waterlogging of low-lying soils in the winter and spring for long durations. Floods that last several months and inundate millions of hectares with 1 m or more of water occur several times a century (Paruelo and Sala, 1990). Consequently, most soils have an aquic moisture regime. Salts move upward by capillary rise from the shallow fluctuating groundwater table driven by high evapotranspiration in the summer (Lavado and Taboada, 1988). Because much of the groundwater has a high pH and contains Na_2CO_3, sodic soils, mostly Natraqualfs, are abundant (Alconada *et al.*, 1993). Neither rainfall nor mineral weathering contribute significantly to soil salinity or sodicity (Lavado, 1983; Lavado *et al.*, 1983).

Most soils (80%) in the Pampa Deprimida have an aquic soil moisture regime with Natraquolls (28 000 km²) and Natraqualfs (11 000 km²) being dominant over continuous areas of hundreds of thousands of hectares. Natralbolls and Argialbolls are also present (INTA, 1990a). Large areas of soils with a mollic A and a natric B horizon occur in this region, but not under an aquic soil moisture regime. Because soil taxonomy does not cater adequately to such soils, Godagnone *et al.* (1991) have proposed a new Great Soil Group, Natrudolls.

The Bajos Submeridionales region occupies about 40 000 km² in northern Santa Fé and southern Chaco provinces. The northern part is characterized by subnormal to concave relief, with slope gradients between 0.1 and 0.2 %. Water flow to the Paraná River is impeded by the physiography which promotes periodic flooding (Morras and Perman, 1977). Parent materials vary from loesslike (west) to alluvial materials from the Brazilian shield (east) with sediments from the Andes in between. The monsoonal climate is subtropical and subhumid. The predominant soils are associations and complexes of Natraquolls and Natraqualfs, including Natrudalfs, Natralbolls, etc. Gley horizons occur frequently (INTA, 1990b). As in the Pampa Deprimida region, sodification arises from the discharge of very saline shallow groundwaters (Morras and Candioti, 1982).

The Oeste Bonaerense (several million hectares) located in the northwest of the Buenos Aires and parts of Santa Fé, Cordoba, and La Pampa provinces are characterized by a series of longitudinal dunes which impede the general drainage of the area. The soils, which have developed in the interdunal depressions, have a higher salt content than those described above. The soils have formed from loesslike material but coarser than that of the other two regions under a subtropical/temperate and semiarid climate with floods being more episodic (INTA, 1990a). Salt is supplied by deep thaptonatric horizons (Altamore *et al.*, 1983) and eolian sources in dry summers (Lavado and Reinaudi, 1988). Natraquolls, Natraqualfs, and some Salorthids occur in the depressions with Natrustalfs and Natrustolls on midslope positions (INTA, 1990a).

Brazil

In northeastern Brazil, sodic soils are widespread in Bahía, Ceará, and Pernambuco States (Fig. 2.4). Throughout northeastern Brazil, where the climate is semiarid and warm with a rainfall of 600–1100 mm, sodic soils have developed on Precambrian migmatites and gneisses, intruded by acid igneous bodies, on the Cambro-Ordovician Formaçao Estancia or in Holocene sandy clays in landforms with slightly undulating relief (EMBRAPA, 1977). A characteristic xerophytic vegetation formation of open scrubland to open woodland called *caatinga* generally coincides with the distribution of the semiarid climate and sodic soils.

Sodic soils (Solodic Planosols and others) also occur in the state of Matto Grosso under a seasonally contrasted, humid tropical climate on sandy to sandy clay Holocene alluvial deposits of the Paraguay River at a low elevation on subdued microrelief (2–3 m) subject to annual flooding (EMBRAPA, 1971).

Paraguay

Most of Paraguay is covered with sodic soils under a semiarid tropical (west) to subhumid subtropical climate (east) (Fig. 2.4). Solonetz soils derived from poorly consolidated clastic Cenozoic sediments, many with salic properties, occur over a large portion of the Chaco basin in western Paraguay (FAO, 1964). Low slope gradients and dense subsoil clay pans are responsible for seasonally perched water tables and waterlogging. These soils appear to be similar to those of the Oeste Bonaerense in northern Argentina. In eastern Paraguay, sodic soils developed from Permian shales and Triassic red bed sandstones only occur in topographic lows.

2.4.5 Sodic Soils in Eurasia

In Europe, there are 22 million ha of sodic soils (FAO, 1988), including 7 million ha of Solonetz (FAO, 1991) whereas Asia has an estimated 120 million ha of sodic soils (Abrol *et al.*, 1988), 30 million ha of which are Solonetz, mostly in central Asia (Figs. 2.2 and 2.5) (FAO, 1991). Eurasia is the landmass on which most anthropogenic salinization and sodification have occurred and where the reclamation of saline and sodic soils has been the focus of most research (Glazovskaya, 1984; Szabolcs, 1989). Because of the influence of Russian and central European pedologists, many of the models developed on soils in Eurasia are still used worldwide. Sodic soils occur over a range of climates in Eurasia.

Europe
Much research has been conducted on sodic soils (*szik*) in the Carpathian basin (mean annual temperature 10–11ºC, mean annual precipitation 520–580 mm) in Hungary, Romania (Oprea *et al.*, 1971; Obrejanu and Sandu, 1971), Serbia (Adam *et al.*, 1988), and Slovakia (Hrasko, 1971). Most of the sodic soils formed in the Quaternary period in Hungary are found either in the Trans-Tisza region or in the Danube-Tisza interfluve at elevations above 100 m with microrelief of about 0.5 m (Tóth and Rajkai, 1994). In the Danube-Tisza rivers interfluve, mostly saline (Solonchak) soils have formed on calcareous sands of the Danube, while in the Trans-Tisza region, the dominant Solonetz soils have formed on sediments intermingled with loesslike material of silt or clay texture. In both areas, the dominant salt is Na_2CO_3, with accessory amounts of Na_2SO_4 and sodium chloride (NaCl).

Despite previous disagreements concerning the factors involved in the formation of sodic soils in the Carpathian basin (Jenny, 1941), the present consensus is that hydrological conditions, namely, groundwater level and composition, have played the dominant role in the genesis of these soils. The occurrence of salt-affected soils closely matches groundwater discharge areas, and capillary rise from saline groundwater seems to be the means by which salt accumulates. Three major sources of salts, (a) released by rock weathering (feldspars), especially volcanic

tuffs, in the surrounding mountains and transported into the groundwater of the lowlands, (b) Pliocene evaporites from which salt seeps upward (Erdélyi, 1979), and (c) salt in surface waters, have been identified.

Catenary relationships in the sodic soils of Hungary were recognized early, giving rise to popular and scientific terms distinguishing soil categories according to topographic position (Treitz, 1924). de'Sigmond (1927a) and Magyar (1928) distinguished the dry (seldom waterlogged) and wet types of salt-affected soils while Kreybig and Endrédy (1935) demonstrated that salt-affected soils tend to occur at the same elevation at the same latitude. The sodic soils of the Danube-Tisza interfluve have the following catenary sequence downslope: Calcareous Meadow Soil → Sodic Meadow Solonetz → Sodic Solonchak → Solonetz → Peaty Meadow Soil (Rajkai and Molnár, 1981), while in the Trans-Tisza Region, the sequence is Meadow Chernozem (Haplustoll) → deep Meadow Solonetz (Natrustoll) → shallow Meadow Solonetz (Natrustalf) → crusty Meadow Solonetz (Natraquept) → Meadow Soil with saline subsoil (Haplaquoll) (Magyar, 1928; Bodrogközy, 1965; Tóth *et al.*, 1991; Tóth and Rajkai, 1994). The terms *deep*, *shallow*, and *crusty* describe the thickness of the eluvial A horizon, as in the Russian terminology. Quantitative prediction of soil properties can be made on the basis of vegetation (Tóth and Rajkai, 1994) and catenary relationships (Kertész and Tóth, 1994).

The genesis of Ukranian sodic soils is similar to that of the Carpathian basin. Around the Black Sea (mean annual temperature 7–8 ºC, mean annual precipitation 500–550 mm), Solonetzic Soils, Solonetz, and, in closed depressions, Gleyic Solods occur while at Stavropol, Vertic Solonetzic Chernozems have formed on Tertiary marine clays (Hitrov, 1988).

Most of the Transcaucasian plain in Georgia and Azerbaijan (mean annual temperature 14ºC, mean annual precipitation 200–250 mm) is covered by sodic soils (Solonetz) with salinity increasing and alkalinity decreasing from northwest to southeast as precipitation decreases (Ostrikova, 1991).

Northern and Central Asia
Sodic phases of Kastanozems are commonly associated with Solonetz throughout the desert steppe region of central Asia under a climate characterized by low rainfall with cold winters and warm summers (FAO, 1978). An extensive belt of sodic soils occurs west of the Caspian Sea covering Kazakhstan, Turkmenia, Uzbekistan, Tadzhikistan, and Kirghizia (Figs. 2.2 and 2.5). Rozanov (1961) noted a genetic connection between Solonetz and

weathering granite in the Tien Shan Mountains, which consist mostly of Precambrian to Devonian crystalline rocks, with a widespread occurrence of granitoids (FAO, 1978).

The Caspian lowland at the northern end of the Caspian Sea (mean annual temperature 6 °C, mean annual precipitation 150–250 mm) consists of Quaternary moraine deposits which together with rivers produce the salts. In Kazakhstan, sodic soils have formed on saline and gypsic Tertiary deposits with very little microrelief (0.3–0.5 m) (Eskov, 1991). Solonchak-Solonetz occur on microhighs with Solonetz on slopes and leached Meadow Chestnut Soils in depressions (Dementyeva and Motuzov, 1988).

South of the Tien Shan Mountains, in the basin comprising the Xinjiang-Gansu-Qinghai provinces in China (mean annual temperature 7–9 °C, mean annual precipitation <350 mm), which is hydrologically closed from the surrounding area, most of the soils are saline with SO_4 and Cl anions dominant but some Solonetz soils, including Mg-Solonetz, occur. On the Huang and Huai river plains in eastern China (mean annual temperature 12.5 °C, mean annual precipitation 580 mm), saline soils have developed on the alluvial sediments salinized by ground- and seawater (SO_4 and Cl) (Inanaga, 1991). Sodification can be observed (Renpei, 1988) in depressions where HCO_3^- accumulates with the vegetation pattern reflecting pH changes (Tóth *et al.*, 1994). In northeastern Manchuria (mean annual temperature 4 °C, mean annual precipitation 440 mm), sodic soils are found on the low river and lake terraces of the Amur River and its northeast-flowing tributaries. The dominant salts in the subsurface waters and soils are NaCl and Na_2SO_4.

The northernmost sodic soils in Eurasia are found in the Russian Republic of Yakutia (mean annual temperature -8 °C, mean annual precipitation 130–250 mm) on the Lena River plain in central Yakutskaya derived from saline Devonian, Cambrian, and Jurassic materials. The arid climate coupled with permafrost limits the migration of salts, which accumulate in depressions, giving rise to Solonchak and Solonetz soils containing $NaHCO_3$. At more elevated positions, solodized soils occur. The sodic soils of the west Siberian forest-meadow-steppe region (mean annual temperature -1.6 °C, mean annual precipitation 324–354 mm) consist of Peaty Solonchak soils on the flat plains, Na_2CO_3-rich Solonchak soils in depressions at slightly higher elevations, and Solonetz soils on the high ground (Bazilevich, 1965). The Solonetz soils encircle the ridges covered by leached Chernozems. Dominant anions in the sequence of soils from north to south are CO_3^{2-}, HCO_3^-, SO_4^{2-}-CO_3^{2-}, Cl^--SO_4^{2-}, and SO_4^{2-}-Cl^-.

Southern Asia

In the Middle East and southern Asia, under arid, semiarid, and subhumid climates, saline and sodic soils are widespread in poorly drained environments of the alluvial plains of the Euphrates, Indus, and Ganges rivers. Sodic soils have been described in Syria (Al-Saleh, 1991), Iraq (Kadry, 1969), Iran, Pakistan (Sandhu and Aslam, 1980), India (Kanwar, 1969; Bhargava, 1977, 1979; Bhumbla, 1977; Murthy *et al.*, 1982), and Bangladesh (Dent, 1992). The distribution of saline and/or sodic areas in the Indo-Gangetic alluvial plains is determined mostly by groundwater movement and climatic zonation (Bhumbla, 1977; Bhargava, 1979; Murthy *et al.*, 1982). Throughout India and the other areas, the increase in irrigated agriculture has caused salinization and/or sodification arising from (a) poor land preparation, (b) seepage from unlined canals, (c) over-irrigation and the subsequent percolation of water, (d) inefficient water management, (e) inappropriate cropping systems, (f) nonconjunctive use of surface and groundwater, and (g) impeded drainage and lack of outlets (Yadav, 1993). In Pakistan, tube well irrigation with sodic groundwater has contributed to an increase in sodic soils (Sandhu and Aslam, 1980).

2.4.6 Sodic Soils in Africa

In Africa, sodic soils, which cover 27 million ha (0.9% of the total land), occur under warm, semiarid to subhumid climates (FAO, 1991), including 13 million ha of Solonetz soils found mainly in Chad, Nigeria, Somalia, Sudan, coastal Tunisia, and the Kalahari basin. The Solodic Planosols mainly occur in Chad, Burkina Faso, Tanzania, and southern Africa, with minor areas in Senegal and Niger (Fig. 2.6).

West Africa

The geochemical evolution of surface and groundwaters and their effect on pedogenesis and mineralogy in the polders of Lake Chad can give rise to alkaline sodic soils (Cheverry, 1974). In the interdunal depressions bordering the lake, surface waters are concentrated by evaporation. While the surface waters are initially dilute but relatively high in Ca^{2+} and HCO_3^-, they become enriched in HCO_3^-, CO_3^{2-}, and Na as they are concentrated leading to precipitation of $CaSO_4$, Na_2SO_4, $CaCO_3$, and Na_2CO_3 (calcite, gaylussite, and trona) on the surface. Shallow groundwaters are relatively rich in SO_4^{2-} and Ca^{2+}. Bocquier (1964, 1968, 1971) described some typical catenas, incorporating sodic soils, developed on granitic rocks in semiarid West Africa between 14 and 10° N latitude. Essentially eluvial-illuvial, the downslope catenary sequence is: ferruginous leached soils → hydromorphic soils → Planosols → Solonetz → Vertisols. These soil types cover various proportions of the slopes according to latitude. Nevertheless, the actual soil pattern may differ widely from this classical sequence due to local conditions of drainage and parent materials. The Na input from weathering relative to that from other sources is unknown. Sodic soils also occur in the Sahelian Oudalan Province of northern Burkina Faso, where weathering *in situ* governs their development (ORSTOM, 1968; Boulet, 1970, 1978; Leprun, 1977; BGR, 1980; BUNASOL, 1991). The excess Na responsible for the sodification is derived from the crystalline bedrock, which has inclusions of calc-alkaline granite that vary in size and are irregularly distributed in the bedrock causing the distribution of sodic soils to be somewhat patchy and irregular (Ducellier, 1963; Hottin and Ouedraogo, 1975, BGR, 1980). As shown in Table 2.5, calc-alkaline granites have a rather high ratio of Na to Ca + Mg. As a result, despite the large areas of sodic soils shown on maps, they occur as a complex of individuals on a microscale, sometimes with similar morphology but with little or no Na influence. The remaining soils belong to the *sols bruns subarides* or *sols ferrugineux tropicaux* groups according to the French CPCS system (1967). Lateral water movement can spread Na from its original source to soils in lower slope positions where it accumulates by evaporation.

Southern Africa

Sodic soils are found throughout southern Africa occurring widely in the Republic of South Africa (van der Merwe, 1956; Beater, 1957, 1959,1962; van der Eyk *et al.* 1969; de Villiers, 1962; MacVicar *et al.* 1977; Schloms *et al.*, 1983), Swaziland (Murdoch and Andriesse, 1964; Murdoch, 1964), Lesotho, Botswana, and Zimbabwe (Blair Rains and McKay, 1968; Purves and Blyth, 1969; Thompson and Purves, 1978; Stocking, 1979; Verbeek, 1989) but are confined mainly to numerous scattered small localized areas, whereas in northern Namibia and southern Angola, there are much larger tracts covered by these soils (Cass,

TABLE 2.5 Ratios of Na and Na + K to Ca + Mg in various rocks of the Oudalan Province in Burkina Faso [Ducellier, *Mém. BRGM*, 10 Paris, (1963)]

Ratio	Alkaline granite	Calc-alkaline granite	Amphibole plagioclase granite	Amphibole schist
Na/(Ca + Mg)	6.7–10.3	1.5–4.9	0.21–0.76	0.13–0.56
Na+K/(Ca + Mg)	11.5–14.9	2.40–7.66	0.28–1.15	0.15–0.83

1980). The regions where they occur seem to have a mean annual rainfall of less than 800 mm. However, the main factor responsible for their formation is the occurrence of parent materials with high amounts of Na-releasing weatherable minerals (Beater, 1957, 1959, 1962; van der Eyk *et al.* 1969; Thompson and Purves, 1978). Their most frequent occurrence, therefore, does not necessarily coincide with the regions of lowest mean annual rainfall. For example, sodic soils occur more extensively in the Zambezi Valley (Zimbabwe) and near Estcourt (South Africa), where the mean annual rainfall is appreciably higher than in the drier lowveld areas. In these higher rainfall areas, the sodic soils are formed even on uplands in undulating terrain and are developed from the Karroo formation members that are rich in Na-releasing weatherable minerals. In many parts of southern Africa, sodic soils are commonly found on granitic parent materials, but usually occur at the lower end of a catena where the surface horizons (A and E) are predominantly sandy. A very thin albic horizon (very pale, bleached) usually occurs between the A horizon and the underlying impermeable B horizon. The conditions leading to the formation of the abrupt, sharp-line nature of the upper boundary of the B horizon is brought about by lateral movement of water across the surface of the B horizons (Purves and Blyth, 1969). When there are low amounts of iron oxides and oxyhydroxides present (<1%), the clay fraction of the sandy granite-derived soils is easily dispersed at ESP values as low as 3 when these soils experience aquic conditions (Thompson and Purves, 1978). The albic horizon is diagnostic of a marked degree of lateral water movement and redoximorphic conditions. In effect, surplus soil water moving within the solum across the surface of the sodic horizons removes clay to give rise to what amounts to a subsurface erosion pavement (Thompson and Purves, 1978).

Thompson and Purves (1978) have noticed that there tends to be a correlation between the depth of sandy surface horizons and the degree of sodicity in the underlying impermeable horizons. Soils with an appreciable depth of surface sand tend to be weakly sodic, with ESP less than 15 in the sodic horizon, while those with the shallower surface sands tend to have strongly sodic subsoils. The pH values of the upper part of the weakly sodic horizons are not necessarily high if the exchange complex is not fully saturated; occasionally pH values as low as 6.0 are encountered. However, all the strongly sodic horizons are markedly alkaline in reaction, with pH values greater than 7.5.

Sodic horizons also occur in soils with appreciable clay contents, mainly on colluvia and alluvia (van der Eyk *et al.*, 1969). Where the clay content of the surface horizon is sufficient to give rise to sandy clay loams or heavier textures, the upper boundary of the underlying sodic horizon is almost invariably much less apparent, and the change may even be a gradual one.

Saline sodic soils are found mainly in areas of low rainfall in pans and in some of the more extensive diffuse drainage depressions, such as those that occur in part of the lower Sabi Valley (Zimbabwe) and Karroo (South Africa) and on eolian and lacustrine deposits around the Makgadikgadi salt pans (Blair Rains and McKay, 1968). Sodic soil phases are found in the clayey central depressions of sandy alluvial islands in the Okavango delta and in large duricrusted pans (e.g., Thale, Ngwako pans) in the sandveld (Verbeek, 1989). At its Quaternary maximum, Lake Palaeo-Makgadikgadi was second in size to Lake Chad (Thomas and Shaw, 1991) with obvious parallels in the genesis of saline and sodic soils in the two areas.

2.5 CONCLUSIONS

The preceding discussion suggests that sodic soils can occur over a wide range of climates, but generally under seasonally contrasted conditions. Among the environmental conditions that promote the genesis of sodic soils are the presence of shallow saline groundwater, textural discontinuities during the deposition of eolian, glacial, alluvial, or colluvial sediments, the occurrence of perched water tables within 1 m of the surface, low slope gradients, and endorheic or impeded drainage. Salts dissolved in groundwater as well as the weathering of Na-containing minerals, such as albite, in granitic or sedimentary deposits are the sources of Na. The classical theory of Solonetz formation is clearly limited in its applicability on a worldwide basis. More generally, it appears that restricted water movement and soil moisture dynamics that fluctuate between aquic and ustic or aridic, within 1 m of the surface, in a landscape where a source of Na exists, promote the formation of natric horizons or sodic properties. Soils can become sodic as a result of anthropogenic processes that change their water balance, such as irrigation and drainage of saline soils, and land clearing.

Many soils that would not fit classical sodic soil specifications exhibit sodic behavior and cover a large but undetermined area of the world. Because of their agricultural importance, more attention needs to be devoted to these soils in the future.

3

PROCESSES INVOLVED IN SODIC BEHAVIOR

P. Rengasamy and M.E. Sumner

3.1 INTRODUCTION

Salt-affected soils comprise saline and sodic varieties and various combinations thereof, all of which differ greatly in behavior. While sodic soils may also be saline in certain circumstances, the discussion to follow will be confined to a discussion of the processes involved in sodic behavior while fully recognizing that salt content can play a vital role in determining soil properties. Both saline and sodic soils are among the least productive soil types, which even in undisturbed ecosystems can damage the environment due to the release of salts and suspended colloids. Where such soils have been brought into cultivation, environmental contamination is often exacerbated (Chapter 9). The mechanisms responsible for this behavior are entirely different in the two soil categories. sodic soils have extremely poor physical condition, leading to an inadequate balance between water and air regimes within the soil. This imbalance stems from restricted water acceptance and transmission properties, which result in the soil being too wet or dry for much of the time and leads to poor root development and crop growth (Shainberg and Letey, 1984; Quirk, 1986). In addition, sodic soils are difficult to cultivate and have poor load-bearing characteristics. The lack of structural stability in these soils promotes seal and crust formation at the soil surface, resulting in soil erosion, which is a significant contributor to the pollution of water bodies (Shainberg and Letey, 1984; Fitzpatrick *et al.*, 1994; Chapter 9). On the other hand, saline soils usually have good physical properties suitable for crop growth provided that sodium (Na) does not reach very high levels (Shainberg and Shalhevet, 1984). However, the high salt content creates an osmotic potential effect which can reduce water availability to crops and, consequently, can take a heavy toll on yield. In both saline and sodic soils, toxicity due to specific ions may also limit crop produc-

tion. For example, high Na concentrations in sodic soils can be toxic while elevated salt concentrations can dissolve potentially toxic heavy metals from primary and secondary minerals (Chapter 9).

Two processes, swelling and dispersion, are responsible for the physical behavior of sodic soils during wetting. On the other hand, in saline soils swelling is minimal and clay dispersion is absent due to the high electrolyte concentrations present. Both swelling and dispersive behavior are governed by the balance between attractive and repulsive forces, arising from intermolecular and electrostatic interactions between solution and solid phases in the soil. The distinction between saline and sodic soils arises because these forces vary depending on whether the soil solution is concentrated (saline) or dilute with a high proportion of Na to divalent ions sufficient to cause swelling and dispersion (sodic).

In the past, soil scientists have used a model involving Lifshitz-van der Waals, ion correlation, hydration, and electrical diffuse double-layer forces generated between colloidal clay minerals suspended in water to explain sodic soil behavior (Sumner 1993a; Quirk 1994) . However, in natural soils, clay particles which are bound together with silt and sand particles into aggregates of various sizes are confined and not readily suspended in water. Consequently, for clay in soil to become dispersed, forces other than those which operate in colloidal suspension must be overcome. In addition, most investigations of dispersion and flocculation have used pure clay minerals rather than soil clay systems as models for soil behavior (Shainberg and Letey, 1984; Sumner, 1993a). Despite the fact that soils are often categorized on the basis of the dominant clay mineral present (e.g., smectitic soil), soil clay systems which are complex heterogeneous intergrowths of different clay structures intimately associated with organic and biopolymers do not behave in the same way as their pure clay mineral counterparts. Thus soil colloids are unique in their behavior and are not appropriately represented by pure clay mineral systems (Sumner, 1993a).

Consequently, in order to understand sodic soil behavior, all the processes that occur during the initial wetting of dry aggregates, resulting in swelling to the final stage of aggregate disintegration, and leading, in turn, to dispersion of soil clays when completely wet, must be taken into account. Thus classical theories of colloidal behavior in suspension such as the Derjaguin-Landau-Verwey-Overbeek (DLVO) theory (van Olphen, 1977) may not satisfactorily explain the entire phenomenon of sodic behavior. In the discussion to follow, the mechanisms involved in the disintegration of an aggregate, on wetting, into its constituents with the release of dispersed clay will be explored.

3.2 MECHANISMS INVOLVED IN STRUCTURAL CHANGE

When Na is involved in the association between clay particles, soils tend to form massive structure without any hierarchical arrangement of different particle size ranges into micro- and macroaggregates (Barzegar *et al.*, 1994). The stability of different aggregate sizes and their associated pore systems depend on the relative magnitudes of the various forces arising from interactions between aqueous and solid phases in the soil. As the soil water content varies, there is a concomitant change in total potential (energy level), which is comprised of component potentials such as osmotic (due to interaction with dissolved components) or, matric (due to interaction with solid phase) potentials (Bolt and Miller, 1958). When a dry soil aggregate comes into contact with water, these interactive forces lower the potential energy of the water molecules, resulting in the release of energy which is partly consumed in the structural transformation of the aggregate, with the remainder being dissipated as heat (Rengasamy and Olsson, 1991). Slaking, swelling, and clay dispersion are the major mechanisms by which the massive structure of sodic soils disintegrates during rain or irrigation, resulting in the observed poor physical condition.

The potential energy of interacting water molecules in a soil aggregate is changed by:

1. intermolecular force fields whose intensity depends on the surface area available for water adsorption,
2. hydration of cations and anions adsorbed by outer sphere complexation on soil particles,
3. electrical fields induced by positive and negative charges on soil colloids, and

4. chemical interactions such as hydrogen bonding or inner sphere complexation between water molecules and soil particles.

Because slaking, swelling, and dispersive behavior of soil aggregates are not influenced by nonpolar solvents (Murray and Quirk, 1982) (Table 3.1), the polar (electrical) nature of water molecules is crucial in determining the above patterns of behavior. Despite the importance of the polarity of a liquid described by its dielectric constant in sodic behavior, solvation reactions of the liquid appear to be equally important in clay swelling and dispersion (MacEwan, 1948; Graber and Mingelgrin, 1994).

Soil aggregates are composed of several structural units linked together principally by colloidal (clay and organic matter) particles (Greenland and Hayes, 1978). The surface atoms and electric charges on the colloidal particles and their associated cations and anions which are held by outer sphere complexation contribute to the stability or breakdown of the units when exposed to water molecules. Based on this assumption, the changes, which are likely to take place in the structural behavior of sodic and nonsodic aggregates on wetting, will be compared. The magnitude and direction of energy changes taking place during wetting of an aggregate are illustrated schematically in Fig. 3.1, involving aggregate slaking and swelling, spontaneous and mechanical clay dispersion, and flocculation of dispersed clay by electrolytes. The individual stages involved in these processes will now be discussed.

3.2.1 Initial Wetting of Dry Aggregates

Clay particles in dry aggregates (Stage 1, Fig. 3.1) are bound together by inorganic and organic compounds involving several mechanisms and types of bonding (Table

TABLE 3.1 Slaking and spontaneous dispersion of Alfisol aggregates (2-4 mm) at two levels of sodicity in various solvents

Solvent	Dielectric constant (25 °C)	Slaking % < 2mm		Dispersed clay % of total clay	
		ESP 1	ESP 20	ESP 1	ESP 20
Water	78.5	67	80	0	26
Ethanol	24.3	24	12	0	6
Benzene	2.3	0	0	0	0
n-Hexane	1.9	0	0	0	0

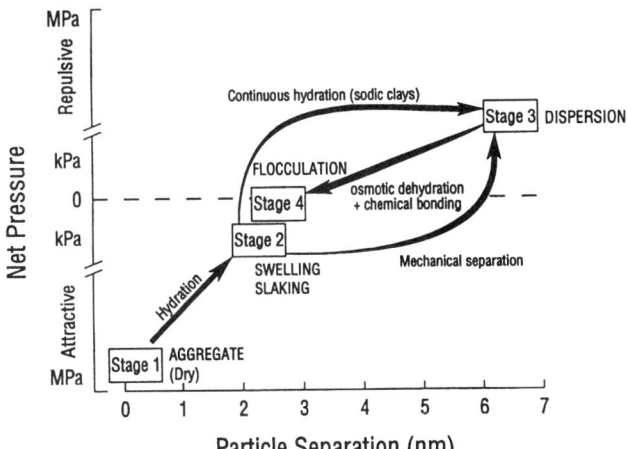

Figure 3.1 Schematic illustration of processes that take place and intensity of attractive and repulsive forces involved when a dry aggregate of soil is wetted.

3.2), which produce strong attractive pressures of the magnitude of megapascals (Rengasamy and Olsson, 1991). The water stability of an aggregate depends on the strength and persistence of these linkages which, in turn, are functions of the type of bonding. Bond strength in the presence of water, generally, decreases in the order: covalent, hydrophobic, Lifshitz-van der Waals, coordination complexing, hydrogen, and finally ionic bonds. However, the inherent mechanical strength of each type of bond may depart from this order. For example, the mechanical strength due to ionic bonding is greater than that due to Lifshitz-van der Waals forces. In contrast to covalent bonds, ionic bonds are readily solvated by water molecules, for example, in pure Na compounds and Na-clay linkages where ionicity dominates over covalency. In fact, any given heteronuclear bond found in natural systems has a mixture of covalent and ionic character (Huheey *et al.*, 1994). The degree of covalency in a bond involving metal cations is characterized by the Misono softness parameter Y derived from ionization and ionic potentials (Misono *et al.*, 1967). This parameter is calculated as follows:

$$Y = 10(I_z/I_{z+1})(r_i/z^{1/2}) \qquad [3.1]$$

where

r_i = ionic radius of metal ion
z = its formal charge
I_z = z^{th} ionization potential ($M^{(z-1)+} \rightarrow M^{z+}$)

The Misono factor Y, which expresses the tendency of a metal ion to form a dative π-bond, corresponds to its

"softness" or the ability to enter into electron-acceptor and electron-donor (EAED) or hard-soft acid-base (HSAB) reactions. Consequently, the tendency to form covalent bonding and complexes increases in the order: Na^+, K^+, Mg^{2+}, Ca^{2+}, Fe^{3+}, Hg^{2+}, for example. For more details on bonding models, the reader is referred to Huheey *et al.* (1994).

3.2.2 Hydration Reactions

With further wetting, solvation or hydration (when the liquid is water) forces, which are distinct from the DLVO or Lifshitz-van der Waals forces, come into play. Hydration forces arise from solvent effects in which the molecular structure of the solvent is altered (Ninham, 1985). Both DLVO and Lifshitz–van der Waals forces are changed by the solvent interaction, even when solvent molecules themselves are not transformed. Solvation forces in water differ from those in other liquids by exhibiting a long-range monotonic component, in addition to a superimposed oscillatory one. The monotonic component can be repulsive or attractive, long (>4 nm) or short range (<1.5 nm), and appears to depend on both the chemical nature and the dynamic state of the surface (Israelachvili, 1991).

TABLE 3.2 Types of bonding and mechanisms involved in the linkages between clay particles in an aggregate [Rengasamy and Olsson, *Aust. J. Soil Res.* 29:935-952, (1991)]

Type of bond	Mechanism
van der Waals forces	bonding between permanent or induced polar units
Ionic bond	cation and anion exchange, protonation, cation bridging
Hydrogen bond	water bridging, H-bond with oxygens of the silicate sheet, proton mediated H-bond
Coordination complex	ion-dipole interaction, ligand exchange, inner- and outer-sphere complexation
Covalent bond	O-H, Al-O, Si-O bonding, inner- and outer-sphere complexation
Hydrophobic bond	interaction between the hydrophobic portion of the clay surface and non-polar groups of organic molecules

Because, in clay systems with a uniform charge (positive or negative), monotonic hydration forces are repulsive (the pressure being in the megapascal range), the initial attractive force between clay particles in the aggregate decreases markedly on wetting, and as the hydration continues, the distance between these particles increases. This is reflected in reduced aggregate strength with increasing water content (Dexter, 1988b). Although very weak, the net force is still attractive and the clay particles are held together by hydrated cations. If the clay particles are Ca or Mg saturated, further wetting does not increase the interparticle distance beyond 2–3 nm, whereas in monovalent cation saturated clays, particles continue to separate. This type of swelling due to divalent cation hydration called *crystalline swelling* (Stage 2, Fig. 3.1) occurs in clay particles even at high electrolyte concentrations. However, two regions of limited swelling in divalent ion saturated smectites have been identified by Slade and Quirk (1991). The change in basal spacing from 1 to 1.55 nm is not affected by the concentration of the wetting solution, whereas the ionic strength (salt concentration) below which the transition from 1.5 to 1.9 nm (an osmotic process) takes place, depends on the charge density and origin (tetrahedral or octahedral) in smectites. Thus smectites with high charge and tetrahedral substitution require a lower electrolyte concentration for this transition to occur than low charge octahedrally substituted analogues. This phenomenon has even been observed in two Na smectites (Nibost and Drayton montmorillonites) with high charge arising mostly from tetrahedral substitution, which did not swell beyond 1.6 nm even in distilled water (Slade *et al.*, 1991).

3.2.3 Comprehensive Hydration Mechanism for Soil Clays

The following working hypothesis of hydration reactions in soil clays is proposed: The attractive forces between clay particles depend on the nature of bonding mediated through the cations (Na, K, Mg, Ca, Fe, and Al) commonly found in soils. Generally, a bond is considered to be covalent or ionic as long as the bond in question is predominantly one type or the other. However, in hydration reactions, it is convenient to be able to characterize the bond as intermediate. In EAED interactions, if the cation has no polarizing effect on the clay surface atoms, the bond is essentially ionic. When the cation and the surface atoms are mutually polarizing, the resultant bond is *polar covalent*. The degree of ionicity (or covalency) of such bonds depends on the nature of the cations as characterized by the

Misono factor, with Na contributing more to ionicity than Ca. When the polarization by the cation is sufficient, a covalent bond is formed.

These three types of bonds between clay surfaces and cations are illustrated in Fig. 3.2. The linkage between Al hydroxy cations and clay surfaces is dominated by covalency because of the high polarizing power of Al (Fig. 3.2a). Thus Al and Fe hydroxy species are specifically adsorbed or inner spherically coordinated to the clay surfaces. Hence, hydration of Fe and Al in such linkages becomes difficult. Calcium forms polar covalent bonds (Fig. 3.2b) in which the hydration of Ca is limited and determined by the polarity of the bond. For Na, the linkage of clay surfaces is ionic (Fig. 3.2c), and hence, hydration is extensive.

In addition, clay charge characteristics modify the EAED interactions and the nature of bonding. If the anion were large and "soft" enough, the cation should be capable of being polarized, with the extreme case being the penetration of the anionic electron cloud by the cation giving rise to a covalent bond. Because the polarizability of the anion is related to its "softness", that is, to the deformability of its electron cloud (Huheey *et al.*, 1994), the large size and high charge of the anion favor this interaction. Here the clay particle is considered as a superligand with heterogeneous charged sites having different polarizing abilities per unit of charge depending on location. These factors also determine whether cations link clay particles by ionic, polar covalent, or covalent bonding. Thus the ionicity of the clay-cation-clay bonds is determined by both the type of cation and the nature of clay ligand. In the case of soil clays with net positive charge, bonding will depend on the nature of the anions present. More detailed accounts of surface complexation have been presented by Sposito (1984) and Stumm (1992) among others, to whom the interested reader is referred.

These attractive forces should not be confused with those of Lifshitz-van der Waals interactions, which are not involved in the hydration reactions of clay particles. Forces arising from water interactions with clay particles can be attractive when the interaction is hydrophobic or repulsive due to polar interactions (not electrostatic). For soil clays, these polar interactions, which are based on EAED or HSAB interactions between the polar solvent (water) and the polar moieties (charged clay particles), are repulsive and functions of ionicity of the bonds between particles. These interactions can exhibit energies that may be up to two orders of magnitude greater than those commonly encountered in the components of traditional DLVO energy balances (van Oss *et al.*, 1988).

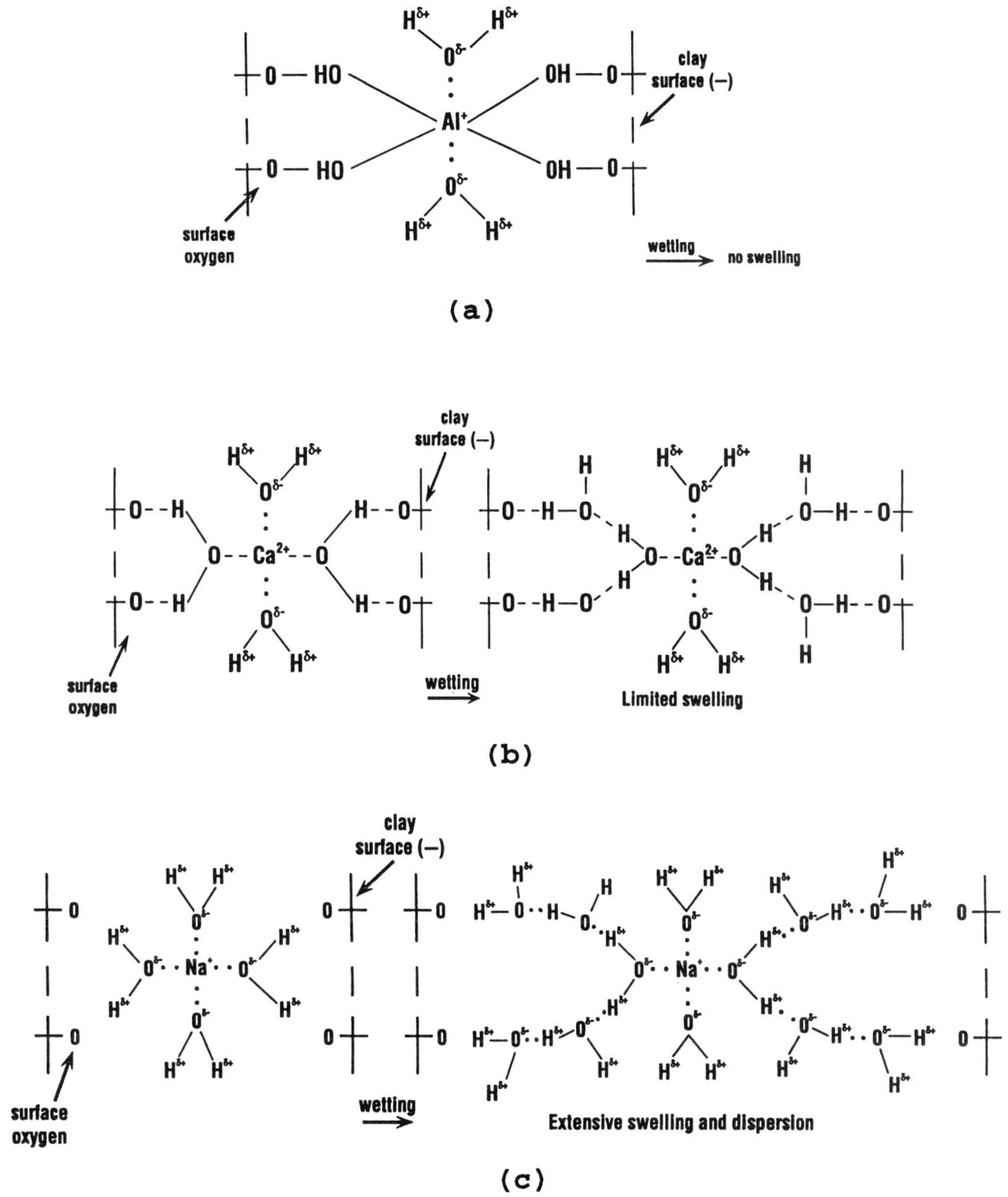

Figure 3.2 Schemetic representation of the effect of wetting and nature of bonding between cations and clay surfaces on swelling and dispersion: ([Al(OH)₂]⁺ ion linking clay particles by covalent bonding (—) (b) Ca²⁺ aquo ion linking clay particles by polar covalent bonding (– –) and (c) Na⁺ aquo ion linking clay particles by ionic bonding. Water molecules are linked to cations by hydrogen bonding (• •).

Earlier in Russia, Derjaguin and Churaev (1974) considered hydration forces as "the structural component of disjoining pressure," and, more recently, hydration forces causing repulsive interactions between mica surfaces separated by less than 5 nm have been extensively implicated by Pashley and Quirk (1984) and Pashley (1985). They concluded that, in dilute suspensions, the forces observed were generally in good agreement with DLVO theory, but in more concentrated suspensions (clay particle separations less than 5–10 nm), short-range non-DLVO forces (hydration forces) arose. These short-range repulsive forces were clearly related to the type of cation adsorbed at the mica surface. Although the Lifshitz-van der Waals forces were certainly present in concentrated suspensions, they were completely overshadowed by the repulsive hydration forces. Physical and chemical forces other than electrostatic repulsive and Lifshitz-van der Waals attractive forces have been considered responsible for deviations from DLVO theory for colloidal particles separated by short distances (less than 5 nm) (Israelachvili and McGuiggan, 1988). Low (1979, 1991) has also considered that the repulsive force in macroscopic swelling of montmorillonites is not of the DLVO variety but a structural force arising from the modified structure of water molecules close to clay surfaces.

The origin of non-DLVO forces lies in solvation reactions where polar solvents react with polar surfaces. Clays do not swell or disperse in nonpolar solvents (MacEwan, 1948; Murray and Quirk, 1982, Graber and Mingelgrin, 1994) (Table 3.1). Solvation reactions are controlled by EAED interactions and not merely by the polarity of the solvent as measured by the dielectric constant (Mayer and Gutmann, 1975; Christenson and Horn, 1985; van Oss *et al.*, 1990).

Only charged clay particles exhibit swelling and dispersion, which depend on the ability of water molecules to solvate the cations or anions involved in clay-clay bonding and are related to the EAED interactions mentioned. It is important to note that the extents to which free (solution) versus adsorbed cations and anions are hydrated are different. For example, the hydration of cations as shown by the hydration numbers follows the order $Ca^{2+} > Mg^{2+} > Na^+ > K^+$ (Table 3.3), whereas the extent of swelling of smectites caused by the hydration of cationic clays in pure water at zero matric potential is in the order $Na^+ > K^+ > Mg^{2+} > Ca^{2+}$ (Table 3.4).

As mentioned earlier, hydration reactions during limited crystalline clay swelling lead to the separation of individual crystals in a series of discrete steps, corresponding to 0, 1, 2, 3, and 4 layers of water molecules, depending on EAED interactions during hydration. Initial separation up to a distance of about 1.6 nm is independent

TABLE 3.4 Interlayer swelling of various cation-saturated smectities in pure water

Clay	Cation	Interlayer spacing d_{001} (nm)
Montmorillonite[a]	Na⁺	>13.0
	K⁺	>6.0
	Mg²⁺	1.92
	Ca²⁺	1.89
Wyoming bentonite[b]	Mg²⁺	1.86
	Ca²⁺	1.85
Otay smectite[b]	Mg²⁺	1.90
	Ca²⁺	1.85

[a]Data from Norrish (1954).
[b]Data from Slade and Quirk (1991).

of ionic strength, but further separation to 2 nm (Fig. 3.1) appears to be determined by both ionic strength and EAED interactions (Slade *et al.*, 1991). During extensive hydration, as is observed in sodic clays, separation of clay particles beyond 7 nm depends on both ionic strength and EAED interactions. When the ionic strength of the equilibrium solution is decreased, water molecules enter between the clay particles. The magnitude of pressure developed during macroscopic swelling, usually known as clay dispersion (Stage 3, Fig. 3.1), depends on the difference in chemical potential of water in the equilibrium and inner (i.e. in the space between the particles) solutions. Once the clay particles are completely dispersed (i.e., separated into distinct individual entities), electrostatic repulsive forces, as predicted by DLVO theory, come into play at which stage Lifshitz-van der Waals attractive forces are negligible.

As the difference in the chemical potentials of water in inner and outer solutions approaches zero as a result of increasing ionic strength in the outer solution, the clay particles begin to approach each other. The electrostatic repulsive pressure (DLVO) is balanced by the increasing osmotic pressure, at which stage Lifshitz-van der Waals attractive pressures become dominant. At this stage, *flocculation*, or *coagulation* (used synonymously), occurs in clay or soil suspensions in water (Stage 4, Fig. 3.1). On drying, the flocculated clays are increasingly attracted to each other by EAED interactions, resulting in far greater attractive pressures than predicted by Lifshitz–van der Waals interactions. This association is called aggregation, or *flocculation plus* as suggested by Bradfield (1936).

TABLE 3.3 Thermodynamic values for the hydration of ions at 25°C

Ion	Enthalpy (kJ mol⁻¹)	Entropy (J mol⁻¹ K⁻¹)	Hydration number
Na⁺	-406	-60.25	4
K⁺	-322	-102.51	4-6
Mg²⁺	-1922	+117.99	6-8
Ca²⁺	-1577	+55.23	9-12

TABLE 3.5 Factors influencing hydration of clay particles

Enthalpy effects	Variations in cation bonding with ionic potential and Misono's factor
	Ligand field effects (clay is considered a ligand)
	Steric and electrostatic repulsions between clay particles in aggregates
	Effects related to conformation of clay-organic complexes
	Other coulombic forces involved in the aggregation of clay particles
	Change in bond strength due to heterogeneous charge distribution
Entropy effects	Concentration and size of clay particles
	Molecular structure and size of associated organic polymers
	Changes of hydration associated with hydrophobic components
	Hydration of clay-cation complex

When a soil aggregate is wetted causing the water content to increase, clay particles will not separate and become dispersed if the electrolyte concentration of the wetting solution is sufficient to balance the electrostatic repulsive pressure (DLVO).

The hydration reactions in soil clays are influenced by both enthalpy and entropy effects (Table 3.5). Enthalpy of hydration will depend on the nature of bonding between cations and clay surfaces, ligand field effects, and steric and electrostatic interactions. Factors related to clay and associated organic polymers such as concentration and particle size and the presence of hydrophobic components will introduce entropy effects. However, when many of these factors are relatively constant, the nature of bonding between cations and clay surfaces will be the major determinant.

3.2.4 Slaking of Soil Aggregates

Despite the fact that the clay particles are separated by water molecules (Stage 2, Fig. 3.1), the net pressure, which is in the kiloPascal range, is still attractive. If the clay particles are Ca- or Mg-saturated, further wetting does not increase the interparticle distance beyond 2–3 nm. Hydrostatic and pneumatic pressures in the range of kiloPascals associated with pore filling are sufficient to break this

weak attractive force. As a result, the linked units of clay particles in a microaggregate become separated, which is known as *aggregate slaking*. Unless these clay linkages are weakened by hydration reactions, the low pressures associated with entrapped air or raindrop impact (in the range below kiloPascals) will fail to cause disruption.

The rate of wetting is critical in the initial hydration involving EAED reactions. Stronger attractive bonding between clay particles through water/cation bridging is maintained during slow rather than rapid wetting, when entropy plays an important role. When saturated with divalent cations, aggregate integrity is maintained and slaking is minimized. However, in sodic aggregates, dispersive breakdown is not prevented by slow wetting under suction (Table 3.6). On the other hand, under saline conditions (high EC) where hydration is controlled due to the osmotic effect (Section 3.2.2), aggregate breakdown is minimized, irrespective of whether the clay is Na- or Ca-saturated (Table 3.7). However, when the aggregates are initially Na-saturated or the solution in which they are sieved has a high SAR, severe slaking takes place at low EC values, with the effect being less marked for Ca- and Mg-saturated systems.

3.2.5 Spontaneous Dispersion

Spontaneous dispersion often takes place without energy inputs when sodic clay is placed in water of very low electrolyte concentration. When the bridging between clay particles involves Na and the EC is low, the interparticle distance continuously increases, with continued wetting to beyond 7 nm (Fig. 3.1). The proportion of clay particles separated in this way depends on the number of Na ions involved in the clay linkages and hence on the SAR of the

TABLE 3.6 Effect of mode of wetting and crop rotation on spontaneously dispersible clay from aggregates of an Alfisol [Barzegar *et al.*, *Soil Tillage Res.* 32:329-345, (1994)]

	Dispersed clay (% of total clay)	
Crop rotation	Rapid wetting	Suction (2 kPa) wetting
Wheat-fallow	1.7	1.4
Wheat-wheat	1.3	1.0
Wheat 2 years and pasture 4 years	0.8	0.7
Permanent pasture	0.3	0.3
Virgin soil	0.2	0.4

TABLE 3.7 Average mean weight diameter of homoionic[a] wet sieved aggregates (2–4 mm) in solutions of different sodium adsorption ratio (SAR) and electrical conductivities (EC)

| | Solution composition | | Mean weight diameter (mm) | | |
| | | | Cationic form of aggregates | | |
Soil	SAR	$EC_{1:5}$ (dS m^{-1})	Ca	Mg	Na
Alfisol	0	0.1	0.50	0.48	0.08
	0	4.0	0.62	0.60	0.57
	30	0.1	0.12	0.12	0.04
	30	4.0	0.60	0.58	0.57
Vertisol	0	0.1	0.85	0.78	0.02
	0	4.0	0.96	0.92	0.72
	30	0.1	0.16	0.14	0.02
	30	4.0	0.96	0.90	0.72

[a]Homoionic aggregates prepared prior to immersion in solutions.

soil-water system (Rengasamy and Olsson, 1991). At water contents lower than saturation, limited but more than crystalline swelling and incomplete separation of clay particles take place, with the interparticle distance depending on the water content. Similar mechanisms are involved in interlayer and interparticle separations in smectites and other clays (kaolinite and illite) (Quirk, 1986).

3.2.6 Mechanical Dispersion

Hydrated clay particles which have undergone limited separation can be pushed further apart by applying external mechanical pressure in the range of Pascals to kiloPascals such as from raindrop impact. At distances of separation greater than $2/\kappa$, where κ is the thickness of the electrical double layer in nanometers (see Equation [3.3]), electrostatic repulsive forces predominate over attractive forces. Hence, on continued wetting, which reduces electrolyte concentration and increases double-layer thickness, the particles become progressively more separated, finally reaching the stage of dispersion. Thus, a Ca or Mg clay aggregate can be dispersed when uniformly remolded (energy input) at or above a critical water content, defined by Emerson (1983) as the *water content for dispersion*. At lower water contents, attractive forces dominate in spite of the mechanical repulsive pressure introduced by remolding.

In subplastic soils (Norrish and Tiller, 1976), the attractive pressures between clay particles have to be overcome by higher levels of mechanical energy involving sonification or grinding. These attractive forces increase with the extent of covalent bonding between particles caused by cementing agents, thereby decreasing the potential for solvation by water molecules. This is analogous to the difference in solvation and subsequent dissolution of $CaCl_2$ and $CaCO_3$. In the case of $CaCl_2$, solvation is complete and hence it dissolves, whereas for $CaCO_3$ the much lower solvation renders it insoluble. Similarly, Na-saturated Drayton montmorillonite did not swell on wetting (Norrish and Tiller, 1976) because the high tetrahedral charge (Slade *et al.*, 1991) promoted strong inner-sphere coordination between Na and the clay making hydration more difficult. On the other hand, when lithium (Li)-saturated, the clay hydrated and dispersed readily.

3.3 THE ELECTRICAL DOUBLE LAYER

In an aqueous suspension, the charge on clay particles is neutralized by hydrated ions of opposite charge. In sodic soils, clay surfaces usually always carry a net negative charge, which is neutralized by a diffuse cloud of ions in which the concentration of cations increases and that of anions decreases as the surface is approached. The electrical double layer consists of the surface charge and the surrounding ion swarm as illustrated in Fig. 3.3.

Various theories (Gouy-Chapman, Stern) and modifications thereof (multiple layer models) have been proposed in an attempt to predict the behavior of the electrical double layer present in dispersed colloidal systems. The

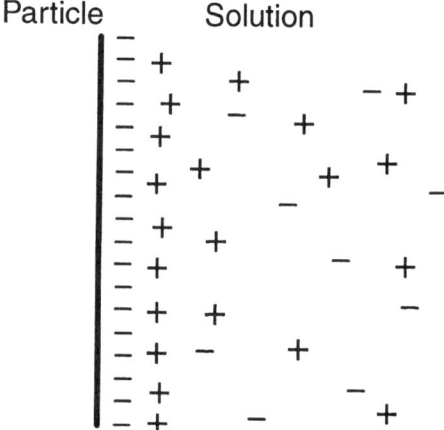

Figure 3.3 Schematic distribution of charges in the diffuse electrical double layer according to Gouy-Chapman theory [van Olphen, *An Introduction to Clay Colloid Chemistry*, 2nd, (1977)].

classical DLVO theory (Gregory, 1989) explains colloidal stability (dispersion and flocculation) on the basis of the opposing forces of attraction (Lifshitz-van der Waals forces) and repulsion (electrostatic or coulombic forces). All have failed to predict behavior quantitatively, except under very limited equilibrium conditions. The DLVO theory has been found to be unsatisfactory in clays with divalent cations, where diffuse double-layer formation is restricted due to stacking or aggregation of particles (Swatzen-Allen and Matijevic, 1974). This failure has been assumed to be caused by factors such as partial blocking of negative surfaces by adjacent particles, incomplete double-layer formation on some or all surfaces, and specific adsorption (inner sphere complexation) of cations with the negative sites on the clay (Rengasamy *et al.*, 1984a). However, once clays become dispersed, either spontaneously or mechanically, the different forces operating in colloidal suspensions can be described adequately by these models, discussed in detail by van Olphen (1977), McBride (1989), Gregory (1989), and Iwata *et al.* (1995). The Gouy-Chapman theory will now be followed to present mathematical expressions describing the electrical double layer.

The relationship between the electrical potential ψ and the charge σ at the particle surface, and the nature and concentration of the equilibrium solution are given by

$$\sigma = (2\epsilon nkT)^{1/2}/\pi \sinh(ze\psi_0/2kT) \qquad [3.2]$$

where

σ = surface charge on particle
ψ_0 = surface potential
n = electrolyte concentration in equilibrium solution
z = valence of the counter ions (cations)
e = electronic charge
ϵ = dielectric constant of solvent
k = Boltzmann's constant
T = temperature in °K

The effective thickness $1/\kappa$ of the double layer is given by

$$1/\kappa = (\epsilon kT/8\pi \ z^2 e^2 n)^{1/2} \qquad [3.3]$$

Theoretically the double layer is infinite in thickness due to the exponential nature of the relationship between ion concentration and distance. The approximation made here is equivalent to assuming that the ion cloud is concentrated as a planar charge at a distance from the surface, analogous to the situation for an electrical condenser.

TABLE 3.8 Effect of valence and concentration of counterions on effective thickness of double layer [Sumner and Stewart, *Soil Crusting: Chemical and Physical Processes*, pp. 1-32, (1992)]

Valence	Effective thickness of double layer (nm) Concentration (mol L^{-1})				
	10^{-5}	2×10^{-5}	10^{-4}	2×10^{-4}	10^{-3}
1	100	67.7	30	21.4	10
2	50	33.8	15	10.7	5
5	33	22.5	10	7.1	3.3

Equation [3.3] shows that for negatively charged systems, the thickness of the double layer is inversely proportional to z (cation valence) and $n^{1/2}$ (Table 3.8). By increasing either the valence of the cations in the soil solution and/or their concentration, the double-layer thickness is reduced so that the clay suspension flocculates.

3.3.1 Permanent Charge Systems

In clays where the charge arises from isomorphous substitution as in smectites, for example, the surface charge must be constant, and consequently, according to Equation [3.2], changes in electrolyte composition and concentration (z and n) cause the surface potential ψ_0 and the

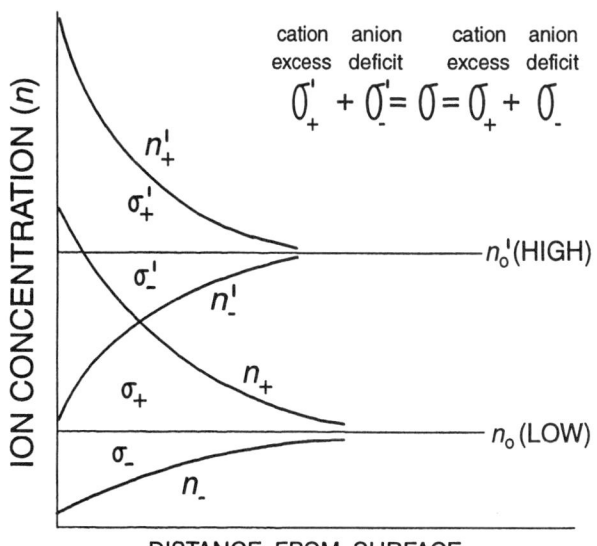

Figure 3.4 Distribution of counterions and co-ions in electrical double layer for a permanent charge surface at two electrolyte concentrations [van Olphen, *An Introduction to Clay Colloid Chemistry*, 2nd, (1977)].

distribution of ions in the diffuse layer to change, as illustrated in Fig. 3.4. On increasing the electrolyte concentration from n_0 to n_0' (or the valence) of the cations, the double-layer thickness is reduced, and the proportion of the constant surface charge neutralized by the excess counterions σ_+ and the deficit of co-ions σ_- change to σ_+' and σ_-', respectively, but the total areas $(\sigma_+ + \sigma_- = \sigma)$ and $(\sigma_+' + \sigma_-' = \sigma)$ remain constant.

3.3.2 Variable Charge Systems

In variable charge systems such as kaolinites and oxides of Fe and Al, the charge is dependent on pH, electrolyte concentration, and counter ion valence. When n or z are changed (Equation [3.2]), the surface charge increases or decreases while the surface potential ψ_0 remains constant. The distribution of ions in the double layer under these conditions is illustrated in Fig. 3.5. The balance between H^+ and OH^- (pH), which are the potential-determining ions, controls the polarity and magnitude of the surface charge. When the pH is constant, electrolyte composition and concentration determine the areas σ and σ' which are not equal (Fig. 3.5), but the relative proportions of the total surface charge σ neutralized by the diffuse layer cations and anions $(\sigma_+/\sigma_- = \sigma_+'/\sigma_-')$ remain constant. Thus except for a difference in the magnitude of the charge, variable charge systems react in essentially the same way to changes in n and z as do permanent charge systems, namely,

increasing n and z reduce the effective thickness of the double layer but increase the total surface charge.

3.3.3 Mixed Charge Systems

In systems containing mixtures of permanent and variable charge (e.g., inorganic and organic colloids), the surface charge σ, the surface potential ψ_0, and the effective thickness of the double layer $(1/\kappa)$ will all change simultaneously with changes in solution composition and pH of the suspension. In addition, at low pH, some particles (sesquioxides) can develop positive charge and interact with negatively charged particles, leading to reductions in net charge and $1/\kappa$ and, resulting in flocculation. The way in which positive and negative charges change with pH and electrolyte concentration for permanent, variable, and mixed charge systems is illustrated in Fig. 3.6. When soil clays comprise only permanent charge minerals, the surface charge is independent of pH and electrolyte concentration, as illustrated by the horizontal line. If the clay has only variable charge sites, the surface charge and its sign are entirely governed by pH and electrolyte concentration. At pH values above pH_0 (the pH at which variable, positive, and negative charges are equal), the charge becomes increasingly negative. Increasing the electrolyte concentration at a given pH value above pH_0 also increases the negative charge. The reverse is true below pH_0 where the system becomes increasingly more positively charged. At pH_0 for mixed charge situations, the system has a net negative charge because of the contribution from the permanent negative charge component. Therefore, a point of zero net charge (PZNC), at which the total net charge (variable plus permanent) is zero, has to be defined. The PZNC will always be at a pH value below pH_0 for soil clays containing closely associated clay minerals and organic polymers. A more detailed discussion of the PZNC concept is presented in Uehara and Gillman (1981).

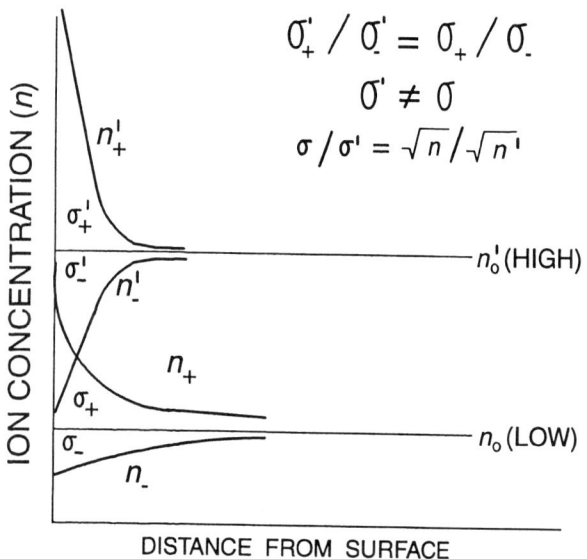

$$\sigma_+' / \sigma_-' = \sigma_+ / \sigma_-$$
$$\sigma' \neq \sigma$$
$$\sigma / \sigma' = \sqrt{n} / \sqrt{n'}$$

Figure 3.5 Distribution of counterions and co-ions in electrical double layer for a variable charge surface at two electrolyte concentrations [van Olphen, *An Introduction to Clay Colloid Chemistry*, 2nd, (1977)].

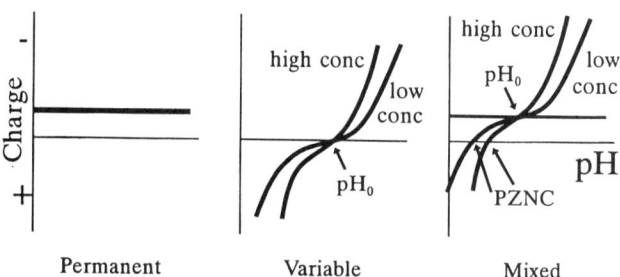

Figure 3.6 Variation in charge with pH and electrolyte concentration for permanent, variable, and mixed charge systems [Sumner and Stewart, *Soil Crusting: Chemical and Physical Processes*, (1992)].

3.3.4 Net Charge and Clay Dispersion

Because repulsion between particles increases with an increase in $1/\kappa$ in a system of uniform polarity, the nature and concentration of the cations control the behavior in negatively charged systems and anions in their positively charged analogues. In mixed charge systems, charge reversal can be effected by changing pH and/or adsorption of organic or inorganic ligands (e.g., phosphate on positively charged sites) by innersphere complexation to the surface (specific adsorption) (Sumner, 1992). Thus adding phosphate to a soil promotes repulsion between particles because any positive charges present become negative, reducing the attractive forces between sites of opposite charge. In terms of sodic soils, in which there are elevated levels of Na, expansion of the double layer and hence repulsion between particles are promoted.

The electrostatic interaction (predicted by DLVO theory, Section 3.4) in dispersed clays depends on the net charge of the system. For example, Chorom *et al.* (1994) demonstrated a significant relationship between the net negative charge of soils (CEC at pH 8.0) and their tendency to disperse (Section 3.7). However, the slope of the relationship was a function of soil type and pH, indicating that the actual charge present in the field was not reflected well by the CEC measurement at pH 8.0. This inappropriate measure of charge could have affected the dispersive potential (Rengasamy and Naidu, 1994). To resolve this situation, the effect of net negative charge associated with sodium

$\sigma_{Na}{}^{-}$ at the natural pH and electrolyte levels in various soils belonging to Alfisol, Aridisol, and Vertisol Orders on spontaneously dispersible clay was investigated (Fig. 3.7). The spontaneously dispersed clay was highly correlated with $\sigma_{Na}{}^{-}$ irrespective of the soil type, indicating that the magnitude of the net negative charge is a determinant of dispersion.

Net negative charge also influences the electrokinetic properties of the clays. For example, the electrophoretic mobility of clay particles, or zeta potential ς depends on the net charge available for outersphere complexation of cations, the mode of attachment of Na on clay surfaces in most cases. On the other hand, divalent-ion clays always exist as domains or quasicrystals, where a number of individual particles are linked by innersphere coordination (Shainberg and Letey, 1984). As a result, these clays have lower electrophoretic mobilities than sodic clays (Chorom and Rengasamy, 1995) (Fig. 3.8). While ς potentials of Na-kaolinite, illite, and smectite increased linearly with pH, those of the Ca-treated clays only increased up to pH 7 and then decreased sharply due to the precipitation of $CaCO_3$, which caused the clay particles to aggregate.

3.4 REPULSIVE AND ATTRACTIVE FORCES IN CLAY-WATER SYSTEMS

The classical DLVO approach to colloidal stability (i.e., stability of dispersed clays) is based on opposing forces of attraction due to van der Waals interaction and repulsion

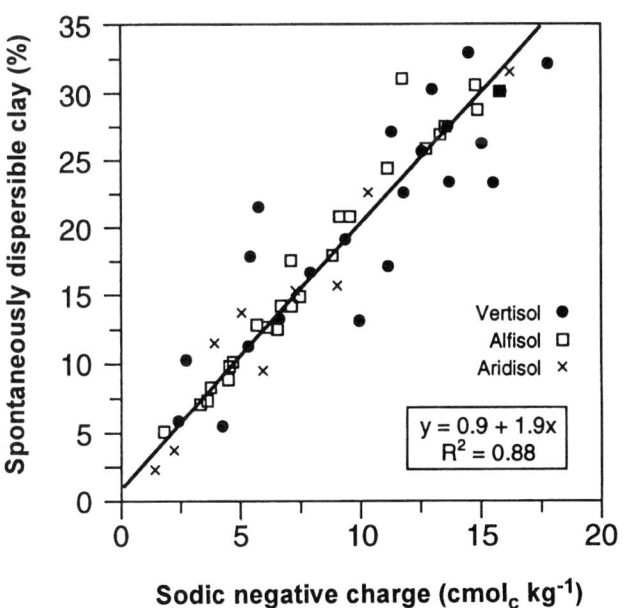

Figure 3.7 Effect of negative charge saturated with Na on spontaneously dispersible clay in a range of soils.

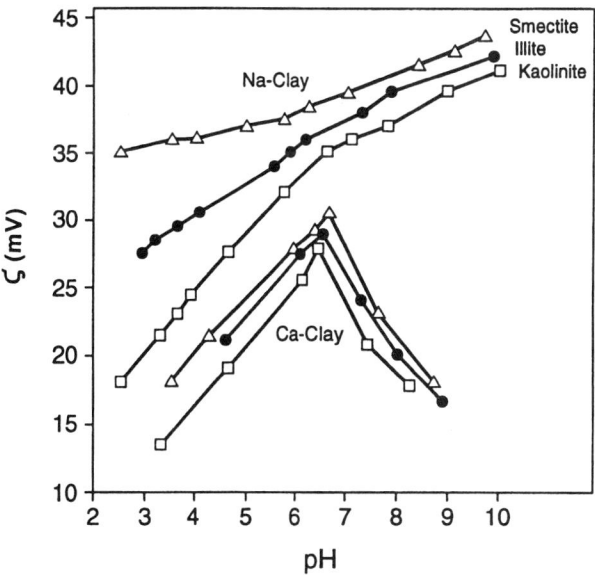

Figure 3.8 Effect of pH on Zeta (ς) potentital of Na- and Ca-saturated clays [Chorom and Rengasamy, *Eur. J. Soil Sci.* 46:657-665, (1995)].

arising from electrostatic forces in the double layer. Although DLVO theory predicts the behavior of completely dispersed particles after the development of the diffuse double layer, failure in divalent-ion saturated clays, even in suspension, can be attributed to particle associations (as domains or quasi crystals) which reduce the effective net charge in the double layer. Readers interested in a detailed discussion of DLVO theory are referred to van Olphen (1977) and Gregory (1989).

3.4.1 Repulsive Forces

As can be seen from Equation [3.3], the thickness $1/\kappa$ of the double layer is a function of counter ion concentration and valence. When a clay suspension is diluted, double layers on adjacent flat particles begin to interact, setting up a repulsive force between the particles with the repulsive (swelling) pressure being given by

$$\text{Swelling pressure} = 2nk\text{T}[\cosh(ze\psi_d/k\text{T}) -1] \qquad [3.4]$$

where

ψ_d is the potential midway between the plates.

As double layers begin to overlap, ψ_d increases, giving rise to greater repulsion. Similarly, as n and z of the counter ions are reduced, repulsion increases. Although it does not follow from Equation [3.4], repulsion is also directly proportional to particle size (Gregory, 1989). Because of the complex relation between $_d$ and d, the repulsive energy V_r can be evaluated from Equation [3.5] by assuming weak interaction between particles,

$$Vr = (64nk\text{T}/\kappa)\left\{\frac{\exp(ze\psi0/2k\text{T})}{\exp(ze\psi0/2k\text{T})}\right\} \qquad [3.5]$$

The repulsive energies between clay particles suspended in electrolyte solutions of varying concentration and valence obtained in accordance with Equation [3.5] are illustrated schematically in the upper portion of Fig. 3.9. At low electrolyte concentrations, the repulsive force is high, particularly at large particle separations, and as the concentration increases, the decay in repulsive force becomes more marked and of shorter range.

3.4. 2 Lifshitz-van der Waals Attraction in the Condensed State

At the point when clay particles flocculate, the electrostatic repulsive forces are balanced by the attractive forces

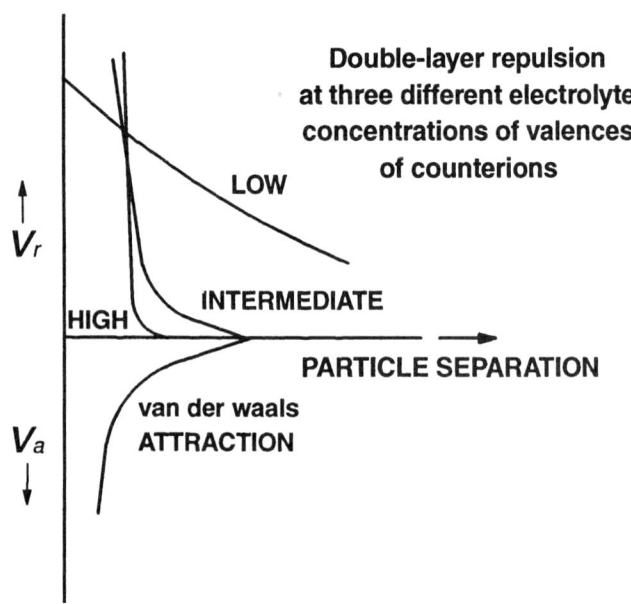

Figure 3.9 Schematic representation of variation in repulsive and attractive forces between colloidal particles of like charge with distance from particle surface [van Olphen, *An Introduction to Clay Colloid Chemistry*, 2nd, (1977)]

known as van der Waals forces. The Lifshitz theory of condensed media interaction, which is based on spontaneous electric and magnetic polarizations of the atoms in the particles in the close condensed state, has been used to derive van der Waals forces under various conditions (Mahanty and Ninham, 1976). These attractive forces alone do not cause flocculation. However, when the clay particles are brought together by electrolytes (or removal of the solvent molecules between the clay particles by other than osmotic processes), these attractive forces begin to operate.

For a pair of atoms, this attractive energy is inversely proportional to the sixth power of the distance separating the atoms, whereas for large spherical particles, it varies with the second power. For most aqueous colloidal systems, van der Waals forces are independent of the ionic strength of the bathing medium (Gregory, 1989). For clay particles where the distance between particles is small compared with their thickness, the attractive energy V_a is given by the simplified equation

$$V_a = -(A/48\pi d^2) \qquad [3.6]$$

where

$A = $ Hamaker constant
$d = $ half distance between the particles.

The way in which V_a varies with the distance between particles is illustrated in the lower portion of Fig. 3.9.

3.5 CRITICAL FLOCCULATION CONCENTRATION

As the net negative charge associated with Na, σ_{Na}^-, or the SAR of the solution, increases, the electrostatic repulsive force in the double layer and its thickness $1/\kappa$ increase. By increasing the electrolyte concentration or introducing multivalent cations, $1/\kappa$ is reduced to a distance where V_r becomes equal to V_a and the clay particles flocculate. The electrolyte concentration at this stage, where the total potential energy per unit area of particle surface becomes zero, is defined as the critical flocculation concentration (CFC) which, for a symmetrical electrolyte, is given by

$$CFC = K[\tanh(ze\psi_0/4kT)]^4/A^2z^6 \qquad [3.7]$$

From this equation, CFC is a function of $1/z^6$ which is popularly known as Schulze-Hardy rule. Although this rule explains the effect of valence on flocculation, quantitative predictions deviate from experimental results because flocculation by multivalent ions involves a charge reduction on clay surfaces and EAED interactions in addition to the reduction in the thickness of the double layer. Critical flocculation concentrations for a number of homoionic clays by chloride solutions of the same cation (Table 3.9) illustrates the major role that valence plays. The order of flocculating power is Ca > Mg > K > Na. However, even with a given valence, flocculating power differs (e.g., K > Na). The constancy of the relative flocculation concentration (RFC) for a given cation system, irrespective of the origin of the clay, suggests that the average RFC may have some predictive value within a

given range of soils. Compared to Na = 1, the flocculation power (RFC_{NaCl}/RFC_{XCl}) of the other cations would be K = 1.8, Mg = 27, and Ca = 45.

Flocculation by electrolytes appears to be a combination of an osmotic phenomenon and EAED interactions. The osmotic pressure of the electrolyte in solution causes the removal of water molecules between the clay particles (dehydration), thereby reducing their separation distance. The cationic effects cause EAED interactions, resulting in particle bonding and reduction in net charge available for outersphere complex formation. On the one hand, the Misono factor and the ionic potential can be related to the EAED interactions, as illustrated in Fig. 3.10 (Sposito, 1989), while on the other, the reduction in double layer thickness is related to z^2 (Equation [3.3]). Combining these functions, the following relationship has been used to calculate the flocculation power of the cations:

$$\text{Flocculating power} = 100(I_z/I_{z+1})^2 \, z^3 \qquad [3.8]$$

The flocculating powers (dimensionless) of Na^+, K^+, Mg^{2+}, and Ca^{2+} calculated from this equation are very close to those determined experimentally (Table 3.10). However, before this relationship can be used generally, further validation is necessary.

3.6 HYDROPHOBIC ATTRACTION

A significant proportion of soil surfaces may have hydrophobic character caused by the adsorption of nonpolar organic molecules, often a feature of nonwetting sands. The ordering of water molecules on such surfaces tends to be limited by the inability to form hydrogen bonds. In water suspensions of materials exhibiting this character,

TABLE 3.9 Critical flocculation concentrations (CFC) for homoionic soil clays in electrolyte with the same anion

Clay form	Electrolyte	Critical flocculation concentration (CFC) (mmol$_c$ L^{-1})				Relative flocculation concentration (RFC)[a]				Average RFC[b]
		Alfisol			Ultisol	Alfisol			Ultisol	
		Australia	S. Africa	Niger	Georgia	Australia	S. Africa	Niger	Georgia	
Na-clay	NaCl	14.0	5.6	6.8	8.5	46.7	56.0	34.0	42.5	44.8
K-clay	KCl	8.5	3.4	4.2	5.2	28.3	34.0	21.0	26.0	27.3
Mg-clay	MgCl$_2$	0.6	0.2	0.3	0.3	2.0	2.0	1.5	1.5	1.75
Ca-clay	CaCl$_2$	0.3	0.1	0.2	0.2	1.0	1.0	1.0	1.0	1.0

[a]Calculated as CFC_{XCl}/CFC_{CaCl2} where X = Na, K, or Mg.
[b]Averaged over the four clays.

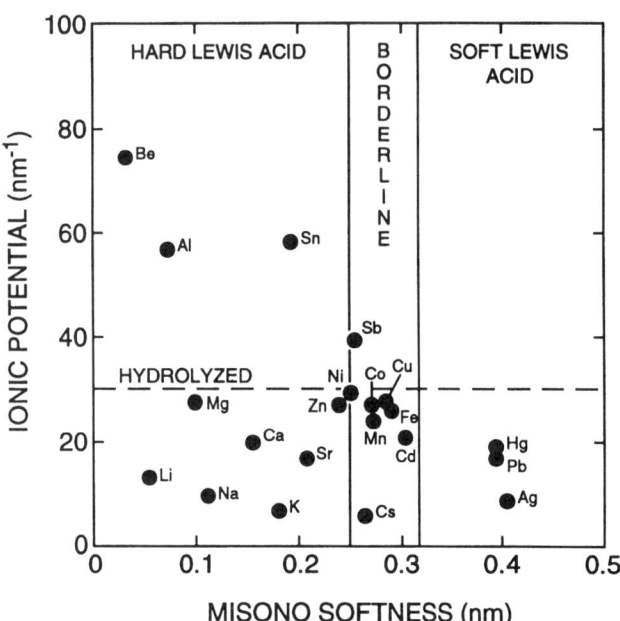

Figure 3.10 Classification of metals according to ionic potential and Misono softness [Sposito, *The Surface Chemistry of Soils,* (1984)].

the particles associate among themselves in such a way that contact with water molecules is minimized. Hydrophobic forces of attraction (in the range of 200–400 kPa) can be stronger than van der Waals forces and can operate at distances of separation of about 80 nm (Gregory, 1989). The adsorption of hydrophobic substances on soil clays leads to a reduction in net charge, making this mechanism even more important.

3.7 DISPERSIVE POTENTIAL

The improvement of the structure in sodic soils requires the application of appropriate amendments in order to promote clay flocculation. However, the CFC values obtained in dispersed systems are not directly applicable to soil at field moisture contents where clay particles occur as aggregates. The repulsive forces present in dispersed clay systems are far greater than those in the aggregated clay particles. Hence, the critical concentration required for clay to disperse from soil aggregates is always lower than the CFC obtained for suspensions (Chorom *et al.*, 1994). A threshold electrolyte concentration (TEC), however, is necessary for each level of sodicity to maintain favorable soil structure in sodic soils. For a large group of Australian Alfisols, Rengasamy *et al.* (1984b) obtained a linear relationship between $SAR_{1:5}$ and $EC_{1:5}$ (measured in a 1:5 soil-water system), reflecting the TEC for preventing clay dispersion from soil aggregates. Thus, for a given SAR, the line A–B in Fig. 3.11 predicts the TEC necessary to

prevent dispersion from sodic Alfisols. However, TEC is not only a function of SAR and EC, it also varies with the net negative charge, as discussed earlier. With increasing negative charge, the slope and the intercept of the line A–B increase toward a line C–D, shown schematically in Fig. 3.11. For a given soil, increases in pH, acidic functional groups on biopolymers and adsorbed organic ligands, and exposure of particle surfaces due to fragmentation of aggregates, all cause an increase in the negative charge (Rengasamy and Olsson 1991; Sumner 1993a).

In order to derive a single parameter that will combine the effects of SAR and EC, Rengasamy *et al.* (1991) proposed the use of *Dispersive Potential*, which is derived from the electrolyte concentration and composition at which the tendency of soil aggregates to disperse spontaneously is prevented. This potential P_{dis} is defined as the difference in osmotic pressure between the concentration required to flocculate (or prevent dispersion) from aggregates P_{tec} and the ambient solution concentration P_{sol}.

$$P_{dis} = P_{tec} - P_{sol}, \text{ for } P_{sol} < P_{tec} \qquad [3.9]$$

P_{tec} and P_{sol} are estimated using the valence factor of the Schulze-Hardy rule for divalent ions and from the osmotic pressure due to individual ions using the equation

$$P_{osm} = (\Sigma C_i z_i)RT \qquad [3.10]$$

where

C_i = concentration of ion I
R = gas constant

The dispersive potential of soils has been found to be related to many sodic soil properties, such as clay dispersion and modulus of rupture (MOR) (Rengasamy *et al.*, 1992). Because factors such as pH and net charge greatly influence the dispersive potential (Chorom *et al.*, 1994)

TABLE 3.10 Relative flocculating power of cations

Ion	Calculated (Equation [3.8])	Derived experimentally
Na^+	1.18	1.00
K^+	1.86	1.70
Mg^{2+}	28.16	27.00
Ca^{2+}	42.99	43.00

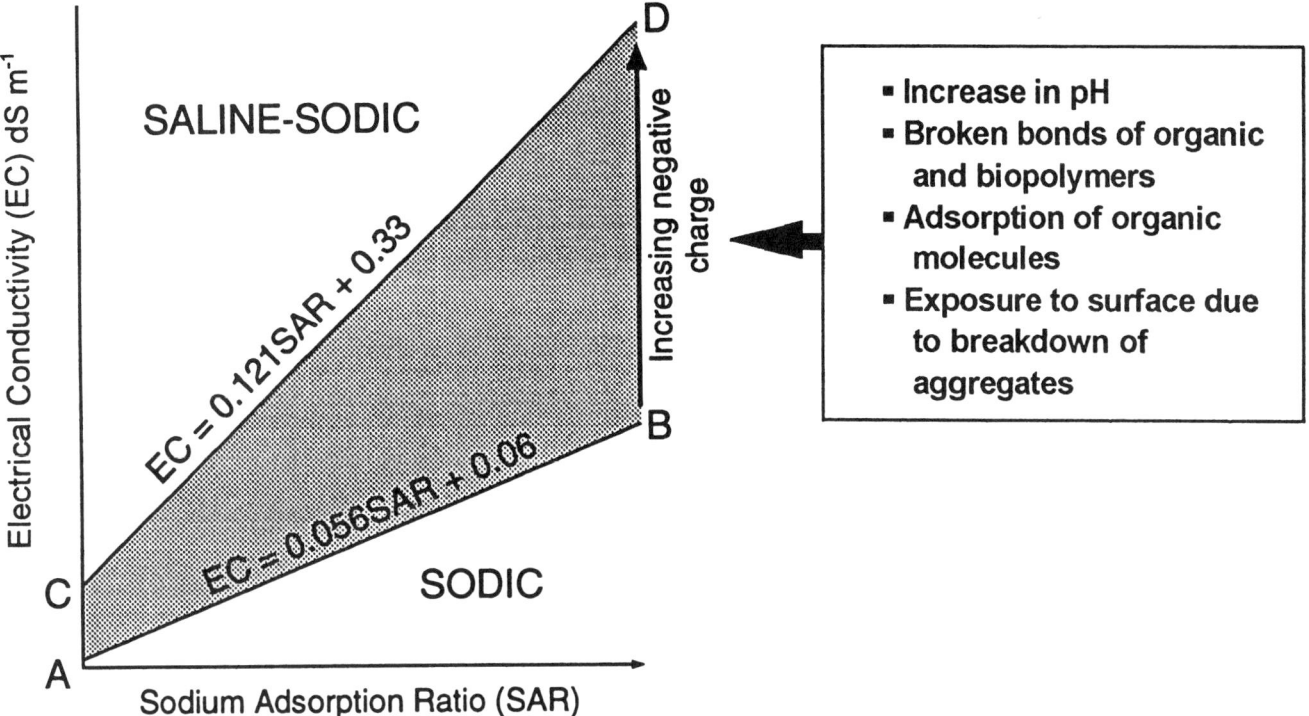

Figure 3.11 Schematic illustration of factors affecting threshold electrolyte concentration (TEC) [Rengasamy and Olsson, *Aust. J. Soil Res.* 29:935-952, (1991)].

and because the use of the Schulze-Hardy rule is not applicable for calculating the cationic effect on soil clays, this concept must be refined further to be of practical use. The use of the flocculating power of cations discussed earlier instead of the Schulze-Hardy rule will avoid treating cations of the same valence equally. Many workers have found the effects of divalent ions on soil clay swelling and dispersion and soil hydraulic properties to be different (Emerson, 1983; Rengasamy *et al.*, 1986). For example, Na/Mg-saturated soils have higher dispersibilities and lower HC than Na/Ca-saturated soils and the CFC values for Mg saturated clays are higher than for Ca saturated clays. Similarly, among monovalent cations at comparable concentrations, Na saturated soils are more highly dispersive than their K counterparts. Levy and van der Watt (1990) reported that exchangeable K cannot be grouped together with either Ca or Na as ions having similar effects on hydraulic properties. These differences between the various cations can be accounted for by their flocculating powers discussed earlier.

3.8 CONCLUSIONS

The poor physical condition of sodic soils is caused by the processes of swelling and dispersion of clay particles originally bound in the form of soil aggregates. The disintegration of aggregates on wetting and the subsequent dispersion of clay particles from the aggregates cannot be explained quantitatively by the classical DLVO theory or its modifications, which are based on the attractive Lifshitz–van der Waals and the repulsive electrostatic forces present in a dispersed colloidal suspension.

The swelling and dispersion of soil clay particles on wetting the aggregates proceed through a number of different stages. Initial hydration of cation-saturated clays leads to swelling while continuous (extensive) hydration of highly sodic clays results in the liberation and spontaneous dispersion of clay particles from the aggregates. Initially, the dry aggregates are strong with high attractive pressures between clay particles, but hydration reactions lead to repulsive forces, which reduce the attraction between clay particles, rendering the swollen wet aggregates weak. In divalent ionic clays, the clay particles can be further separated beyond the swollen stage by mechanical forces applied externally.

The extent of hydration depends on the nature of bonding between the cations and the clay surfaces. Highly ionic bonding, as is the case with Na-saturated clays, leads to extensive hydration, whereas in Ca-saturated clays with polar covalent bonding, limited hydration occurs. The nature of the bonding or the degree of covalence in an ionic bond is determined by the "hard-soft" nature of the cation and the clay surface. The Misono factor and the ionic potential of the cations are implicated in these interactions.

Once the clay particles are separated beyond 7 nm, they remain dispersed in suspension. At this stage, the electrical double layer develops and the stability of the suspension is controlled by electrostatic repulsive forces, which depend on the net particle charge. Flocculation of clay particles is brought about by the addition of electrolyte, which as a result of its osmotic effect, causes dehydration of the clay-water system, thereby reducing the distance of separation between particles. At this stage, the bonding of clay particles by cations nullifies the electrostatic repulsive forces. The degree of aggregation of the clay particles depends on the cations and their ability to form polar covalent or ionic bonds. Thus the flocculating power of cations can be derived from their Misono factor (tendency to form a dative π-bond), ionic potential (controlling the nature of bonding), and the valence of the cations (controlling the osmotic pressure).

The mechanisms to explain sodic behavior discussed in this chapter provide a framework in which the swelling of sodic soils and the release of dispersed clay can be assessed so that systems to manage and reclaim these soils can be developed, as discussed further in the chapters to follow.

4

Organic Matter, Sodicity, and Soil Structure

P. N. Nelson and J. M. Oades

4.1 INTRODUCTION

Sodicity is generally associated with poor soil structure while, on the other hand, organic matter is generally associated with good soil structure. Although there is a large body of literature on the effects of organic matter and sodicity on structure individually, the interactions between organic matter and sodicity are much less well understood (Emerson, 1983; Emerson et al., 1986; Mullins et al., 1990; Churchman et al., 1993). In this chapter, the term structure will be defined (Section 4.2), followed by a brief discussion of the nature of soil organic matter and how it influences structure (Section 4.3). The effect of sodicity on soil structure is discussed in Chapters 5 and 6 and will not be further discussed here. Interactions between organic matter, sodicity, and soil structure will then be examined (Sections 4.4, 4.5, 4.6), and in the final Section (4.7), biological amelioration of sodic soils will be discussed.

4.2 SOIL STRUCTURE

Soil structure has been defined as the size, shape, and arrangement of particles and pores in soil (Marshall and Holmes, 1988). To avoid confusion, it is neccessary to consider soil structure across nine orders of magnitude and to specify clearly what scale is being considered. Terms describing pores and particles across the range of scales are illustrated in Fig. 4.1, covering arrangements of atoms and molecules at the scale of nanometers to pores and aggregates with dimensions of the order of centimeters.

A particular structural arrangement is rarely stable as particles move under mechanical stresses, or stresses created by shrinking and swelling caused by drying and wetting. Desirable structures for specific purposes can be

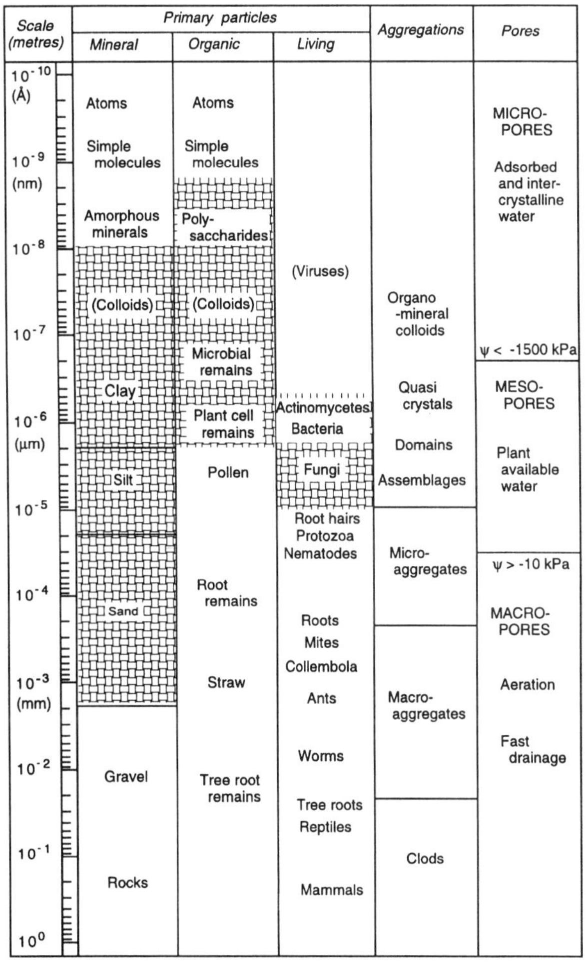

Figure 4.1 Scale in soil structure. In surface mineral soils approximately 90–99% of soil mass is mineral, 1–10% is organic and 0.01–0.5% is living. Shading shows primary particles that predominate in each category [derived from Waters and Oades, *Advances in Soil Organic Matter Research: The Impact on Agriculture and the Environment*, Royal Society of Chemistry, Cambridge, UK, (1991)].

defined. For example, for crop growth, a good structure involves the aggregation of soil particles into compound units with a range of diameters which maintain a range of pore sizes promoting infiltration of water, aeration, and root growth and providing a good physical environment for root systems and associated organisms. Thus a soil with good structure exhibits aggregation, crumbliness, or friability. The stability of this structure to withstand stresses created by drying, wetting and cultivation is very important. The formation of soil structure involves the physical forces of shrinking and swelling created by changes in water status, freezing and thawing, tillage, or movement of particles by water or large biota such as roots and worms. The larger the shrink-swell capacity of soils, the greater the tendency to regenerate macroaggregates on wetting and drying.

This review will concentrate onsoils in which the clay fraction is dominated by layer lattice silicates. Iron oxides and organic matter are minor but important constituents, intimately associated with the clay and other soil components. In these soils, structure may be considered a hierarchy (Tisdall and Oades, 1982). Aggregation (and pores) at the larger scales are dependent on aggregation at smaller scales. Thus the formation and stability of microaggregates and macroaggregates depend on dispersion and flocculation at the scale of colloidal particles. Degradation of soil structure involves at least two processes, referred to as *slaking* and *dispersion* (Chapter 3). Slaking refers to the breakdown of macroaggregates into microaggregates upon wetting, whereas dispersion refers to the release of individual clay particles. Both slaking and dispersion usually result in closer packing when the soil dries, resulting in crusting, hardsetting, high bulk density, low friability, and high strength. Slaking alone leads to the disaggregation of macroaggregates and loss of larger pores, which limits infiltration of rainfall or irrigation water but soils can still be managed, albeit carefully. When individual clay particles become detached from the aggregates, dispersion begins and creates very undesirable structures, which are not stable. Dispersion may occur spontaneously if the exchangeable sodium (Na) percentage (ESP) is high, and in these circumstances, soils should not be exposed to rainfall or be disturbed. Dispersion of nonsodic soils may occur by mechanical means, such as raindrop impact and cultivation. Structural problems created by slaking and dispersion influence plant growth and soil management through their effect on water infiltration, plant available water holding capacity, aeration, root penetration, seedling emergence, runoff, erosion, and the timing of tillage and sowing operations (Chapter 8). Important soil factors influencing slaking and dispersion include particle size

distribution, hydration, ESP, electrical conductivity (EC), mineralogy, specific surface area, organic matter, and cementing materials such as calcium carbonate ($CaCO_3$) and oxyhydroxides of iron (Fe), aluminum (Al), and silicon (Si). (Kemper and Koch, 1966; Shainberg and Letey, 1984; Goldberg *et al.*, 1988). Many measurements, reviewed by Dexter (1988a) and Kay (1990), may be used to determine various aspects of soil structure.

Sodicity influences soil structure at the scale of clay microstructure through its effect on dispersion and flocculation. The nature of sodicity and its influence on the physical properties of the soil have been reviewed by Bresler *et al.* (1982), Shainberg *et al.* (1989), Gupta and Abrol (1990a), Rengasamy and Olsson (1991), Sumner (1993a), and So and Aylmore (1993), as well as in Chapters 5, 6, and 8. Organic matter consists of many different components, which influence structure at different scales and in different ways, as discussed in the following section.

4.3 ORGANIC MATTER AND ITS GENERAL ROLE IN SOIL STRUCTURE

What follows is a brief discussion on the nature of organic materials and their role in the formation and stabilization of soil structure. For a detailed treatment of the topic, the reader is referred to reviews by Oades (1984, 1989, 1993), Emerson *et al.* (1986), and Ladd *et al.* (1996).

Approximately 1–10% of the soil mass is organic matter, of which approximately 1–5% is microbial biomass. While organic matter makes up only a small proportion of the soil mass, it is intimately associated with inorganic particles and plays an important role in the stabilization of soil structure. In soils which undergo wetting and drying, the structural role of organic matter is greatest in loams and least in clays with high shrink-swell potential. In clays, shrink-swell stresses during wetting and drying are the dominant factor in structural changes.

The form in which organic matter is added, and the way in which it is distributed in the soil, have important consequences for its subsequent role. It may be added as crop residues, litter, large roots, fine roots, exudates, animal manure, or other materials, and distributed on the surface or at varying depths. It may be subsequently redistributed by tillage or the action of fauna, such as earthworms.

Different components of organic matter act at different scales of structure (Fig. 4.1). At large scales, plant roots and mycorrhizal and saprophytic fungal hyphae enmesh macroaggregates, thus inhibiting slaking and dispersion. Plant roots and fungal hyphae are transient binding agents in that they are only present when plants are growing and fresh organic materials are being introduced to the soil. They are also easily disrupted by desiccation or mechanical disturbance. At smaller scales, mucilages and colloidal organomineral complexes play an important role in binding microaggregates through a variety of mechanisms, and their effect tends to be more persistent (Fig. 4.2).

In addition to the complex organic materials described, many other organic molecules are found in soil. A small and transient pool of organic matter comprises the simple organic acids, amino acids, and sugars produced by plant roots and microorganisms. They are transient in soil because they are readily decomposed, but their concentrations are maintained by continual production. While perhaps not directly involved in soil structure, they influence chemical and biological processes. Some highly complexing simple organic acids such as citrate influence structure directly by increasing dispersion and decreasing water stability (Section 4.6.2).

4.3.1 Transient and Temporary Bonding Agents

Living roots and particulate organic matter such as dead roots, straw, and seeds are important components of organic matter. These materials are frequently termed the "light" or "macroorganic" fraction, and may be separated from soil using sieving and density techniques following dispersion. Living roots slough off cells and exude a variety of compounds, including mucilages which consist largely of neutral polysaccharides, but they also contain acidic functional groups (e.g., from uronic acids) (Oades, 1978). Microorganisms and fauna colonize roots, the sur-

rounding nutrient-rich soil, and fresh additions of particulate organic matter, with fungi being particularly important in the initial stages of colonization. The microorganisms metabolize the more readily decomposable cell components and substrates, including the mucilages, and excrete various organic compounds and mucilages in turn. Successions of various organisms occur during decomposition of organic additions. For example, when the more readily decomposable substrates have been depleted, fungi are decomposed (Molope et al., 1987). Electron microscopy has shown that inorganic particles adhere to, and become embedded in, mucilages (Foster, 1981,1988). Thus the plant roots and debris, surrounded by microbial colonies and mucilages, become a core for stable aggregates (Oades and Waters, 1991). As readily decomposable substrates such as carbohydrates are consumed, the aggregates become less stable, and clay becomes more easily detached. Eventually, when all of the readily decomposable substrate is depleted, only the chemically resistant remains of the organic particles with no attached inorganic material survive (Golchin et al., 1994). Golchin et al. (1994) separated aggregates from this sequence and characterized their stability using density separation and ultrasonic disruption. Using ^{13}C nuclear magnetic resonance (NMR) spectroscopy, they found that the more stable aggregates contained less decomposed organic matter, whereas, in the less stable aggregates, the relative proportion of carbohydrate had decreased and aliphatic carbon (C) had increased.

4.3.2 Persistent Bonding Agents

Another important component of organic matter is polyanionic colloidal material, often referred to as humic substances. This material includes plant remains and microbial products, components of which are persistent in soil due to their chemical recalcitrance and association with inorganic materials (Stevenson, 1992). Sodium hydroxide (NaOH) or Na pyrophosphate ($Na_2P_2O_7$) have often been used to extract this material from soil, along with many other types of organic matter. Sodium and high pH promote dispersion and dissolution, while phosphates aid in dissolution by complexing metal cations such as calcium (Ca), Fe, and Al. Extraction dissolves organic matter attached to inorganic materials as well as discrete organic particles. The extracted materials are therefore not a particularly good reflection of the nature of colloidal organic matter as it exists in soil. Inferring the properties of organic matter in the soil system based upon extracts should therefore be undertaken with caution. However, a 60-year-old tradition of characterizing alkaline extracts

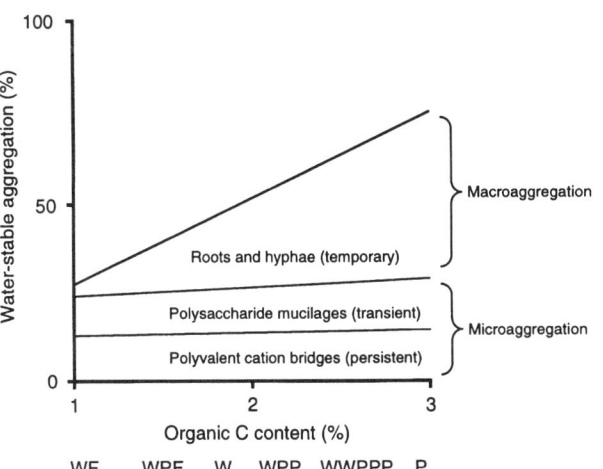

Figure 4.2 Role of organic matter in stabilizing structure in soils with different organic C contents in a long-term trial with different periods of wheat (W), fallow (F), and pasture (P) [Tisdale and Oades, *J. Soil Sci.* 33:141-163, (1982)].

has provided a great deal of information on their properties. The organic matter in these extracts which precipitates upon acidification has been termed *humic acid*, while the material that remains dissolved at low pH is *fulvic acid*. The soil organic matter that is not removed by extraction has been termed *humin*. These extracts contain considerable amounts of inorganic materials and known biochemicals, including acidic polysaccharides. The terms *humic acid* and *fulvic acid* have also been used to define the organic matter in the extracts which cannot be identified as known compounds, or the organic matter remaining after purification. Purification removes the inorganic materials with which humic acid is intimately associated in soils. Humic acid and fulvic acid contain large proportions of colloidal anionic organic matter.

Unlike neutral polysaccharides, anionic materials are mobile and may move below the zone of organic matter inputs. They range in size from molecules with molecular weights of only a few hundred to particles with diameters on the order of micrometers. Solubility, oxygen content, and charge density tend to increase with decreasing molecular size. The chemical structure of colloidal organic materials consists of a flexible backbone of aliphatic and aromatic polymers with a variety of functional groups, and their behavior is strongly influenced by their many acidic functional groups such as carboxyl (-COOH). Data compiled by Schnitzer (1978) showed carboxyl contents of 150–1120 $cmol_c\ kg^{-1}$, alcoholic OH contents of 20–950 $cmol_c\ kg^{-1}$, and phenolic OH contents of 30–570 $cmol_c\ kg^{-1}$ for humic and fulvic acids. Other important functional groups include carbonyl, amino, imidazole (aromatic ring -NH), and thiol (-SH). When dissolved in water, their acidic functional groups become more or less dissociated. As most of the acidic functional groups dissociate between pH 2 and 6, organic materials usually bear a net negative charge in soils. In water, the negative charges within the polymer repel each other, and the polyelectrolyte assumes a stretched-out form. At low concentrations, neutral to alkaline pH, or low ionic strength, they tend to behave as flexible linear colloids. In contrast, they behave as spherocolloids at high concentration, low pH, and high neutral salt concentrations (Ghosh and Schnitzer, 1980). Upon the addition of salts (or in the presence of salts in the soil), cations are attracted to the negatively charged functional groups, which causes a reduction in the intramolecular coulombic repulsion of the polymer chain, resulting in a smaller, more rigid configuration. This effect, termed the *Fuoss effect*, reduces the amount of water held within the molecule and explains why it behaves

as a colloid like a clay particle, obeying double-layer theory (Ong and Bisque, 1968). The effect of different cations on the configuration of organic polymers is discussed in Section 4.4.3.

Depending on their properties, and the amount and nature of charged surfaces and other ions, anionic organic materials also become adsorbed to inorganic surfaces by electrostatic and other forces. They are attracted to the positive charges on inorganic colloids, especially oxides, and decrease their positive charge (Oades, 1989). They thereby reduce the point of zero charge (PZC) of the soil, thus raising the critical flocculation concentration (CFC) and the tendency to disperse at a given pH (Shanmuganathan and Oades, 1983a; Oades *et al.*, 1989) (Chapter 3). The molecules also link with polyvalent cations, either complexing them or becoming linked to negatively charged inorganic colloids via cation bridges. In the presence of sufficient polyvalent cations, these linkages are relatively persistent, being resistant to microbial decomposition, chemical extraction, and physical disintegration. In this case, the high cation exchange capacity (CEC) associated with organic matter enhances soil structural stability (Theng, 1982; Oades, 1984). However, the high negative and low positive charge densities in soils associated with the presence of anionic organic matter also have a tendency to increase clay dispersion (Section 4.6.2). Basic functional groups containing nitrogen (N) such as amines, which can become positively charged, also exist in organic matter, and tend to be strongly sorbed to clay surfaces. Uncharged alkyl and aromatic moieties may also be involved in bonding by adhering to clay particles through hydrophobic and van der Waals interactions. Alkyl C makes up a large proportion of the organic matter in the clay fraction of soils (Golchin *et al.*, 1994).

4.4 SODICITY INFLUENCES THE AMOUNT AND NATURE OF ORGANIC MATTER

Sodicity influences plant production and the amount of organic matter that enters the soil, and the loss of organic matter by mineralization, erosion, and leaching. As a consequence, sodicity influences the amount and nature of soil organic matter. Sodic soils are susceptible to erosion, and the loss of surface soil in this manner can cause significant losses of organic matter. The influence of sodicity on erosion has been discussed elsewhere (Chapter 5), and will not be discussed further here. The effects of sodicity on organic matter are discussed in the following sections.

4.4.1 Organic Matter Content

Examination of the C contents of the Australian Great Soil Groups showed that sodic soils (Solonetz, Solodized Solonetz, Solods, Sodic Red Brown Earths, and Solonized Brown soils) had lower organic C contents than all other soils except sands and arid zone soils (Spain *et al.*, 1983). The sodic Great Soil Groups had C:N ratios less than 12, indicating a high degree of decomposition. Soil groups dominated by oxides of Fe and Al, or those containing $CaCO_3$, generally had higher organic C contents (>2%) and C:N ratios (> 15). In a number of studies, negative correlations have been found between organic matter content and ESP. For example, within a group of 28 Red Brown Earths in South Australia, Grierson (1975) found a significant inverse correlation between organic C content and ESP (Fig. 4.3). Data from long-term trials and naturally varying soils in Western Australia (Aylmore and Sills, 1982) and India (Banerjee, 1959; Chander *et al.*, 1994) fit similar relationships. The low organic matter contents of sodic soils are due to low input rates and high rates of loss.

4.4.2 Inputs of Organic Matter

The inputs of organic matter to soil generally depend on on-site plant productivity and the retention or the harvesting or removal of plant shoot materials. Reasons for poor plant growth, which have been reviewed several times

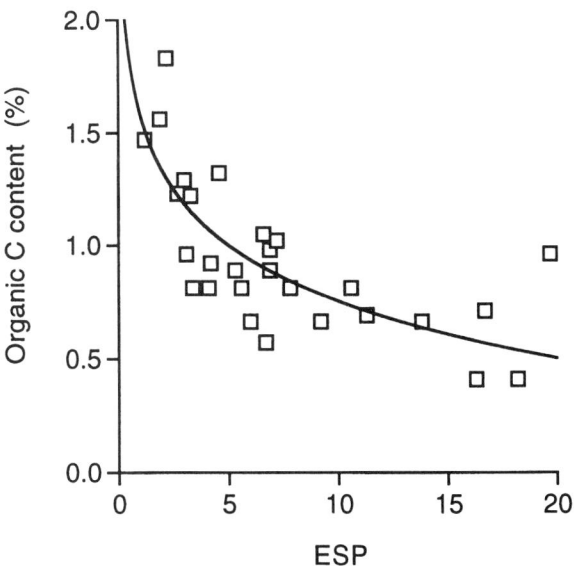

Figure 4.3 Organic C content and exchangeable sodium percentage (ESP) of 0–0.05 m layer of Red Brown Earths from midnorth of South Australia [Grierson, M. Agric. Sc. Thesis, University of Adelaide, Australia, (1975)].

(Gupta and Abrol, 1990a; Naidu and Rengasamy, 1993; Chapters 5, 6, and 7), can be separated into effects of sodicity on soil structure and nutrient availability.

Up to a critical ESP for a given soil and plant, growth is limited primarily by soil structural factors, which include the effects of high soil strength on seedling emergence and root penetration. High strength becomes particularly important as the soil dries due to uptake of water by roots. Limited root penetration and the effect of reduced aeration on root growth curtails nutrient and water uptake, and may also increase susceptibility to root diseases. High runoff from sodic soils may reduce the total amount of water held in the soil, while the water retained may be less available to plants due to being held at a lower potential. Structural factors also affect tillage operations and make their timeliness and effectiveness more critical and difficult to achieve. Rice (Oryza satura) is the only crop where the structural deterioration due to the presence of Na may be beneficial.

In addition to influencing the uptake of water and nutrients by roots due to physical factors, sodicity also affects the forms and amounts of nutrients available. The small amount present and the low inputs of organic matter can result in restricted supplies of sulfur (S), phosphorus (P), and, particularly, N to plants. Deficiencies of N and also zinc (Zn) are common, and their application improves crop tolerance to sodic conditions. At high ESP and low P_{CO2}, Ca may be deficient (Chapter 7). A loss of nutrients by the erosion of dispersed colloids also reduces their availability to plants. Phosphorus is relatively available in alkaline sodic soils, while potassium (K) is normally adequate. For some plants such as sugar beet (Beta vulgaris), which have a requirement for Na, yield can increase to a certain extent with increasing sodicity (Draycott, 1993). However, when exchangeable Na is present in large amounts, it becomes toxic to most plants. Other common toxicities include boron (B), molybdenum (Mo), selenium (Se), and bicarbonate (HCO_3^-), especially where pH is high. Plants vary in their tolerance to sodicity, which is discussed in Section 4.7 and Chapters 7 and 8.

4.4.3 Solubility, Dispersion and Adsorption of Organic Matter

Solubility and Dispersion of Organic Molecules
High solubility of organic matter in the presence of Na, particularly in combination with high pH, low electrolyte concentration, and complexing anions such as pyrophosphate ($P_2O_7^{2-}$), oxalate (COO^-)$_2$, flouride (F), EDTA, and carbonate (CO_3^{2-}), is well known (Greenland, 1965b). The

TABLE 4.1 Solubility of Na and Ca salts of simple carboxylic acids

	Solubility (Max. conc. in cold water M)	
	Na salt	Ca salt
Monocarboxylic acids		
Formic	1.36	0.13
Acetic	14.5	2.8
Propionic	10.4	2.4
Lactic	>8.9	0.1
Gluconic	2.71	0.08
Benzoic	4.58	0.08
Salicylic	6.94	0.09
Cinnamic	0.535	0.006
Oleic	0.329	0.001
Polycarboxylic acids		
Oxalic	0.276	0.0001
Succinic	0.79	0.01
Tartaric	1.26	0.0001
Citric	2.66	0.0015

Na salts of simple organic acids are more soluble than their Ca counterparts by up to several orders of magnitude. The difference in solubility increases with the size of the organic anion and is greater for poly- than for monocarboxylic acids (Table 4.1). Larger organic materials act as colloids, so CFC is a more meaningful way of describing solubility. The Na salts of polyacidic organic colloids ionize readily, thus enhancing double layer repulsion and dispersion. Organic complexes of polycations, especially Al in acidic soils and Ca in neutral-alkaline soils, ionize less readily. Due to cross-linking and bridging with inorganic materials, polyvalent metal-organic complexes become less readily solvated and do not disperse readily (Chapter 3). Ong and Bisque (1968) prepared metallorganic colloids with the alkali-soluble fraction of a peat and various metal cations. The CFC values at pH 7 for the Na and Ca organic colloids were 598 mM NaCl and 7.2 mM CaCl$_2$, respectively. In the presence of Fe^{3+}, the respective CFC values were reduced to 51 and 0.5 mM. The effects of cation valence were in agreement with the Schultz-Hardy rule and DLVO theory [CFC $\propto$ 1/z^6], with small differences between cations having the same valence. Within groups, the CFC of mono- and divalent cations was related

to their hydrated ionic radii. Trivalent cations, because of their high ionic potentials, tend not to occur as simple species. The more rigid organic moieties with closely spaced negative charges sequester highly charged cations such as Ca and do not swell or become dispersed in the presence of Na. The more flexible materials, with more widely spaced carboxyl groups, would have lower relative selectivity for Ca (but probably still higher than clays), and may be the moieties which tend to disperse in a mixed Na/Ca system.

The degree to which organic materials are solubilized or dispersed depends not only on charge characteristics, but also on other properties. Large organic molecules with high contents of non-polar structures such as aliphatic chains or aromatic rings, or strong associations with inorganic colloids tend to dissolve less readily. When organic matter does disperse or dissolve, it does so not only as pure organic matter, but also as organomineral complexes. The mineral part of the complexes ranges from single metal atoms through amorphous (hydr)oxides of Fe, Al, and Si to (hydr)oxide or aluminosilicate crystals (clay particles). The divisions between dissolved, colloidal, and suspended particles are usually operationally defined. Suspended particles are sufficiently large to settle under the influence of gravity. The diameters of dissolved particles range from 10^{-10} to 10^{-9} or 10^{-8} m, whereas colloidal particles range from 10^{-9} or 10^{-8} to approximately 10^{-7} m (Fig. 4.1). The important point is that both dissolved and colloidal particles are mobile in water. Particulate organic matter in soils has been less extensively studied than colloidal organic matter, but components of it too are probably negatively charged and susceptible to the influence of Na.

Solubilization of Metal Complexes

Trace metals in soil are to a large extent complexed by organic matter. Metal-organic complexes predominate in the solid phase, but may also occur in solution, depending on their size and charge characteristics, and on other soil properties (Kabata-Pendias and Pendias, 1991; Moraghan and Mascagni, 1991; Stevenson, 1991, 1992; Harter and Naidu, 1995). Metal cations with a high affinity for organic ligands, such as the transition metals, readily displace Na from exchange sites, especially those with a tendency to form coordinate linkages. However, there may still be Na associated with other sites on the same molecule. Therefore, the dissolving or dispersing action of Na on organic molecules and organomineral complexes can increase the concentration of organically complexed metals in solution (Sholkovitz and Copland, 1981). Higher metal concentrations in solution increase their mobility, and perhaps their mineralizability. Depending on the stability of the complexes, metals may be released by low pH or mineralization, both of which tend to be enhanced in the rhizosphere.

Associations between Organic Matter and Inorganic Particles

Most soil organic matter is associated with mineral particles, and the types of interactions depend on the properties of both materials (Greenland, 1965a,b; Hayes, 1980; Theng, 1982; Oades, 1989). The nature of the exchangeable cations is an important property of the surfaces, and their effect on adsorption is greatest in clays with high specific surface area. Strongly polarizing cations (with high ionic potential) such as Ca^{2+} usually possess an inner hydration shell and an outer coordination sphere. For small organic molecules to interact with the surface, they must either (a) displace cations by an exchange process, or (b) displace H_2O molecules from the outer coordination sphere for strongly polarizing cations, and probably from the inner sphere for monovalent cations, or (c) they must interact with the cation by neutralizing its charge.

Important properties of the organic molecules are polarity, polarizability, charge density distribution, dissociation constants of acidic groups, solubility, size, and shape (Chapter 3). When considering the effect of exchangeable cations on adsorption, the charge characteristics of the organic molecule are the most important properties. Soil organic materials may be positively or negatively charged or neutral. Positively charged organic molecules containing amino groups such as proteins (including enzymes) adsorb to clay by cationic exchange reactions, and their adsorption is influenced by the nature of the saturating cations. Sodicity enhances adsorption of organic cations because exchangeable Na is more easily replaced than exchangeable Ca. The adsorption of uncharged molecules depends on the energy needed to displace water molecules from the exchangeable cations as well as on the surface area available for adsorption. Sodicity increases the adsorption of large neutral polymers such as polysaccharides by increasing the available surface area of clay through dispersion. This has been shown in clay suspensions, especially for expanding clays such as montmorillonite (Greenland, 1963; Parfitt and Greenland, 1970), but it has not been demonstrated in soil. Adsorption of cationic and nonionic polymers is enhanced by low EC (Ben-Hur *et al.*, 1992).

While exchangeable Na may increase adsorption of positive or neutral organic molecules, the opposite is true for organic anions, which constitute a large proportion of soil organic matter. Cation bridges play a major role, linking minerals and polyanions. Links form through a water bridge for strongly hydrating cations such as Ca or through direct ion dipoles for monovalent cations. The strength of adsorption of polyanionic soil organic matter to clays is proportional to the ionization potential of the cation and is hence greater for Ca than Na (Fig. 4.4) (Evans and Russell, 1959; Theng and Scharpenseel, 1976). Adsorption is reduced at high pH (Mortensen, 1962) and enhanced by low pH and high ionic strength (Theng, 1982; Ben-Hur *et al.*, 1992). Sodium may reduce both the affinity of the surfaces for anionic organic matter, i.e., the slope of the adsorption isotherm, and the amount of anionic organic matter adsorbed at saturation (Ceppi *et al.*, 1993).

Implications

High mobility of organic matter and organomineral complexes in sodic soils results in loss from surface layers by leaching (Jacquin *et al.*, 1979). Leaching is evident in Solonetz soils, where organic matter may accumulate and stain the top of the B horizon. Following loss from the soil by leaching, dissolved organic matter and organomineral complexes may have a negative impact on water quality. High concentrations of dissolved organic matter have been measured in streams draining sodic soils (Naidu *et al.*, 1993a), and Skene and Oades (1995) demonstrated a positive relationship between the dissolved organic C

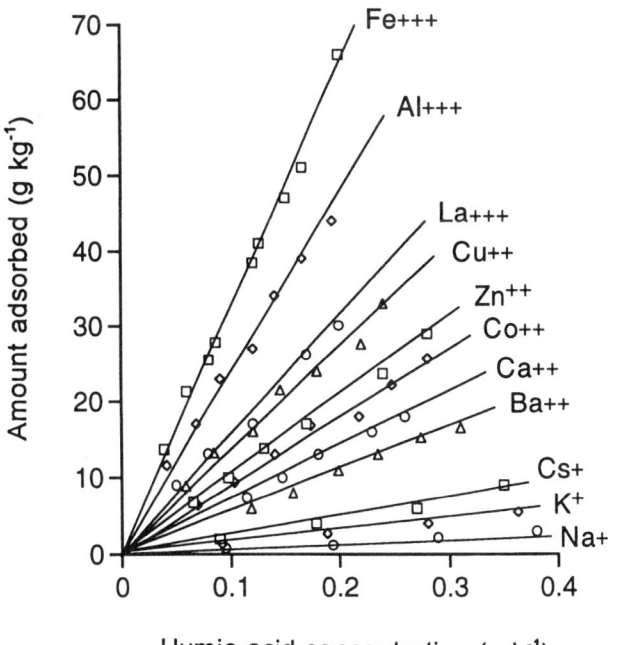

Figure 4.4 Adsorption isotherms for humic acid on montmorillonite saturated with different cations [Theng and Scharpenseel, *Proc. Int. Clay Conf.* pp. 643-653, (1976)].

concentration and sodium adsorption ratio (SAR) of stream waters in South Australia. In alkaline Very Sodic soils (Sod$_H$) (Chapter 1) in areas where evaporation is greater than precipitation, organic matter moves to the soil surface and accumulates as black "slick spots."

4.4.4 Retention or Mineralization of Organic Matter

Aside from losses by erosion or leaching, organic matter added to soil is either retained or mineralized to CO_2 and returned to the atmosphere. Sodicity may influence the decomposition of organic matter both directly and indirectly, but most effects lead to reduced retention of organic matter in soils. Because organic matter and clay contents are positively correlated (except at very high clay contents), dispersion and eluviation of clay lead to coarser textured A horizons, less capable of retaining organic matter over time.

Sodicity also influences the structure of soils, which plays a role in microbial and faunal activity and trophic interactions. For example, pore size distribution influences species composition by determining aeration and the movement of micro- and mesofauna. The effects of soil structure on microbial activity have been reviewed by Ladd *et al.* (1993, 1995).

Over the last 80 years, evidence has accumulated which shows that the presence of exchangeable Na stimulates mineralization and increases losses as dissolved organic matter from the soil (Fig. 4.5), which may be due to greater solubility of the Na than Ca salts of organic materials (Sokoloff, 1938). Oxygen uptake was greater in soil amended with Na than with Ca humate (Juste and Delas, 1970). The assumption is that Ca blocks functional groups which represent sites of initial decomposition (Juste *et al.*, 1975) and also cross-links flexible polymers to create dense rigid molecules, which are more stable to both chemical and biological attack (Oades, 1988, 1989). Sodium has the opposite effect, dispersing and dissolving organic materials, making them more accessible for microbial decomposition. Other polyvalent metals stabilize organic materials in the same manner as Ca and, in addition, can be toxic to microorganisms and inactivate enzyme systems. The positive influence of Na on decomposition could be expected to be greatest for small or colloidal anionic substrates and least for particulate, uncharged substrates.

Incubation of ^{14}C-labeled plant materials in soils with different Ca and Na status has also shown a relative stabilizing effect of Ca versus Na. For example, when ^{14}C-labeled barley straw was incubated in a soil with ESP values of 6 and 52, mineralization of the ^{14}C was initially greater in the soil with an ESP of 6. However, after 2 months, the trend was reversed, so that after 200 days, the ^{14}C mineralized was greatest at an ESP of 52 (53% vs. 46%). The initial retarding effect of the high ESP on mineralization was thought to be due to poor aeration of the soil sample (Abdou *et al.*, 1975). The anaerobic conditions caused by sodicity could decrease mineralization in the field through their negative impact on microbial activity. In both laboratory and field experiments with ^{14}C-labeled glucose and straw, respectively, the addition of Ca salts decreased the mineralization of the ^{14}C (Muneer and Oades, 1989a,b). Interestingly, the addition of $CaCO_3$ initially increased the mineralization of C, presumably due to a positive effect of the pH change on microbial and faunal activity (Baldock *et al.*, 1994), but, subsequently, the rate of mineralization decreased compared to a soil with no $CaCO_3$ added; after 6 months, the mineralization of C was depressed by additions of $CaCO_3$ or $CaSO_4$ (Muneer and Oades, 1989a,b). The addition of Na salts has generally enhanced the rate of C and N mineralization, particularly with salts such as Na_2CO_3 which increase pH (Laura, 1973, 1976, 1991; Bruno and Pietramellara, 1992).

Although sodicity may increase the rate of mineralization, salinity has the opposite effect due to its osmotic influence on microbial activity. Thus when a Saline Sodic soil (Sod$_H$ Sal$_H$ or Sod Sal) (Chapter 1) is ameliorated and both sodicity and salinity are reduced, mineralization of organic matter may be greater in the ameliorated than in the original soil (Malik and Haider, 1977). In an experiment examining the interaction between sodicity and salinity, sodicity increased the mineralization of ^{14}C in a soil amended with ^{14}C-labeled clover residues, whereas salinity reduced mineralization irrespective of sodicity (Fig.

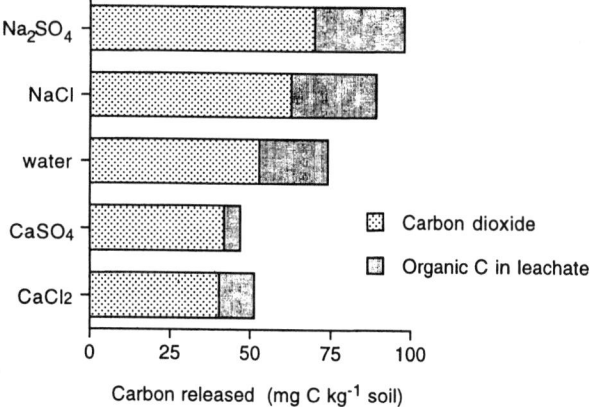

Figure 4.5 Influence of leaching with water or solutions of Na or Ca salts (0.05 mol$_c$ L^{-1}) on loss of organic C from a sandy loam (0.13% organic C) during 45 d incubation [Sokoloff, *J. Agric. Res.* 57:201-216, (1938)].

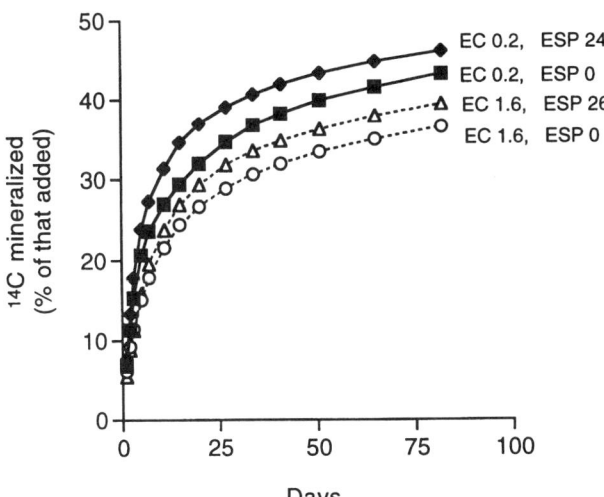

Figure 4.6 Influence of exchangeable sodium percentage (ESP) and electrical conductivity ($EC_{1:5}$ dS m^{-1}) on mineralization of ^{14}C-labeled *Trifolium subterraneum* shoots [Nelson *et al., Soil Biol. Biochem.* 28:433-441, (1996)].

4.6). The positive effect of sodicity on the mineralization of native soil C was reduced by salinity (Nelson *et al.,* 1996). In conclusion, sodicity may increase the mineralization of organic matter by making it more accessible to microorganisms, but may also decrease mineralization due to the effects of anaerobic conditions on microbial activity.

4.5 ORGANIC MATTER INFLUENCES SODICITY

Organic matter influences sodicity through (a) its CEC and the selectivity of exchange sites for Ca over Na, and (b) its ability to donate protons. As a consequence, organic matter may inhibit the degree to which a soil becomes sodic under the influence of sodic water, or it may reduce sodicity when added to a sodic soil. These effects are discussed in the following sections.

4.5.1 Cation Exchange Capacity of Organic Matter

Organic matter contributes 25–90% of topsoil CEC in mineral soils (Stevenson, 1994) and, therefore, has a direct bearing on sodicity. The relative contribution of organic matter to the CEC of soils is greatest in soils with low inorganic CEC, i.e., low clay content, or low-charge minerals such as kaolinite. Its contribution is least in soils with high inorganic CEC, i.e., high clay content, or highly charged minerals, such as vermiculite or montmorillonite, or at low pH, where acidic organic groups are not dissociated. Because of complexing, the CEC values of organic matter and clay are not additive. As a rule of thumb, each

weight percent of organic matter contributes about 3 cmol$_c$ kg^{-1} to the CEC of neutral, permanent-charge soils (McBride, 1994) and about 1 cmol$_c$ kg^{-1} in variable-charge soils (Oades *et al.,* 1989).

The CEC of organic matter originates mostly from carboxylate (-COO$^-$) (~55% of CEC) but also from phenolate, enolate, and imide (=NH) groups (Broadbent and Bradford, 1952; Schnitzer and Skinner, 1963), with negative charge being directly proportional to pH. The actual CEC at a given pH equals the potential CEC, or total acidity, multiplied by the degree of dissociation. The dissociation of organic acids is discussed further in Section 4.5.5. The CEC of soil organic matter ranges from 60 to 300 cmol$_c$ kg^{-1} (Leinweber *et al.,* 1993; Stevenson, 1992) with values up to 1400 cmol$_c$ kg^{-1} of total acidity being recorded for humic and fulvic acids. Organic matter has a greater CEC than most soil minerals (Fig. 4.7). Even at a given pH, it is not possible to give an exact value for the CEC of organic matter, as it depends on associations with inorganic materials. Negatively charged sites on organic matter interact with positive sites on inorganic colloids, especially oxides, thereby reducing the CEC of organic matter and the amount of positive charge in the soil (Oades *et al.,* 1989). Because of its complexation of metal ions, principally Fe^{3+} and Al^{3+}, soil organic matter has a weaker acidic character than could be expected from its carboxyl content (Coleman and Thomas, 1967). In acid mineral soils, where Al^{3+} is soluble, the adsorption of Al species is so strong that CEC is effectively reduced, and organic matter contributes little to CEC (McBride, 1994). In neutral to alkaline soils, Ca could be expected to have a similar effect, though not as great, because Ca has a lower ionization potential than Fe or Al. Hence, for a given organic matter content, its contribution to CEC would be greatest in sodic soils. For a more detailed discussion of the cation exchange properties of organic matter, the

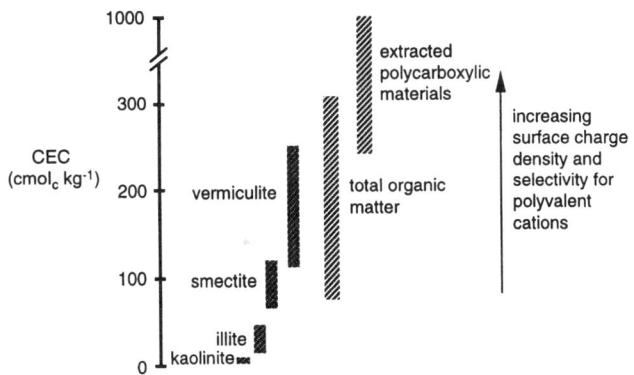

Figure 4.7 Cation exchange capacity (CEC) and selectivity of soil clay minerals and organic matter.

reader is referred to Bloom (1981), Sposito (1989), and Stevenson (1992). In the presence of polyvalent cations, the high CEC associated with organic matter enhances soil structural stability (Section 4.3). However, in the presence of exchangeable Na, a high negative charge increases dispersion (Section 4.5.5).

4.5.2 Selectivity of Organic Exchange Sites

The selectivity and bonding strength of all cation exchange sites in soil are greater for divalent than for monovalent cations. Calcium is preferred to Na because of its higher ionic potential. On layer minerals, bonding energies are, on average, twice as high for divalent than for monovalent cations (Marshall, 1964). The preference for magnesium (Mg) is slightly less than that for Ca (Hunsaker and Pratt, 1971; Curtin *et al.*, 1994). However, ion-exchange equilibria vary considerably with the nature of the exchange material (Wiklander, 1964). Schachtschabel (1940) found that organic materials adsorbed more Ca than clay minerals, and he was the first to put forward the idea that cations exist in different proportions on different types of colloids in soil. The selectivity of exchangers can be expressed as the relationship between the proportion of ions on the exchange phase and those in solution. A number of approaches have been taken for quantifying selectivity, and they have been reviewed by Marshall (1964), Thomas (1977), Bresler *et al.* (1982), Shainberg and Letey (1984), and Sposito (1989). None of the parameters used to express selectivity for ions of unequal charge are entirely satisfactory for all soil conditions because of the problems of measuring cation activity in the exchanger phase, and the variety of cations and exchange sites involved. Therefore, simple expressions of selectivity like the Gapon constant (Equation [4.1]) are still useful (Tucker, 1983).

$$ESR = K_g \times SAR \qquad [4.1]$$

where

ESR = exchangeable Na ratio (exch. Na/exch. Ca)
K_g = Gapon constant

Reasons for Selectivity

An important characteristic of exchanger material is the exchange capacity or, more importantly, its surface charge density (SCD). Ions are arranged unevenly on soil surfaces depending on their SCD. The ratio of adsorbed divalent to monovalent cations on exchange materials increases in proportion to their SCD (Eriksson, 1952). The

CEC and the charge density of organic matter are considerably higher than those of most clay minerals in soil (Fig. 4.7). Although it should be kept in mind that the SCD of clay minerals is not necessarily correlated with CEC, the ratio of adsorbed Ca or Mg to adsorbed Na or K generally decreases in the order organic matter > montmorillonite > illite > kaolinite (Wiklander, 1964). Schachtschabel (1940) leached various exchangers with a mixture of Ca and NH_4 acetate (both 0.05 mol_c L^{-1}). Muscovite and biotite took up only 5–6% Ca, feldspar 19%, kaolinite 54%, montmorillonites 56–63%, and humic matter 92%. The relative selectivity of organic matter for Ca over Na increases with increasing pH due to increasing CEC (Pratt *et al.*, 1962; Gupta *et al.*, 1984b). As selectivity is related to cation charge, the trivalent metal cations Fe^{3+} and Al^{3+} form highly stable complexes with organic matter and, in effect, block exchange sites, thus reducing CEC. An important interaction is CEC with ionic strength. The replacing power of polyvalent cations relative to that of monovalent cations is greater at low than at high solution activities, according to Schofield's Ratio Law. This effect becomes more marked as CEC increases (Wiklander, 1964).

Although important, differences in the SCD of different exchanger materials do not fully explain the high relative selectivity of organic matter for Ca over Na (Pratt and Grover, 1964), which is also related to the localization of charge and the molecular structure. Studies of synthetic organic cation exchangers have shown that the degree of selectivity for Ca over Na is proportional to their degree of cross-linking (Helfferich, 1962). In clays, negative charges tend to be diffuse, especially in 2:1 minerals, in which charge originates mainly from the octahedral layer such as in montmorillonite. This type of exchanger surface is termed a weak field exchanger (McBride, 1989). Organic exchange sites also behave as weak field exchangers due to their overall negative charge. However, external acidic groups have a discrete negative charge when ionized, and may behave as strong field exchangers. Strong field exchangers have a higher relative selectivity for cations with high ionization potential such as Ca because displacement of hydration shells results in strong coordinate linkages being formed. These coordinate linkages have also been termed *inner sphere complexes*, and the molecule that combines with the metal ion is commonly referred to as the ligand. Organic groups containing O, N, and S have unshared pairs of electrons that can form coordinate linkages with Ca. Sodium, on the other hand, tends to form diffuse ion swarm associations, or outer sphere complexes with organic matter (Sposito, 1989). The stability constant of complexes provides a measure of affinity. In general the affinity of organic functional groups, measured as stability constants, for metal ions decreases in the order: enolate (-O⁻) > amine (-NH₂) > azo (-N=M-) > ring N (-N-) > carboxylate (-COO⁻) > ether (-O-) > carbonyl (C=O) (Chaberek and Martell, 1959).

In addition to being discrete, some of the negatively charged groups on organic matter are able to move in order to attain configurations with maximum stability, unlike the negative charges on clay, which are fixed in space. Where two or more negative charges lie close together on organic matter, chelates may form, conferring very high stability on the complex, and greatly enhancing its selectivity for more highly charged cations. Chelation can be considered to render cations nonexchangeable. In general, chelation occurs through oxygen or nitrogen atoms and takes place only when 5- or 6-membered rings containing the metal atom can be formed. The stability of chelates depends on the properties of the chelator and metal cation. Neither Ca nor Na form complexes with organic matter as readily as some metals, but of these two, Ca forms much more stable complexes than Na, especially where two or more negative charges on the ligand allow chelation. Complex stability is proportional to the ionization potential of the cation and generally decreases in the order Al > Fe > Mg > Ca >> Na > K (Chaberek and Martell, 1959). The presence of metals with higher ionization potentials than Ca will reduce the amount of Ca that is chelated. Of the 1028 complexes with organic ligands reported by Martell (1964), 205 Ca complexes and only 8 Na complexes were recorded. The Na complexes had stability constants 1–9 orders of magnitude lower than the corresponding Ca complexes under the same conditions (Table 4.2). The differences in dissociation constants between the Na and Ca complexes were proportional to the number of acidic groups on the ligand. Some of the properties of the chelator which influence stability include steric effects and ring size. Stability increases with increasing polydentate nature and increasing basicity of the chelator, with five-membered rings being the most stable (Chaberek and Martell, 1959). Although the extent to which Ca is chelated in soil is unknown, the existence of chelating groups, especially in root exudates, is well documented (Stevenson, 1992). Any conditions which reduce the activity of Ca in solution, such as high ESP, high pH (Gupta *et al.*, 1984b), or the presence of complexing anions such as carbonate or phosphate, favor the adsorption of Ca by exchange sites having high affinity.

Due to the complex nature of organic matter, the importance of specific organic sites or structures for selectivity cannot be quantified (Stevenson, 1992). Stability constants are very difficult to measure for complex organic materials. A range of values may occur on one molecule, which are affected by prior complexation with other groups on the same molecule, and pH. Stability constants are lower for complex materials than for simpler compounds (Stevenson, 1992). Modeling approaches which have been used to describe the strength of binding at different sites are discussed by Stevenson and Fitch (1986) and Stevenson (1991, 1992). Because of hysteresis in cation exchange, the order in which cations are introduced influences the extent of adsorption (Marshall, 1964). In reviewing the

TABLE 4.2 Stability constants of some organic complexes of Na and Ca under corresponding conditions [data from Martell, *Stability Constants of Metal-Ion Complexes*, The Chemical Society, London, UK, (1964); and Pettit and Powell, *Stability Constants Database*, International Union of Pure and Applied Chemistry/Academic Software, Timble, UK, (1993)]

Ligand	No. of carboxyl groups	Stability constant (pK)	
		Na complex	Ca complex
Acetic acid CH₃•COOH	1	-0.27	0.57
Salicylic acid HO•C₆H₄•COOH	1	-0.5	0.63
Succinic acid HOOC•(CH₂)(CH₂)•COOH	2	0.47	1.45
Malonic acid (HOOC)•CH₂•COOH	2	0.57	1.64
Malic acid HOOC•CH₂CH(OH)•COOH	2	0.30	1.95
N-(o-Sulfophenyl)iminodiacetic acid	2	0.98	4.57
N-(o-Carboxyphenyl)iminodiacetic acid	2	0.98	5.06
Uramil-N,N-diacetic acid	2	3.32	8.77
Citric acid HOOC•CH₂C(OH)(COOH)•CH₂COOH	3	1.03	4.91
Nitrilotriacetic acid N(CH₂COOH)₃	3	2.15	8.17
Ethylenediamine-N,N,N′,N′-tetra acetic acid	4	1.66	10.59

literature on hysteresis in cation exchange on 2:1 clay minerals, Verburg and Baveye (1994) found that it was greatest in heterovalent exchange and where the system had undergone drying. Although there is much less information on hysteresis of exchange on organic materials, it would be at least as great as for clay minerals. Due to their ionic nature, Na-organic matter complexes retain their solubility upon drying and rewetting. When Ca-organic matter complexes are dried, their molecular structure rearranges, causing entrapment of Ca, so that upon rewetting, the solubility of the complex and the exchangeability of the adsorbed Ca are reduced (Toth, 1964).

Significance of Selectivity in Soil

In soils, the overall selectivity of exchange sites for Ca over Na is proportional to organic matter content where it varies due to cultivation history, depth, or treatment with H_2O_2 (Pratt and Grover, 1964; Poonia and Talibudeen, 1977; Poonia *et al.*, 1984; Curtin *et al.*, 1995a). The relationship between selectivity (expressed by the Gapon constant K_g) and soil organic matter content is illustrated diagrammatically in Fig. 4.8. The intercept depends on the selectivity of the inorganic exchange sites present. A K_g of around 0.01475 (Richards, 1954) applies to a wide range of irrigated soils partly because the organic C content of arid-zone soils generally lies between 0.3 and 1%. Poonia and Talibudeen (1977) established a variation in organic

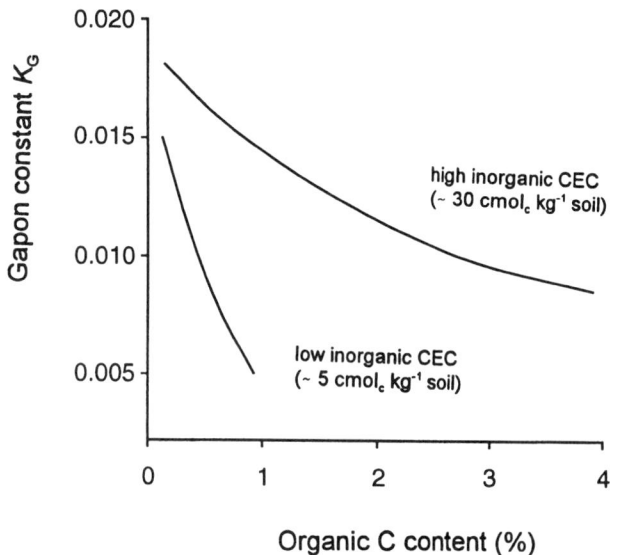

Figure 4.8 Relationship between Gapon constant K_g (exchangeable sodium ratio divided by sodium adsorption ratio of equilibrium solution) and organic C content in neutral soils [derived from Poonia and Talibudeen, *J. Soil Sci.* 28:276-288, (1977)].

matter content by sampling soils from different depths and from plots in India and England receiving different amounts of farmyard manure. In each set of soils, selectivity was a function of organic matter content and total SCD. In a group of 16 Canadian soils with organic C contents ranging from 4 to 92 g kg^{-1}, K_g was negatively correlated with organic C content ($r = -0.90$), pH ($r = -0.90$), and CEC ($r = -0.95$) (Curtin *et al.*, 1995a). Raising the pH of a soil with high organic C content also decreased K_g (Curtin *et al.*, 1995a). Poonia *et al.* (1984) found that soils with high organic matter content had high selectivity for Ca at ESP less than 50, but not at ESP greater than 50. Added manure had a higher selectivity for Ca than natural soil organic matter. Gupta *et al.* (1984a) added an NaOH extract of decomposed farmyard manure to a soil from which the organic matter had been removed by oxidation, and found a similar increase in selectivity. In peat soils containing 45–50% organic matter, selectivity for Ca over Na was found to be much higher than for mineral soils or clay minerals (Naylor and Overstreet, 1969).

Summary and Implications

In conclusion, organic exchange sites tend to hold more Ca and less Na than their clay counterparts, with the difference being moderated by the presence of more highly charged cations such as Fe^{3+} and Al^{3+}. The selectivity of organic exchange sites for Ca over Na increases with (a) increasing charge density or CEC of the organic matter, particularly where adjacent carboxylate groups allow the formation of chelates, (b) increasing solution pH, and (c) decreasing Ca activity, due to decreasing ionic strength, increasing ESP, or complexing anions.

An implication of exchange site selectivity is that resistance to sodification is greater in soils having high rather than low organic matter contents. The effect is proportional to the organic matter content. Another consequence is that the measurement of exchangeable cations using molar displacing solutions may underestimate the effective sodicity in soils having high organic matter contents, by measuring Ca which is not readily exchangeable under soil conditions (Naylor and Overstreet, 1969).

4.5.3 Inhibition of Sodification

Under irrigation with waters having different sodicity hazards, the increase in soil ESP was moderated where farmyard manure was added annually (Chander *et al.*, 1994) or where the soil had a higher organic matter content initially (grassland) (Rengasamy and Olsson, 1993). Similar results have been found in pot experiments where plant residues, manure, or compost were added in combination with saline irrigation water (Sekhon and Bajwa, 1993; Lax *et al.*, 1994). Organic amendments can vary in their inhibition of sodification. In a pot experiment using a calcareous soil irrigated with waters of different quality, Sekhon and Bajwa (1993) found that rice straw was more effective in inhibiting increases in ESP and pH than both green manure (*Sesbania aculeata*) and decomposed farmyard manure. However, the yields of rice, wheat (*Triticum aestivum*), and maize (*Zea mays*) were greatest in the green manure treatment, probably due to N nutrition (even though extra N was given to the rice straw treatment in the rice crop). A gypsum treatment showed that the effects of gypsum and organic amendments were synergistic.

The inhibition of sodification by organic matter is important in maintaining acceptable physical properties in soils irrigated with sodic water. The following mechanisms help to explain this phenomenon, but this relative importance in unknown:

1. Permeability is generally greater in soil with high rather than with low organic matter content (Section 4.6.1). Therefore, under irrigation the leaching fraction (LF) tends to be higher, favoring the more rapid removal of Na salts (Fig. 4.9). This mechanism was proposed by Rengasamy and Olsson (1993) and was also operative

in the experiments of Lax *et al.* (1994), and Chander *et al.* (1994) but not in those of Sekhon and Bajwa (1993), who kept the LF constant in the different treatments.

2. Calcium bound by organic matter is less exchangeable at the EC attained during leaching with sodic water than that bound by inorganic colloids (Section 4.5.2).

3. Where irrigation water has a high HCO_3^- concentration (high residual sodium carbonate concentration, RSC), or the soils being irrigated are calcareous, inhibition of $CaCO_3$ precipitation due to (a) high P_{CO2} in the presence of organic matter (Amrhein and Suarez, 1987; Sekhon and Bajwa, 1993), or (b) organic matter itself (Inskeep and Bloom, 1986) may lead to the maintenance of high solution Ca concentrations. Alternatively, high Ca activities in solution may be maintained by high rates of mineralization of organically complexed Ca compared with slow precipitation of calcite (Amrhein *et al.*, 1993).

4. Organic matter may inhibit exchange reactions by binding aggregates or making them hydrophobic and thus rendering exchange sites innaccessible. This mechanism could be considered analogous to K "fixation" on mica surfaces. Although the K can be considered an exchangeable cation, exchange is severely limited by diffusion in the interlayer region and high selectivity.

In conclusion, when soils are irrigated with sodic water having a given SAR, RSC concentration, and EC, a higher ESP is reached in soils with low than with high organic matter content (Fig. 4.10).

4.5.4 Desodification of Noncalcareous Sodic Soils

Leaching alone can reduce the ESP of noncalcareous sodic soils, especially when they are also saline. However, leaching of sodic soils is restricted by their low permeability. On the other hand, organic matter may increase permeability and leaching by the mechanisms described in Section 4.6.1, and it may also reduce sodicity slightly by increasing the CEC, as already described. Aylmore and Sills (1982) showed that organic matter buildup under continuous pasture reduced ESP compared to continuous cultivation. Srivastava and Srivastava (1993) reported that, following chemical amendment and cropping with rice or wheat, the decrease in ESP was related to an increase in CEC, which was a function of the organic matter content. Because cation exchange sites on organic

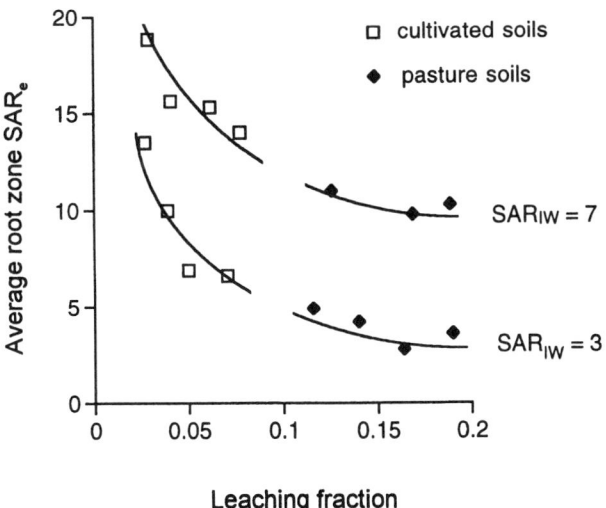

Figure 4.9 Influence of leaching fraction (LF) and sodium adsorption ratio of irrigation water (SAR_{iw}) on sodicity (sodium adsorption ratio of saturation extract, SAR_e) of cultivated and pasture soils [Rengasamy and Olsson, *Aust. J. Soil Res.* 31:821-837, (1993)].

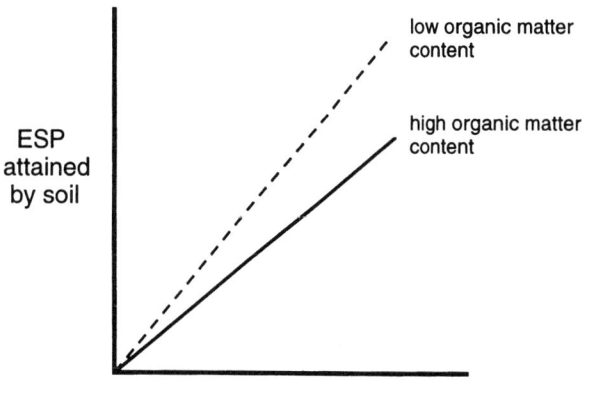

Sodicity hazard of irrigation water

Figure 4.10 Influence of soil organic matter content on exchangeable sodium percentage (ESP) attained by soils under irrigation with water having increasing sodicity hazard.

matter have higher affinity for Ca than those on clay, the overall affinity of soil surfaces for Ca is proportional to organic matter content. Therefore, under a leaching regime, the loss of Na relative to Ca is increased when organic matter is added. However, while the addition of organic matter may reduce ESP slightly, an improvement in soil structure does not neccessarily follow (Section 4.6). The addition of Ca salts (gypsum) with the organic matter offsets the detrimental effects that are possible (Baldock *et al.*, 1994). Singh (1974) showed that incubation with sesbania (*Sesbania aculeata*) residues reduced the ESP of a noncalcareous soil even more than the addition of $CaCO_3$, especially under saturated conditions. When added together, the effects of $CaCO_3$ and organic matter were synergistic. Organic materials contain metal cations, and the K, Ca, and Mg added in organic amendments contribute directly to decreasing ESP. Similarly, deep-rooted plants may take up Ca from deep horizons and deposit it on the surface as litter (Gill and Abrol, 1990; Section 4.7).

4.5.5 Acidification and Desodification of Calcareous (Alkaline) Sodic Soils

Principles of Acidification and Desodification
Alkaline sodic soils contain Ca in quantities which are usually more than sufficient for desodification, but the Ca is present as $CaCO_3$, which is insoluble at high pH. Therefore, the ESP and SAR of soil solution remain high. The pH of alkaline soils is a function of the activity of CO_3^{2-} + HCO_3^-, the partial pressure of CO_2 P_{CO_2}, and the ionic strength, as shown in Equation [4.2] (Masshady and Rowell, 1978).

$$pH = 7.82 - \log P_{CO_2} + \log a - 0.5\ (I)^{0.5} \qquad [4.2]$$

where

$$a = \text{activity of } CO_3^{2-} + HCO_3^-$$
$$I = \text{ionic strength}$$

The presence of Na increases the pH of calcareous soils by increasing the activity of CO_3^{2-} and HCO_3^-. Nakayama (1970) has described the relationship between pH, P_{CO_2} and concentrations of $CaCO_3$, $NaHCO_3$, and Na_2CO_3. The latter two salts are relatively soluble, and their removal by leaching can substantially reduce pH and ESP in alkaline sodic soils. However, because of the high pH and high ESP, permeability is low enough to make leaching impractical. Therefore, the addition of acid is the favored technique for amelioration. In addition to neutralizing the high pH caused by Na_2CO_3, acid dissolves the $CaCO_3$ present, increasing soluble Ca concentrations and thus encouraging the exchange of Ca for Na (Equations [4.3a], [4.3b] and [4.3c]). In order to dissolve the $CaCO_3$, pH must first be reduced to less than 8.5.

$$Na_2CO_3 + HA = 2Na^+ + HCO_3^- + A^- \qquad [4.3a]$$

$$CaCO_3 + HA = Ca^{2+} + HCO_3^- + A^- \qquad [4.3b]$$

or

$$2Na_{(ads)} + CaCO_3 + HA = Ca_{(ads)} +$$
$$2Na^+ + HCO_3^- + A^- \qquad [4.3c]$$

where

HA is an acid.

Sulfuric acid is a commonly used ameliorant. Alternatively, elemental S or pyrite (FeS_2) may be used as a source of S for the production of H_2SO_4 by S-oxidizing bacteria. The amount of strong acid needed to reduce the pH to a reference value is termed the acid neutralizing capacity (ANC). Components that contribute to the ANC depend on the reference pH chosen. In alkaline sodic soils, this should be around the equilibrium pH of $CaCO_3$ (between 8.0 and 8.5). Acidification is defined as a decrease in ANC, but not necessarily a decrease in pH. The ANC of sodic soils may be calculated (van Breemen *et al.*, 1983), but this is complicated by the degree of concurrent leaching. Weaker acids may also be used, but they are important proton sources only when the pH is greater than their pK_a. While the addition of strong acid lowers ANC, weak acids do not directly contribute to change unless the conjugate base and associated cation are removed. Therefore, the removal of Na^+ and HCO_3^- by leaching is necessary for desodification. Protons added occupy the lowest proton energy available, i.e., any acid added reacts with the strongest base in the system until that base is consumed. Additional acid will react with the next strongest base and so on (Equation [4.3a] followed by [4.3b]). The strength of acids may be measured by their degree of dissociation into a proton and its conjugate base. The stronger the acid, the larger the dissociation constant K_a, and the smaller the pK_a (or $-\log K_a$), which is the pH at which 50% of the acid is dissociated. As the addition of acid lowers pH, leaching and the removal of Na_2CO_3 and $NaHCO_3$ becomes easier, and the dissolution of $CaCO_3$ increases.

Influence of Organic Matter

Upon the addition of organic matter to an Alkaline Sodic Soil (Sod$_H$ Sal pH$_H$ or Sod$_H$ Sal$_L$ pH$_H$) (Chapter 1), the same processes occur as in a noncalcareous sodic soil. However, an important additional process is the reduction of pH. The effect of organic matter in this regard is widely recognized and utilized for the amelioration of alkaline sodic soils (Figure 4.11) (Puttaswamygowda and Pratt, 1973; Robbins, 1986; Ahmad et al.,1988; More, 1994). The addition of organic matter in whatever form (roots, root exudates, crop residues, manures) and the microbial and faunal activity that is stimulated by these additions may reduce soil pH and ESP by a number of mechanisms. In the case of crop growth, the more roots, the greater the extent of amelioration (Robbins, 1986). The extent to

which pH is reduced and $CaCO_3$ is dissolved depends on the balance of the reactions in Table 4.3. Net changes in pH occur with the addition and removal of materials.

Production of Carbonic Acid and Organic Acids

In alkaline soils, two major acidification mechanisms are the dissociation of carbonic (H_2CO_3) and organic acids. Carbon dioxide is produced continually by soil biota, including roots, and the greater the input of organic matter, the greater its production. As shown in Equation [4.2], the pH of alkaline soils is inversely proportional to P_{CO2}. This is because CO_2 dissolves in water to produce H_2CO_3,

$$2CO_{2(gas)} + H_2O = H_2CO_3 + CO_{2(aq)} = 2HCO_3^- + 2H^+ \quad [4.4]$$

Although H_2CO_3 is a weak acid (Table 4.4), it may control pH in the absence of stronger proton donors. Dissociation of H_2CO_3, and hence acidification, is high at high pH. In a H_2O/CO_2 system, the concentration of H_2CO_3 depends on P_{CO2}. As H_2CO_3 concentration increases, pH decreases and $CO_{2(aq)}$ is liberated as $CO_{2(gas)}$, so pH does not descend below approximately 5. Where $CaCO_3$ is also present, its solubility increases with P_{CO2} (Table 4.5), and pH does not descend below approximately 8.2. In soil, P_{CO2} may be up to 1 kPa (1% of soil air by volume), and much higher under anaerobic conditions (Ponnamperuma, 1972). The effect of H_2CO_3 on the neutralization of Na_2CO_3 and the dissolution of $CaCO_3$ may be seen by substituting H_2CO_3 for HA (where A is HCO_3^-) in Equations [4.3a], [4.3b], and [4.3c]. The equations are driven to the right by high P_{CO2}, and the removal of HCO_3^- salts by leaching. The quantitative significance of CO_2 for acidification is uncertain (Goertzen and Bower, 1958; Loveday, 1984), although Robbins (1986) found that the most effective organic amendments in a calcareous sodic soil produced the highest P_{CO2}.

Although organic acids are weak, they are stronger than H_2CO_3 (Table 4.4) and can be a significant source of protons in soils when they dissociate. The effect of organic acids on sodicity may be observed by substituting R-COOH for HA in Equations [4.3a], [4.3b], and [4.3c]. Plant roots and microorganisms excrete organic acids to alter the mobility and uptake of various elements, including P, Fe, and Al (Uren and Reisenauer, 1988; Stevenson, 1991). Under anaerobic conditions, microorganisms also produce organic acids (mostly low molecular weight fatty acids) as by-products of decomposition (Ponnamperuma, 1972). Many different acids, including simple aliphatic, sugar, aromatic, phenolic, and more complex polyprotic acids, have been detected in soils (Stevenson, 1991, 1992). The concentrations of simple aliphatic, amino, and aromatic acids are of the order of 0.05 to 5 mM in soil solutions. The main simple organic acids found in anaero-

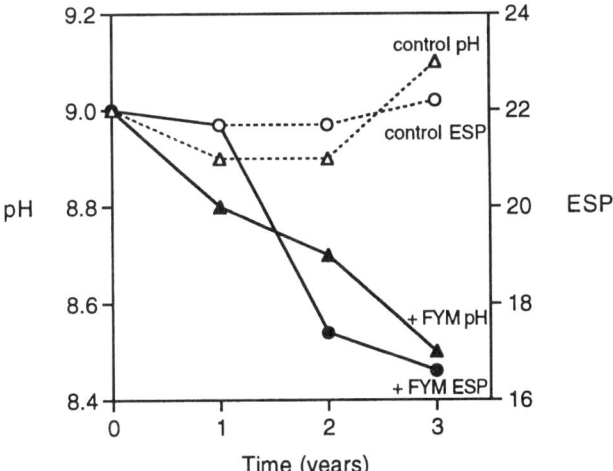

Figure 4.11 Influence of straw and farmyard manure (FYM) at a rate of 50 Mg ha^{-1} on exchangeable sodium percentage (ESP) and pH of a calcareous sodic soil [More, *J. Indian Soc. Soil Sci.* 42:253-256, (1994)].

Sodic Soils

TABLE 4.3 Biological reactions involving transfer of protons and related processes in soil [van Breeman *et al*, *Plant Soil* 75:283-308, (1983)]

Process from left to right	Reaction equation*		Process from right to left
	H^+–indifferent processes		
	Biota/atmosphere		
Photosynthesis	$CO_2 + H_2O$	$= CH_2O + O_2$	respiration
N_2 fixation	$N_2 + H_2O + 2R \cdot OH$	$= 2R \cdot NH_2 + 3/2 O_2$	
NH_3 uptake	$NH_3 + ROH$	$= R \cdot NH_2 + H_2O$	volatilization of NH_3
H_2S uptake	$H_2S + R \cdot OH$	$= R \cdot SH + H_2O$	volatilization of H_2S
SO_2 uptake	$SO_2 + R \cdot OH$	$= R \cdot SH + 3/2 O_2$	
H^+ source	**H^+ transfer process**		**H^+ - sink**
	biota/solution		
Uptake of cations	$M^+ + R \cdot OOH$	$= R \cdot OOM + H^+$	mineralization of M+
Uptake of NH_4^+	$NH_4^+ = R \cdot OH$	$= R \cdot NH_2 + H_2O + H^+$	mineralization of org. N
Mineralization + nitrification of organic N	$R \cdot NH_2 + 2O_2$	$= 2OH + NO_3^-$	uptake of NO_3^-
Mineralization + oxidation of organic S	$R \cdot SH_2 + 3/2 H_2O + 7/4 O_2$	$= R \cdot OH + SO_4^{2-} + 2H^+$	uptake of SO_4^{2-}
Mineralization of P	$R \cdot H_2PO_4 + H_2O$	$= R \cdot OH + H_2PO_4^- + H^+$	uptake of P
	solution or solution/atmosphere		
Dissociation of H_2O	$2H_2O$	$= OH^- + H^+$	protonation of OH^-
Dissociation of CO_2	$CO_2 + H_2O$	$= HCO_3^- + H^+$	protonation of HCO_3^-
Dissociation of organic acids	$R \cdot OOH$	$= R \cdot OO^- + H^+$	protonation of organic anions
Complexation of metal ions (L=org.	$HL + M^+$	$= ML + H^+$	decomplexation of metal ions
Oxidation of H_2S	$H_2S + 2O_2$	$= SO_4^{2-} + 2H^+$	sulfate reduction
Oxidation of SO_2	$SO_2 + ½ O_2 + H_2O$	$= SO_4^{2-} + 2H^+$	
Nitrification of NH_4^+	$NH_4^+ + 2O_2$	$= NO_3^- + H_2O + 2H^+$	
Nitrification of NO_x	$NO_x + 1/4 (5 - 2x)O_2 + ½ H_2O$	$= NO_3^- + H^+$	denitrification
Nitrification of N_2	$N_2 + 5/2 O_2 + H_2O$	$= 2NO_3^- + 2H^+$	denitrification
	solids/solution		
Reverse weathering	$M^n + n/2 H_2O$	$= n/2 M_{2/n} O + nH^+$	weathering
M^{n+}/H^+ exchange	$M^n + nH$ exch	$= M$ exch $= nH^+$	H^+/M^{n+} exchange
Oxidation of Fe^{2+}	$Fe_2 + 1/4 O_2 + 5/2 H_2O$	$= Fe(OH)_3 + 2H^-$	reduction of $Fe(OH)_3$
Oxidation of FeS	$FeS + 9/2 O_2 + 5/2 H_2O$	$= Fe(OH)_3 + SO 2/4 + 2H^+$	reduction of $Fe(OH)_3$ and SO_4^{2-}
Desorption of SO_4^{2-}	exch $SO_4 + 2H_2O$	$= exch(OH)_2 + SO_4^{2-} + 2H^+$	adsorption of SO_4^{2-}

*In soil, the reactions shown are always coupled, so the amount of free H^+ (H_3O^+) is extremely small relative to the amount of transfer. For convenience O_2 is used as an electron acceptor in the reaction equations. For any of the reduction reactions organic matter may be substituted as an election acceptor by combining with the photosynthesis reaction.

TABLE 4.4 Strength of acids

Acid	pK_{a1}	pK_{a2}
Aldehydes and alcohols	12–15	
Monophenolics	8–10	
H_2CO_3	6.4	10.3
"Humic acids"	4–5	
Monocarboxylic acids	4.5	
Dicarboxylic acids	1–4	4–6
HCl, HNO_3	–1	
H_2SO_4	–3	2

bic soils are formic, acetic, propionic, and butyric acids, with acetic being by far the most abundant (Ponnamperuma, 1972). While CO_2 is formed as long as decomposable substrates are present in the soil, organic acids are produced for 3–4 weeks after the addition of fresh organic matter, after which their concentrations fall to zero (e.g., Tsutsuki and Ponnamperuma, 1987). Their concentrations remain significant only with continual additions of readily decomposable organic matter such as occurs under growing plants. Simple organic acids are readily mineralized to CO_2, but they may act as a proton source before being oxidized.

Only the release of organic acids in undissociated form will lower the pH, and then only where the soil pH is greater than the pK_a. The formation of metal complexes is probably a more important source of protons than simple dissociation. Very weak acids such as aldehydes, alcohols, and many phenolics are not very important because their pK_a values are greater than soil pH. Most of the acidic functional groups in soil are associated with this colloidal polyanionic organic matter. The pK_a values of functional

TABLE 4.5 Approximate concentration of Ca^{2+} in water in equilibrium with various atmospheres

Parameter	Atmosphere		
	Air	Soil	Waterlogged soil
P_{CO2} (kPa)	0.03	1	10
$[Ca^{++}]$ (mM)	0.5	1.5	3.4

groups on this material cover the range of the organic groups in Table 4.4, with the mean falling around 4.5. Colloidal polyanionic organic matter is more persistent in soils than simple acids because of its low mineralizability. However, because it exists largely as organomineral complexes, it is not an important source of protons. The pK_a values in Table 4.4 can act as a guide to the strength of acids, but there are pitfalls in their application to polyacidic material in soils, particularly in the solid phase. In polyacids, the dissociation of each acid group weakens the acidity of the subsequently dissociated group.

The production of both CO_2 and organic acids is favored by the addition of fresh, readily decomposable material rather than more decomposed material. For example, using anaerobic incubation, Tsutsuki and Ponnamperuma (1987) found that amendment with rice straw or green manure enhanced the formation of CO_2, methane (CH_4), and acetic acid. In addition, green manure enhanced the formation of isovaleric acid and alcohols, while rice straw enhanced the formation of phenolic acids. Compost amendment, on the other hand, did not enhance the formation of any of these products. The production of organic acids and maintenance of high P_{CO2} is also maximized under anaerobic, such as waterlogged, conditions (Puttaswamygowda and Pratt, 1973; Singh, 1974; Gupta and Abrol, 1990b). Another possible beneficial effect of anaerobic conditions on soil structure is the mobilization of Fe as Fe^{2+} (Puttaswamygowda and Pratt, 1973).

Other Biological Acidification Mechanisms

Excretion of protons from plant roots can be a major acidification mechanism in soils. For example, when measuring acidification in the rhizosphere of maize seedlings, Petersen and Böttger (1991) found that organic acids contributed only 0.2–0.3% to acidification, which was mainly due to proton excretion. Where N nutrition is not quantitatively important, there is a net proton flux from plant to soil, because the uptake of cations far exceeds that of phosphate and sulfate. The form in which N is taken up (as NO_3^- or NH_4^+) has an important influence on proton excretion. Protons may also be excreted to solubilize Ca phosphates in order to obtain P (Hedley *et al.*, 1982).

Mineralization of organic N, P, and S produces acidity, while uptake of their inorganic forms (N as NO_3^-) consumes protons. Processes involving these elements go in one direction or the other, depending on the needs of the organisms, the source of C, and removal of products by volatilization or leaching. Loss of NO_3^- and SO_4^{2-} by leaching favors net mineralization as well as limiting denitrification. Denitrification is an alkalinizing reaction which is favored by anaerobic conditions and the presence of readily decomposable organic matter and NO_3^-. Waterlogging, therefore, reduces pH most effectively when NO_3^- concentration is low. Removal of NO_3^- also promotes uptake of N as NH_4^+ by organisms, especially roots. This encourages H^+ excretion to balance cation and anion uptake. On the other hand, nitrification is an acidifying reaction, and as the primary source of N is N fixation, the presence of legumes increases acidification.

The biological acidification processes described here occur naturally in soils. In areas where low rainfall and heavy texture limit the depth of leaching, acidification of the topsoil is a major reason for alkalinization further down the profile. The leached HCO_3^- accumulates in the subsoil, causing it to become alkaline. Where Ca is present, the HCO_3^- precipitates as $CaCO_3$. Where Na is also present, the subsoil becomes sodic and highly alkaline. Dispersed clay also moves out of the A horizon and accumulates in the subsoil. A large proportion of Australian sodic soils have this type of profile, namely, acid to neutral loamy topsoil and alkaline sodic clayey subsoil.

Summary

The addition of readily decomposable organic matter to calcareous sodic soils, whether by deliberate addition or by the growth of plants, especially legumes, tends to reduce pH and thus cause desodification. The relative roles of different acidification mechanisms vary according to the situation. Leaching is essential for desodification, and the anaerobic conditions associated with waterlogging are beneficial.

4.6 ORGANIC MATTER INFLUENCES THE STRUCTURE OF SODIC SOILS

Aside from influencing sodicity directly, organic matter acts as both a bonding and a dispersing agent in soils with a given sodicity. The balance between these opposing effects depends on the ESP, the nature of the organic materials and their interactions with inorganic colloids, and the influence of mechanical disturbance. Transient organic bonding agents such as roots can have a beneficial effect on the structure of sodic soils, but anionic organic matter may act as a dispersing agent, especially at high ESP. An indirect effect of organic matter additions on structure is through their supply of nutrients to plant roots, which consequently proliferate. The effects of organic matter on the structure of sodic soils are discussed in the following sections.

4.6.1 Aggregation and Bonding in Sodic Soils

The creation and stabilization of large pores following the addition of poorly decomposable organic matter such as rice husks, cereal straw or compost, or crop growth, increase the permeability of sodic soils (Robbins, 1986; Gupta and Abrol, 1990b; Lax *et al.*, 1994). Plant roots increase the number of continuous vertical pores and improve drainage when they shrink, die, decay, or dessicate the soil (Chhabra and Abrol, 1977; Robbins, 1986; Ilyas *et al.*, 1993; Oades, 1993). They may also remove trapped air from larger conducting pores (McNeal *et al.*, 1966b). The extent to which roots affect structure depends on the amount of roots and their distribution. Ilyas *et al.* (1993) found that sesbania (*Sesbania* spp.) and alfalfa (*Medicago sativa*) increased permeability of a sodic soil to depths of 0.25 and 0.8 m, respectively, and that the effect was enhanced by the addition of gypsum. Neither subsoiling nor drainage improved permeability in the same soil.

High organic matter content makes soils less susceptible to the unfavorable influence of exchangeable Na (van Beekom *et al.*, 1953; Aylmore and Sills, 1982). Aylmore and Sills (1982) found that 54% of the variation in modulus of rupture (MOR) in a group of soils was accounted for by ESP, whereas 18% was attributed to organic matter (inverse relationship), but none of the other factors measured had a significant effect. They speculated that the nature of the exchangeable cations at the time of organic matter addition could explain the stability of these soils. Similarly, Loveland *et al.* (1987) found that most of the variation in dispersion (following gentle shaking) in a large group of sodic soils was explained by ESP, whereas organic matter had a significant stabilizing effect.

Aylmore and Sills (1982) measured the MOR of surface soils collected from the wheatbelt of Western Australia with a range of ESP values established by leaching with $CaCl_2$ and NaCl solutions. The Na sensitivity (MOR/ESP) and MOR at any given ESP were inversely correlated with the organic matter content. For those soils with 10–20% clay, the relationship between Na sensitivity and organic matter content was particularly strong (Fig. 4.12), but soils with higher clay contents did not fit the relationship as well, presumably because the shrink-swell potential of the clay would dominate the MOR. Soils from a long-term rotation trial (Merredin) with approximately 33% clay showed decreasing Na sensitivity with increasing organic matter content. Soils from a similar rotation trial but with low clay contents (approximately 7%) had lower organic matter contents and lower Na sensitivities (Fig. 4.12).

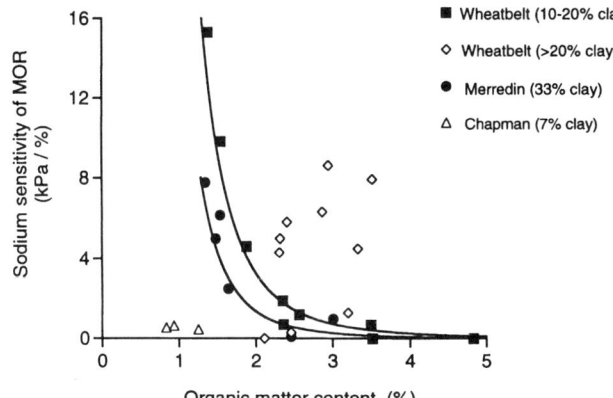

Figure 4.12 Sodium sensitivity of modulus of rupture (MOR) [ratio of MOR to exchangeable sodium percentage (ESP) in soils equilibrated with solutions with different sodium adsorption ratios (SAR)] as a function of organic matter content for a range of soils collected from the wheatbelt of Western Australia and from long-term rotation trials in the same area (Merredin and Chapman)[derived from Aylmore and Sills, *Aust. J. Soil Res.* 20:213-224, (1982)].

In sodic soils, roots, fungal hyphae, other particulate organic matter, and worm casts can stabilize aggregates, preventing slaking and dispersion in the same way as they do in nonsodic soils (Quirk and Murray, 1991) The resistance to slaking conferred by these materials is not affected by Na pyrophosphate (Oades, 1984). They stabilize macroaggregates by enmeshing them and linking particles at the micrometer to centimeter scale. Neutral polysaccharides stabilize clay domains and microaggregate. Due to their nonionic bonding mechanisms, the role of roots, hyphae, and neutral polysaccharides are relatively unaffected by interactions between water, clay, and exchangeable cations, even at very high ESP values. The same is probably true for aliphatic and aromatic materials in soil organic matter which bond to mineral surfaces through hydrophobic and van der Waals interactions. Long-chain organic anions can also act as bonding agents in Na-saturated systems by linking the positive charges on the edges of clay plates (Emerson, 1954) or oxide surfaces. In soils which contain sufficient polyvalent cations, long-chain organic anions can link negatively charged clay particles through cation bridges. Emerson (1954) imposed high ESP values on illitic soil aggregates from an old grassland and a nearby cultivated field with organic C contents of 5.0 and 0.7%, respectively. When leached with progressively more dilute NaCl solutions, the cultivated soil dispersed completely at 0.02 *M*, while the grassland soil aggregates were still intact, even in pure water. The influence of organic matter was attributed to its

effect on the rate of wetting. Once a sodic soil is wet, roots, hyphae, mucilages, and hydrophobic materials must counteract the tendency of organic anions to disperse the clay. Barzegar *et al.* (1996) found that when pea straw was added to soils in which ESP varied from 3 to 36%, wet aggregate stability increased, irrespective of the ESP.

The influence of transient organic binding materials which stabilize aggregates against slaking is susceptible to mechanical disturbance and decomposition. In nonsodic soils, the microaggregates produced by slaking are quite stable and only disperse with considerable input of mechanical energy. However, in sodic soils, once these transient bonds are broken and disaggregation has occurred, the organic matter can act as a dispersant (Emerson, 1977, 1983). The fragile nature of organic binding materials in a sodic soil was shown by Emerson (1954), who compared dispersion in grassland (5.0% organic C) and cultivated (0.7% organic C) soils, both of which he had saturated with Na. The cultivated soil dispersed more readily when undisturbed, but the grassland soil dispersed more readily after disturbance (shaking in suspension). Similarly, Barzegar *et al.* (1996) found that following a 7-day incubation, the addition of pea (*Pisum sativum*) straw to sodic soils (ESP 3–36) decreased spontaneously dispersible clay, but increased mechanically dispersible clay, especially at high ESP. The higher the organic matter content of sodic soils, the more sensitive will they be to mechanical disturbance. Hence, one could speculate that results such as those of Aylmore and Sills (1982) shown in Fig. 4.12 would be reversed if the soils were mechanically disrupted before measurement of MOR, especially at high values of ESP.

Where soils with different organic matter contents have been treated with saline solutions, organic matter reduces the adverse effects of Na on structure. However, where ESP was not measured, the effect may have been due partly to inhibition of sodification and partly due to a beneficial effect on structure at a given sodicity. For example, Black and Abdul-Hakim (1985) examined the effect of sodicity and organic matter on permeability using soils from a rotation trial. Following leaching with solutions having different SAR and EC values, soil permeability from a wheat treatment (2.8% organic matter) was less than that under pasture (3.1% organic matter) at low EC and moderate SAR. Barzegar *et al.* (1994) examined the disaggregation and strength of aggregates from a rotation trial (0.98 to 10.53% organic C) which they had equilibrated at various levels of sodicity and salinity. They found that the stability of aggregates low in organic matter in water was low, and not greatly affected by sodicity. The wet aggregate stability of soils with higher organic matter contents was greater, and was reduced significantly by sodicity. The tensile strength of aggregates was correlated with clay dispersion, and both were inversely related to wet aggregate stability (Fig. 4.13). Although organic matter reduced the impact of sodicity, its effects were not simple and differed with depth. The virgin aggregates from 0.04 to 0.10 m in depth had considerably higher strength, and strength increased more with sodicity, than the pasture aggregates from 0 to 0.1 m in depth, which had similar organic C (2.9%) and clay (20.6%) contents.

Synthetic organic polymers have also been used to improve the structure of sodic soils. The anionic polymers vinyl acetate-maleic acid (VAMA) copolymer, hydrolyzed polyacrylonitrile (HPAN), anionic polysaccharide (APS), and polyacrylamide (PAM) have been found to increase wet aggregate stability, hydraulic conductivity (HC), and plant growth, and decrease MOR in sodic soils (Allison, 1956; Allison and Moore, 1956; Wallace *et al.*, 1986; Helalia and Letey, 1988b; El-Morsy *et al.*, 1991a; Zahow and Amrhein, 1992; Gu and Doner 1993). The beneficial effect of synthetic anionic organic polymers is probably due to their long, flexible linear structure. They create linkages by bonding to positive charges or polyvalent cations on several clay particles (Ruehrwein and Ward, 1952; Theng, 1982; Durgin and Chaney, 1984) and also reduce the rate of wetting (Gu and Doner, 1993). Nonionic polymers, which have the same effects as the natural polysaccharides described (Zahow and Amrhein, 1992), can spread over adjacent soil particles like a coat of paint (Theng, 1982). Positively charged organic polymers have also been shown to decrease clay dispersion and increase infiltration rate (IR) in sodic soils (El-Morsy *et al.*, 1991a,b; Zahow and Amrhein 1992; Chapter 5). The effectiveness of charged synthetic organic polymers in flocculating clay from suspension decreased in the order cationic > nonionic > anionic (Helalia and Letey, 1988a).

In summary, organic matter has a beneficial effect on the structure of sodic soils, but its effect is much greater following displacement of Na by Ca (Muneer and Oades, 1989b; Rengasamy and Olsson, 1991). The beneficial effect depends on the variable and complex nature of the organic matter and all of the other factors governing dispersion. It tends to be greatest when ESP is low, clay content is moderate, EC is high (but not Very Saline [Sol$_H$]), and dominated by Ca or Al, Mg content is low, mechanical disturbance is minimal, and inputs of fresh materials are maximized.

4.6.2 Dispersion in Sodic Soils

Clay dispersion results from the interaction of many soil factors and the amount of energy applied, so soils which do not disperse spontaneously can be made to do so by subjecting them to mechanical stress (Rengasamy *et al.*, 1984; Goldberg *et al.*, 1988; Mullins *et al.*, 1990; Levy *et al.*, 1993b; Chapter 3). Sodic soils are particularly prone to stress-induced dispersion, because when the transient organic bonds are broken, the dispersive effects of anionic organic matter may predominate. As dispersion results from ionic interactions, any ionic soil components, including charged organic matter, have a direct influence on the process. Organic anions are known to increase clay dispersion, especially where variable charge minerals are present (references cited by Greenland, 1965b; Narkis *et al.*, 1968; Emerson and Smith, 1970; Shanmuganathan and Oades, 1983a; Oades, 1984; Durgin and Chaney, 1984; Visser and Caillier, 1988; Oades *et al.*, 1989; Pojasok and Kay, 1990; Frenkel *et al.*, 1992a,b; Gu and Doner, 1993; Kretzschmar *et al.*, 1993). The way in which organic anions increase negative and decrease positive charge on inorganic soil colloids has already been discussed (Section 4.5.1). The dispersing power of organic anions in-

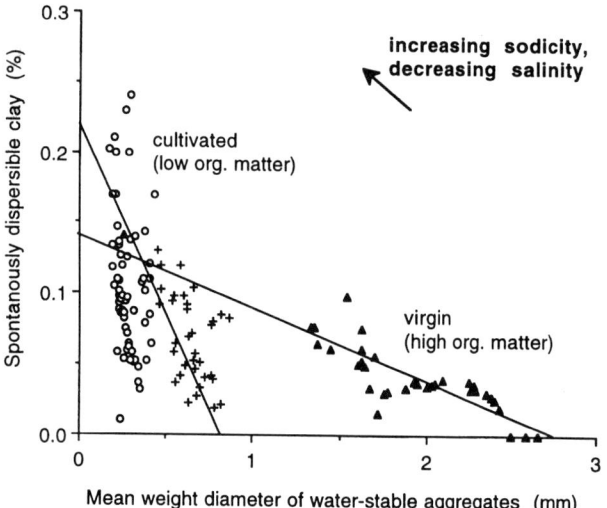

Figure 4.13 Spontaneously dispersible clay, and breakdown in water of 2–4 mm aggregates from a rotation trial on Urrbrae fine sany loam. Aggregates had been equilibrated with solutions with different sodium adsorption ratios (SAR 0–30) and electrolyte concentrations. Virgin soil samples had organic C contents of 2.98 or 10.53% (two depths). Organic C contents were 0.98–1.71% for cultivated soils (o) and for rotations including pasture (+) were 2.02–2.90% [Barzegar *et al.*, *Soil Tillage Res.* 32:329-345, (1994)].

creases with charge density or the number of, and decreasing distance between, carboxyl groups (Durgin and Chaney, 1984; Jekel, 1986; Gu and Doner, 1993). The increased tendency to disperse brought about by increased negative charge density on soil colloids is enhanced at high ESP. The presence of exchangeable Mg or high pH increases the dispersive effect of organic matter on sodic clay (Emerson and Smith, 1970). In addition to increasing the negative charge on soil colloids, organic acids may increase dispersion by lowering exchangeable Ca concentration, particularly when it is already low, by complexation.

Evidence for the influence of soil organic matter on the dispersion of sodic soils has been given by the fact that CFC values for Na-saturated soil clays (montmorillonite, kaolinite, and illite) are higher than those of corresponding pure clays (Goldberg and Forster, 1990; Frenkel *et al.*, 1992a; Gu and Doner, 1993). In addition, removal of soil organic matter considerably decreased the CFC values of Na-saturated soils (Goldberg *et al.*, 1990; Gu and Doner, 1993). Emerson (1954) showed that the NaCl concentration required to flocculate suspensions of a grassland soil with 5.0% organic C was six times higher than that for a nearby cultivated soil (0.7% organic C).

Montmorillonite is more dispersible than kaolinite, but kaolinite is more susceptible to the effects of anion adsorption, and can be made to be as dispersible as montmorillonite. The difference between the effects of anionic organic materials on Ca and Na clays is greatest for montmorillonite, because Ca montmorillonite forms quasi crystals into which organic molecules cannot penetrate. Exchangeable Na also tends to be concentrated on the outside of these quasi crystals (Shainberg and Letey, 1984; Chapter 3). Using smectite, kaolinite, and three soils whose clay fractions were dominated by one or the other of these minerals, Frenkel *et al.* (1992a) showed that the CFC values of Na soils were much higher and much more affected by organic matter than those of Ca soils. Tarchitzky *et al.* (1993) showed similar comparisons between Na- and Ca-montmorillonite suspensions to which they added different amounts of humic acid (Fig. 4.14). These dispersive effects have been attributed to the adsorption of organic matter onto positive charges on inorganic colloids, such as clay edges. Edge-face interactions are thereby disrupted, and high electrolyte concentrations are needed to bring about flocculation by compression of the double layer. Humic substances have smaller effects on Ca clays because, even at low pH, their preferred association is face-face, and CFC values for Ca-saturated humic colloids are similar to those of Ca-saturated clays,

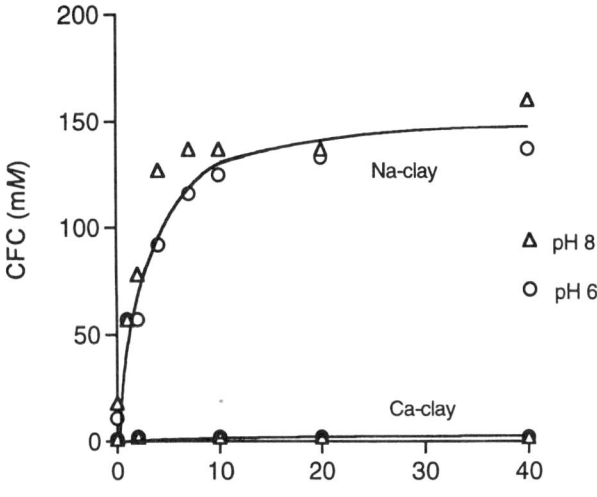

Figure 4.14 Critical flocculation concentration (CFC) of Na- and Ca-montmorillonite suspensions as a function of humic acid concentration [Tarchitzky *et al., Soil Sci. Soc. Am. J.* 57:367-372, (1993)].

reducing their effects on the CFC of the mixture. Experiments in which the electrophoretic mobility of soil particles, before and after the removal of organic matter by peroxidation, was measured have been inconclusive, probably because of artifacts produced by the treatment (Lebron and Suarez, 1992).

Addition of small amounts of anionic polysaccharide or humic acid substantially increased the dispersion of Na-saturated soil or clay in the order humic acid > soil polysaccharide ≥ anionic polysaccharide (Gu and Doner, 1993). Gupta *et al.* (1984a) added farmyard manure to a soil equilibrated with solutions having different SAR values. Manure increased dispersion of fine clay at all levels of sodicity, but the effect was greatest at high SAR (Fig. 4.15). When 5% $CaCO_3$ was present in the soil, dispersion was less, but was influenced in the same way by the addition of manure. When they added manure to soils varying naturally in sodicity (ESP 12, 34, and 64), they found that dispersion was proportional to the rate of manure addition (0–1.5%), except at the highest ESP, where dispersion was complete in all treatments. As a result of these experiments, Gupta *et al.* (1984a) cautioned against the use of manures in soils undergoing sodification. After several years of straw addition (0, 5, and 10 Mg ha^{-1}) in the field, Baldock *et al.* (1994) found that mechanically dispersible clay was proportional to the rate of straw addition, corresponding to th dispersive effects of persistent anionic organic matter. This was despite a positive

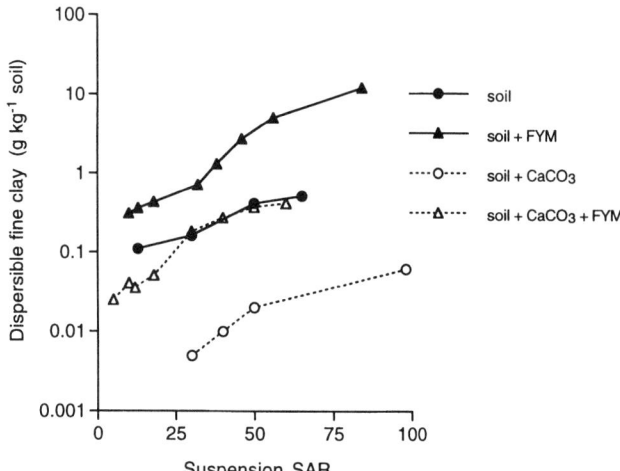

Figure 4.15 Effect of farmyard manure (FYM) addition (1%) on mechanically dispersible fine clay in soil suspensions with or without CaCO₃ at various sodium adsorption ratios (SAR) [Gupta *et al., Soil Sci.* 137:245-251, (1984a)].

effect of straw on water-stable macroaggregation. Gypsum addition (3.4 Mg ha⁻¹) decreased mechanically dispersed clay, but its effectiveness was reduced at high rates of straw addition. The dispersed clay, or dispersive organic materials from straw, caused a reduction in permeability at 0.2 m depth. A similar reduction in permeability at depth has been noted in sodic soils where straw was added, or where crops were grown without gypsum amendment (Kosmas and Moustakas, 1990; Ilyas *et al.*, 1993). These results suggest that the greater the amount of organic matter added, the greater the amount of Ca needed to prevent mechanical dispersion.

In conclusion, anionic components of soil organic matter can increase dispersion, particularly when the ESP is high and the soil is mechanically disturbed.

4.7 BIOLOGICAL AMELIORATION OF SODIC SOILS

The processes involved in the reclamation of sodic soil are reasonably well understood, and the addition of a variety of organic ameliorants is widely recognized and practiced (Table 4.6). The addition of animal manures and plant residues has been applied extensively, but of all the approaches to biological amelioration, the most generally useful involves stimulating sufficient plant growth to initiate the buildup and maintenance of organic matter levels in soil. The ultimate aim of any biological amelioration program should be to produce the maximum amount of biomass possible per unit area and time. Many studies have involved comparisons between the effectiveness of inorganic and organic amendments, and results vary de-

pending on the situation. However, it is clear that initial amendment with Ca-supplying materials such as gypsum, or acid in calcareous sodic soils, followed by plant growth and fertilizers is the most efficient means of reclamation. The inorganic chemical and physical aspects of amelioration are covered elsewhere (Chapter 8). As expected from the mechanisms discussed so far, the ameliorative effects of organic matter are greatest in Alkaline, Very Sodic soils [pH_H Sod_H], and this is where biological techniques have been most successful. Much of this research has been carried out at the Central Soil Salinity Research Institute in Karnal, India. In less sodic soils, there is still great scope to assess the ability of different plants to penetrate strongsubsoils. Biological reclamation is most successful in the surface horizons where inputs of organic matter are highest. Reduced ESP in the topsoil is accompanied by increased ESP further down the profile. This reduces the effectiveness of biological amelioration techniques where the topsoil is neutral to acidic and the subsoil is alkaline sodic (Sod_H Sal pH_H or Sod_H Sal_L pH_H), as is often the case. Placement of inorganic or organic amendments at depth, for example, in riplines or slots (Jayawardane and Chan, 1994) (Chapter 8), or sufficient leaching can solve the problem.

Amelioration must involve leaching, for which irrigation and drainage provide the most effective means. The poorer the quality of the irrigation water (or groundwater if it reaches the root zone), the more important is the removal of water from the soil by drainage. For crops other than rice, irrigation poses major difficulties in sodic soils due to low IR and poor drainage. Irrigation must

TABLE 4.6 Examples of various organic materials used for the amelioration of sodic soils

Organic materials	References
Food processing wastes	Chand *et al.* (1977); Singh *et al.* (1980); Lehrsch and Robbins (1994); More (1994)
Animal manures	Kanwar *et al.* (1965); Sharma *et al.* (1983); Robbins (1986); Fuentes Godo *et al.* (1980); Sekhon and Bajwa (1993); Chander *et al.* (1994); More (1994)
Green manures	Uppal (1955); Yadav and Agarwal (1961); Robbins (1986); Ahmad *et al.* (1990); Singh *et al.* (1991); Ilyas *et al.* (1993); Sekhon and Bajwa (1993)
Composts	Lax *et al.* (1994); Avnimelech *et al.* (1990)

therefore be more frequent and in smaller amounts than for nonsodic soils (Dahiya and Anlauf, 1990). Where irrigation water is unavailable, the speed of reclamation is limited by rainfall and the amount of leaching that occurs. Leaching can be maximized by reducing runoff and evaporation and encouraging infiltration. Aside from being slower, reclamation under rainfed cnditions follows the same principles as under irrigation, namely, a source of soluble Ca and maximization of plant productivity. Maximizing plant productivity involves many cultural factors, one of the most important being nutrition.

The choice of plants cultivated depends on their tolerance to sodicity and any climatic, social, and economic factors. Much work has been undertaken to assess the sodicity tolerance of various crop, green manure, and pasture plants, and lists of some tolerant species are compiled in Tables 4.7 and 4.8. Species which tolerate high pH and salinity as well as sodicity are particularly useful. Rice is commonly used during the initial stages of reclamation because of its tolerance to sodicity and salinity, as well as the waterlogged conditions which favor reclamation (Abrol and Bhumbla, 1979). It has been suggested that blue-green algae play an ameliorative role

under waterlogged conditions (Subhashini and Kaushik, 1981; Kaushik, 1990), but their significance has been questioned (Rao and Burns, 1991). In India, following the addition of Ca and leaching, a series of rotations are followed, generally including rice, sesbania and wheat in the first year. Reclamation begins in the summer when microbial activity and soil P_{CO2} are high due to high temperatures (Gupta and Abrol, 1990b). The soils must be cropped continuously and not left fallow.

While cropping generally brings the best economic returns, pastures, green manures and agroforestry can also play an important role in reclamation (Dahiya and Anlauf, 1990; Singh et al., 1994). Pasture grasses vary greatly in their tolerance to sodicity, with some of the useful ones shown in Table 4.8. Kallar or karnal grass (Leptochlon fusca) is tolerant of Very Sodic Alkaline soils and has shown high production and rapid reclamation of salt-affected land. Two major defects are firstly, that it needs large quantities of good quality irrigation water, and secondly, that it is not productive during winter (Aslam et al., 1993). Shrubs such as Atriplex spp. and Maireana spp. appear to be useful under more saline and dryer conditions where waterlogging does not occur (Ahmad and Ismail, 1993; Aslam et al., 1993; Davidson et al., 1993; Rashid et al., 1993). Although grazing pressure is an important factor in the management of pastures, it may not be very important in influencing the effectiveness of reclamation. On a Very Sodic soil, grass plots which had been grazed or left ungrazed for 3 or 12 years showed no significant differences in organic matter content, structural stability, HC or clay dispersion (Lavado and Alconada, 1994). Of the green manure plants, sesbania is the most widely used because of its high tolerance to sodicity. Annual plants often have difficulty penetrating strong soils, and Materechera et al. (1991) found that among a selection of annual crop and forage plants, roots of dicotyledonous species were generally better able to penetrate strong soils than roots of monocotyledonous species. Cresswell and Kirkegaard (1995) reviewed the evidence for plants penetrating and improving the structure of dense subsoils. They concluded that there was not much evidence for amelioration of subsoils by plant roots, but tap-rooted legumes were generally considered superior to grasses. They suggested that deep-rooted perennials such as alfalfa would be the best option.

Trees and perennial shrubs can contribute significantly to the productive use and reclamation of sodic soils (Sharma, 1988; Gill and Abrol, 1990; Yadav and Singh, 1991; Singh et al., 1994). A large variety of species have proved promising under both rainfed and irrigated regimes, mainly for fuel wood and fodder, but also possibly

TABLE 4.7 Relative crop tolerance of soil sodicity/alkalinity [compiled from Abrol et al., FAO Soils Bull. 39, (1988) and Gupta and Abrol, Adv. Soil Sci. 11:223-288, (1990a)]

ESP range[a]	Crop
10–15	Safflower (Carthamus tinctorius), mash (Phaseolus mungo), peas (Pisum sativum), lentil (Lens esculentum), pigeon-pea (Cajanus cajan), urd bean (Vigna mungo), beans (Phaseolus vulgaris)
15–20	Bengal gram (Cicer arietinum), soybeans (Glycine max)
20–25	Groundnut (Arachis hypogaea), cowpea (Vigna sinensis), onion (Allium cepa), pearl millet (Pennisetum milaceum), oats (Avena sativa)
25–30	Linseed (Linum usitatissimum), garlic (Allium sativum), guar (Cyamopsis tetragonoloba)
30–50	Raya (Brassica juncea), wheat (Triticum aestivum), sunflower (Helianthus annus), cotton (Gossypium hirsutum), tomatoes (Lycopersicon esculentum), beet (Beta vulgaris)
50–60	Barley (Hordeum vulgare)
60–70	Rice (Oryza sativa)

[a]Crop yields are 50% of the maximum in these sodicity ranges

for timber, pulp, leaf oils, and tannins (Marcar *et al.*, 1993). A number of these species, including several from the *Eucalyptus* and *Acacia* genera, are shown in Table 4.8. It is important to select provenances with similar soils and climate to the areas being reclaimed (Mathur and Sharma, 1984). The ameliorative effects of trees include an ability to penetrate strong soil layers and an improvement in internal drainage, high water use, production of litter, recycling of Ca and nutrients from depth, and moderation of microclimate. Trees tend to prevent secondary salinization and sodification by transpiring water from depth, and by reducing evaporation from the soil surface due to a reduction in temperature and air movement, and the mulching effect of litter (Gill and Abrol, 1993). In nonirrigated areas, it was often the removal of trees that caused salinization and sodification in the first place. Although trees may recycle Ca from depth, Na may be brought to the surface in the same way, and species vary in their ability to recycle different nutrients. Gill and Abrol (1990) reported that more N, P, K, and S accumulated in litter under *Acacia nilotica* than under *Eucalyptus tereticornis*, and that the opposite was true for Ca, Mg, Na, Fe, and manganese (Mn). They concluded that greater recycling of Na through *Eucalyptus* litter may defer soil amelioration in the long term.

Techniques for ensuring good establishment of trees involve reducing the effects of sodicity and salinity in the vulnerable early stages of growth by mounding, ripping,

TABLE 4.8 Some non-crop plants tolerant of sodicity

Plants	References
Herbaceous and shrubby forage and green manure plants	
Kallar grass, karnal grass, or dhaincha (*Leptochloa [Diplachne] fusca*)	Grewal and Abrol (1989); Gupta and Abrol (1990a); Kumar (1990); Ahmad *et al.* (1990); Gill and Abrol (1993); Aslam *et al.* (1993)
Sesbania (*Sesbania aculeata* and other spp.)	Abrol and Bhumbla (1979); Singh *et al.* (1991); Ahmad *et al.* (1990); Ilyas *et al.* (1993)
Rhodes grass (*Chloris gayana*); bermuda grass (*Cynodon dactylon*); para grass (*Brachiaria mutica*); panicums; shaftal clover; hybrid napier; berseem clover (*Trifolium alexandrinum*)	Gupta and Abrol (1990a)
Sordan (*Sorghum bicolor* x *Sorghum sudanese* hybrid)	Ahmad *et al.* (1990); Robbins (1986)
Echinochloa polystachya	Villafane (1989)
Amshot (*Agropyron* spp.)	Pearson (1960)
Deep-rooted perennial forage plants	
Alfalfa or lucerne (*Medicago sativa*); bahia grass (*Paspalum notatum*); tall fescue (*Festuca arundinaceae*); sweet clovers (*Melilotus* spp.); kudzu (*Pueraria lobata*)	Cresswell and Kirkegaard (1995); Ilyas *et al.* (1993)
Atriplex spp; *Maireana* spp.	Ahmad and Ismail (1993); Rashid *et al.* (1993)
Trees	
Acacia nilotica; *Eucalyptus tereticornis*	Gill and Abrol (1990, 1993)
Mesquite (*Prosopis juliflora*); *Dalbergia sissoo*; *Acacia nilotica*	Desh-Raj and Raj (1990)
Eucalyptus tereticornis; *Acacia stenophylla*; *A. ampliceps*; *A. modesta*; *A. nilotica*; *Casuarina obesa*; *Prosopis chilensis*; *P. siliquestrum*; *P. alba*; *Tamarix aphylla*	Hussain and Gul (1993)
Eucalyptus tereticornis; *E. camaldulensis*; *Prosopis juliflora*; *Acacia nilotica*; *A. auriculiformis*; *Zizyphus* spp.	Khanduja (1987)

mulching, and applying amendments to the soil. A great deal of work on tree establishment has been carried out in India, where the recommended method is to plant the trees in auger holes backfilled with the original soil mixed with gypsum and manure (Gill and Abrol, 1993; Singh *et al.*, 1994). Planting trees on ridges and kallar grass in the intervening trenches increased the rate and depth of reclamation (Grewal and Abrol, 1989). Singh *et al.* (1994) have discussed the choice of species and the cultural and economic aspects of the various agroforestry techniques available for the amelioration of sodic soils in India.

Secondary salinization and sodification from rising groundwaters and irrigation with poor quality water seem to be outstripping reclamation (Gupta and Abrol, 1990b). Therefore, it is just as important to consider prevention of sodicity as reclamation.

4.8 SUMMARY

Relationships between organic matter, sodicity and soil structure, which are not simple, depend to a large degree on particle size distribution, EC, pH, the mineralogy of the clay fraction, the nature of the anions and other exchangeable cations, and the presence of amorphous Fe, Al, and Si oxides.

Sodic soils generally have low organic matter contents. Three reasons for this, in their probable decreasing order of importance, are (a) low inputs resulting from the deleterious effects of poor physical and chemical conditions on plant growth, (b) high losses due to erosion and leaching, and (c) high rates of mineralization because of high substrate availability and low clay contents. Organic matter in soils with high ESP is more decomposed, more soluble, and less strongly associated with inorganic materials than organic matter in soils with low ESP.

Organic matter helps to inhibit sodification of soil under the influence of sodic water by stabilizing larger pores and thus facilitating leaching, and by increasing the overall selectivity of soil exchange sites for Ca. The addition of organic matter under a leaching regime can play a role in the desodification of soils, especially when added in conjunction with Ca. In alkaline (calcareous) sodic soils, organic matter and biological activity acidifies and, thus, partially desodifies the soil. Production of acidity and maintenance of high P_{CO2} are maximized under anaerobic conditions, and a decrease in pH is favored by leaching.

In sodic soils, roots, hyphae, polysaccharides, and hydrophobic organic materials can bind micro- and macroaggregates, preventing slaking and dispersion in the same way as they do in nonsodic soils. The effects of transient binding agents are maximized by continual additions of fresh organic matter and minimum disturbance by raindrop impact, cultivation, or rapid wetting or drying. In soils containing sufficient polyvalent cations, large organic polyanions may contribute to binding through cation bridging at the scale of clay domains and microaggregates. Where inputs of organic matter are low, or the soil is disturbed, the dispersive effect of anionic organic matter may dominate, especially in Very Sodic soils [Sod_H]. Anionic organic matter increases clay dispersion in sodic soils by increasing the negative charge density of colloids and reducing the activity of Ca^{2+} and other polyvalent cations. The presence of exchangeable Mg and high pH increase the dispersive effect of organic matter on sodic clay.

Organic amendments and the cultivation of crops, pastures, green manures, and trees have proven useful in the amelioration of sodic soils, particularly when used in conjunction with inorganic amendments and irrigation.

There are large gaps in our knowledge of how organic matter influences sodicity, and of the role that organic matter plays in maintaining desirable structure in sodic soils under different cropping systems. What we do know is that an understanding of the role of organic matter in sodic soils is essential to their management.

Acknowledgements
The helpful comments of Drs. W. W. Emerson, G. J. Churchman, C. D. Grant, P. Rengasamy, A. R. Barzegar, and Prof. J. I. Hedges were greatly appreciated.

5

Physical Properties of Sodic Soils

G. J. Levy, I. Shainberg, and W. P. Miller

5.1 INTRODUCTION

The presence of exchangeable sodium (Na) ions in the soil causes numerous adverse phenomena, such as destabilization of soil structure, deterioration of soil hydraulic properties, increased susceptibility to crusting, runoff, erosion, poor aeration, etc. The main factor that determines the extent of the deleterious effect of Na on soil properties is the electrolyte concentration in the soil solution. With low soil solution electrolyte concentration, as is the case in humid and subhumid regions or during the rainy season in semiarid regions of the world, even a low sodicity level [exchangeable sodium percentage (ESP) below 5], has a marked and irreversible negative effect on the physical properties of most soils. Conversely, when the electrolyte concentration in the soil solution reaches a moderate level (~10–20 mmol$_c$ L^{-1}), high sodicity levels (10 < ESP < 30) cause only small to moderate changes in the physical and hydraulic properties of soils, which are mostly reversible.

The aforementioned separation between the effect of high and low electrolyte concentration on the susceptibility of soil to sodicity is well illustrated by the following example. In Israel, due to the shortage of good-quality water, some farmers are forced to irrigate with saline and sodic waters [electrolyte concentration of 25–85 mmol$_c$ L^{-1}, sodium adsorption ratio (SAR) of 11–26]. Irrigation of cotton fields for more than 15 years on Calcic Haploxeralf and Typic Chromoxerert soils with these saline-sodic waters has caused neither a decrease in yield nor any deterioration in the physical state of the soil during the irrigation season, in spite of high soil sodicity (Frenkel and Shainberg, 1975) because the electrolyte concentration was high enough to prevent the adverse effect of sodicity imposed by the irrigation water. On the other hand, during the subsequent rainy seasons, the electrolyte concentration in the soil solution at the surface became very low, rendering the soil susceptible to crusting, reduced permeability, and increased runoff and erosion (Keren et al., 1983).

In view of the large areas of arable soils worldwide that are affected by low odicity levels, and the significant effect of low electrolyte concentration thereon, this chapter will concentrate mainly on the impact of low (ESP < 5) and moderate (5 < ESP < 10) sodicity levels on the physical and hydraulic properties of cultivated (disturbed) soils. In Chapter 6, the effect of high levels of sodicity (ESP > 10) and electrical conductivity (EC) > 1 dS m^{-1} on the hydraulic properties, soil wetting, and drainage of subsoils is evaluated. In order to understand the processes and mechanisms responsible for soil deterioration under conditions of low sodicity and salinity levels, the soil response to sodicity (Sections 5.3 to 5.5) is preceded by a review of the hydraulic conductivity (HC) of clay-sand mixtures. This chapter presents the effect of clay mineralogy, surface charge density, and specific surface area on swelling, dispersion, and hydraulic conductivity of the clay-sand mixtures.

5.2 HYDRAULIC CONDUCTIVITY OF CLAY-SAND MIXTURES

Swelling and dispersion of clays are the primary processes responsible for the degradation of the physical properties of soils in the presence of Na. Thus an account of the behavior of clay-sand mixtures in relation to Na and electrolyte concentration is a prerequisite to exploring the response of soils to sodic conditions. Because reference clay mineral systems are not always suitable for use as models of soil clay behavior, they will be used only as a first approximation and will be discussed only in qualitative terms. Deviation from pure clay models will be evaluated when discussing the response of soils to sodic conditions.

5.2.1 Effect of Sodicity on Swelling and Dispersion of Clays in Suspensions

The effect of sodicity on clay swelling and dispersion in suspension has been reviewed extensively (e.g., van Olphen, 1977; Shainberg and Letey, 1984; Shainberg and Levy, 1992; Sumner, 1993a; Chapter 3). Thus only a brief synopsis will be presented here.

Smectite Suspensions

The diffuse double layer at clay surfaces consists of the lattice negative charge and compensating countercations, which reside in the liquid immediately adjacent to the particles. Divalent cations are attracted to the surface with a force twice as great as that of monovalent ions. With an increase in the EC of the solution, the diffusion tendency of the adsorbed cations is diminished, and the diffuse double layer is compressed. The greater the compression of the ionic atmosphere toward the clay surface, the smaller the swelling pressures developed between smectite platelets. Low swelling pressures exist between calcium (Ca)-smectite platelets which condense into tactoids (Blackmore and Miller, 1961) or quasicrystals (Aylmore and Quirk, 1959). Each tactoid consists of several (four to nine) clay platelets in parallel array with an interplatelet distance of 0.9 nm.

The ESP influences both the swelling and the dispersion of clays. When both Na and Ca are adsorbed on smectites, "demixing" of the cations occurs so that most interlayer spaces contain Ca ions, and Na ions concentrate on the external surfaces of the tactoids. This model explains

swelling, dispersion, and flocculation in Na-Ca clay systems (Shainberg and Letey, 1984). Swelling of smectite as a function of ESP in Na/Ca systems was studied by Shainberg *et al.* (1971). Clay swelling under 1.33 kPa (20 psi) consolidation pressure is only slightly affected by increases in ESP up to 15 (no change in moisture retention) (Fig. 5.1). With a further increase in ESP above 20, a very sharp increase in swelling (large increase in moisture retained) was observed. The extent of swelling above ESP values of 50 was similar to that in pure Na smectites. Similar results were obtained by McNeal and Coleman (1966) on soil clays. These data show that smectites disperse but do not swell at ESP values below 15, above which swelling is initiated.

The effect of ESP on the flocculation (and dispersion) of Na-Ca clays was studied by Oster *et al.* (1980), who found that a small increase in ESP had a considerable effect on the dispersion [and critical flocculation concentration (CFC)] of the clay (Fig. 5.1). Similar effects of exchangeable Na on the electrophoretic mobility of Ca smectite have been reported (Bar On *et al.*, 1970).

That clay swelling is only slightly affected by an increase in ESP up to 15 and clay dispersion increases sharply with an increase in ESP both support the demixing model. Under these conditions, Na cations concentrate on the external surfaces of the tactoids; a small increase in exchangeable Na has a large impact on the Na on the external surfaces and a significant effect on clay dispersion and electrophoretic potential. Consequently, in this ESP range, the size of the tactoids and the area of osmotically active external surfaces are not affected by an increase in Na, and thus swelling is unaffected. Only when the ESP exceeds 15, does Na penetrate inside the tactoids causing the packet to disintegrate, and new surfaces are exposed to water adsorption and swelling.

Illite and Kaolinite Suspensions

The dispersion/flocculation value for illite suspensions as a function of ESP follows a simple linear relation (Oster *et al.*, 1980), which suggests that "demixing" in Na-Ca illite is not as pronounced as in Na-Ca smectite. However, illite at low ESP values is more dispersed and does not swell as much as the corresponding smectite, and exchangeable Na only promotes swelling slightly (Oster *et al.*, 1980). The low specific area for illite (120 m^2 g^{-1}) explains the low swelling. Because illite particles have irregular surfaces and the planar surfaces are terraced (Greene *et al.*, 1978), only small attractive forces exist between the particles, resulting in pronounced dispersion which requires high electrolyte concentrations to flocculate the clay.

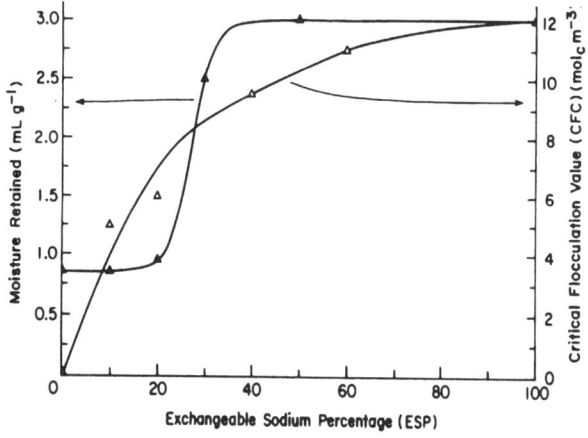

Figure 5.1 Water retention and critical flocculation concentrations (CFC) for Wyoming smectite as a function of exchangeable sodium percentage (ESP) [Shainberg *et al., Soil Sci.* 111:214-219, (1971)].

Pure Na kaolinite is flocculated at pH < 7 (Schofield and Samson, 1954). The attraction between the positive charges on the edges and the negative charges on the planar surfaces in this pH range explains the flocculation which occurs in the absence of salt. Deflocculation of the salt-free suspension takes place on the addition of NaOH. At pH values above 8, the edges become negatively charged, edge-to-face attraction does not occur, and the clay disperses. Because of the low specific surface area of kaolinite (<15 m² g⁻¹) the clay does not swell much and there is a negligible effect of exchangeable Na on clay swelling.

5.2.2 Hydraulic Conductivity, Swelling, and Dispersion in Clay-Sand Columns

The effect of sodicity and electrolyte concentration on swelling, dispersion, and HC in clay-sand mixtures was used to evaluate the dependence of these processes on clay properties (e.g., mineralogy, charge density, specific surface area, specific ion effect, etc.). For the HC determination, clay-sand mixtures containing clay (5%) together with 95% acid-washed quartz sand were packed in 50-mm-diameter plastic cylinders to a bulk density of 1.35 Mg m⁻³ (similar to that of the soil in the plow layer). In order to ensure thorough mixing of the mixture and prevent segregation of the clay and sand during packing, the mixture was wetted slightly and, thereafter, allowed to dry prior to the packing. Saturated "base" level HC was determined by leaching the column with a 0.5 M solution of a desired SAR using a constant head device and measuring the drainage rate. The columns were then consecutively leached with 0.05 and 0.01 M solutions of the treatment SAR until steady-state flow and effluent composition were attained (e.g., Alperovitch et al., 1985; Shainberg et al., 1987b).

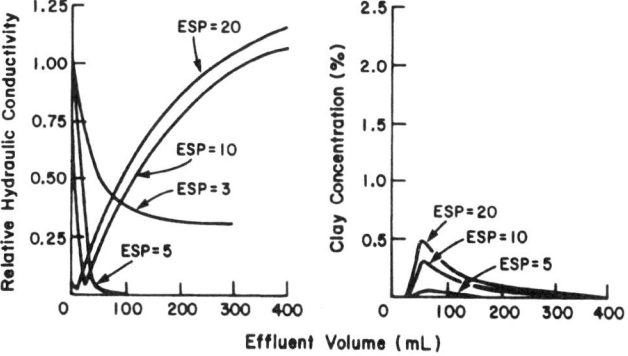

Figure 5.2 Relative hydraulic conductivity (HC) and effluent clay concentration of a Na–Ca Cl⁻ solution [Alperovitch et al., *Clays Clay Min.* 33:443-450, (1985)].

Effect of Electrolyte Concentration and SAR on the HC of Wyoming Smectite-Sand Mixtures

The relative saturated HC values of Wyoming smectite-sand mixtures equilibrated with Na-Ca (0.01 M Cl⁻) solutions and leached thereafter with distilled water are presented in Fig. 5.2. Note that as the ESP of the clay increases, the intercept on the HC axis decreases. For ESP values of 0, 5, 10, and 20, the relative HC at 0.01 M Cl⁻ concentration dropped to 1.0, 0.93, 0.70, and 0.05, respectively. The decrease in HC in 0.01 M Cl⁻ solutions is due to clay swelling, which reduces pore diameters (Alperovitch et al., 1985). Smectite dispersion was not expected at an electrolyte concentration of 0.01 M, which is well above the CFC of the clay at such ESP values (Oster et al., 1980). Swelling of Wyoming smectite increases with an increase in ESP, with a small change between ESP 0 and 10 and a steeper gradient between ESP 10 and 20. These results agree with the swelling data in clay-water pastes presented in Fig. 5.1. The size of smectite packets was not affected at ESP values below 10. As the ESP increases above 10, breakdown of the packets takes place, and the osmotically active surface area (i.e., exposed surfaces) and swelling increase.

When 0.01 M Cl⁻ solutions with SAR of 10 and 20 were replaced by distilled water (DW), the relative HC of the mixture dropped sharply to below 5%, followed by a sharp increase. In addition, the concentration of clay in the effluent peaked at about one pore volume before gradually decreasing (Fig. 5.2). The large percentage of clay in the effluent indicates that dispersion and movement of clay particles over long distances had taken place. Entrainment of clay particles in pore spaces (filtering) led to a sharp drop in HC values. The subsequent increase in HC was explained by a change in flow from that of a solution in a sandy clay matrix to that of a suspension in a sand matrix (Alperovitch et al., 1985; Pupisky and Shainberg, 1979). Removing the swelling clay skin from the sand grains increased the pore radii and thus the HC. When the 0.01 M Cl⁻ solution with SAR of 5 was replaced with DW, the decrease in HC was not followed by an increase in HC because clay dispersion was not sufficiently great to change the mechanism of flow, and, consequently, the column was sealed.

Effect of Smectite Mineralogy on Swelling, Dispersion, and HC of the Clay-Sand Mixtures

The dependence of the HC of four reference smectites on ESP and electrolyte concentration was studied by Alperovitch et al. (1985). The specific surface areas and charge densities of the four smectites were 800, 776, 760,

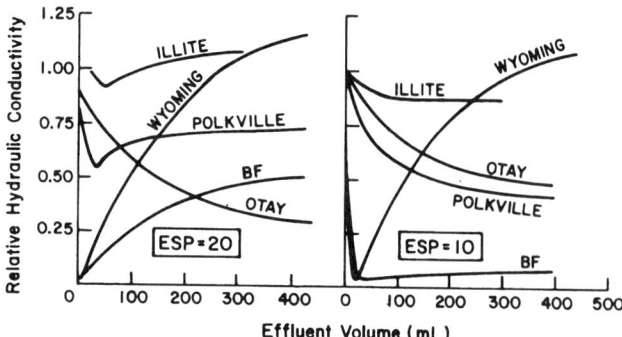

Figure 5.3 Effect of exchangeable sodium percentage (ESP) on relative hydraulic conductivity (HC) of smectite-sand mixtures. (BF = Belle Fourche) [Alperovitch *et al.*, *Clays Clay Min.* 33:443-450, (1985)].

and 552 m² g⁻¹ and 1.13, 1.19, 1.36, and 2.21 × 10⁷ mmol$_c$ cm⁻², respectively, for Upton (Wyoming), Belle Fourche (South Dakota), Polkville (Mississippi), and Otay (California) smectites. The effect of SAR (ESP) on relative HC of the clay-sand mixtures when 0.01 *M* solutions were displaced with DW is illustrated in Fig. 5.3. The effect of smectite mineralogy on clay swelling was estimated from the intercept on the HC axis. Of the four smectites, the order of swelling was Wyoming = Belle Fourche > Polkville > Otay smectites (Alperovitch *et al.*, 1985 and Shainberg *et al.*, 1988), which was similar to that obtained by Low (1980) and depended on specific surface areas and charge densities. Smectites having higher charge densities and lower specific surface areas exhibit greater electrostatic attraction forces between the platelets which are free to swell. Similarly, it was concluded that soils containing smectites with high charge densities will swell less and will be less affected by the sodicity of the electrolyte solutions. The HC of the various Na-Ca clay-sand mixtures varied markedly on leaching with DW. The Belle Fourche and Wyoming smectites swelled to about the same degree, but the latter dispersed more readily (Fig. 5.3). On the other hand, the Otay smectite, having the highest charge density, swelled and dispersed the least, and the change in the HC curve was minimal. The behavior of the Polkville was intermediate between the Otay and the other two smectites.

The effect of smectite mineralogy on the response of clay-sand mixtures to sodic conditions suggests that the mineralogy of the clay fraction in soils affected by Na should be studied quantitatively. In soil studies, not only the smectite content but also the charge density should be determined. Smectites with low charge densities swell and disperse more than those with high charge densities (Chapter 3).

Effect of Exchangeable Potassium on the HC of Smectite-Sand Mixtures

In general, the effect of exchangeable K on soil HC is less clear than that of Na, mainly because results vary or are contradictory. In some studies, exchangeable K had a deleterious effect (Quirk and Schofield, 1955; Levy and van der Watt, 1990) while in others (Ravina and Markus, 1975; Chen *et al.*, 1983), an exchangeable potassium percentage (EPP) of up to 30% improved soil permeability. The effect of exchangeable K on clay behavior appears to depend on clay mineralogy to a greater extent than does that of exchangeable Na. The affinity of soil clays for K increases in the order smectite < vermiculite < illite in parallel with increasing charge density. Shainberg *et al.* (1987a) observed that the affinity of the four smectites (Upton, Belle Fourche, Polkville, and Otay) for K in Na-K and Na-Ca solutions also increased with an increase in charge density. As the charge density on the clay increases, van der Waals and electrostatic attraction forces between the clay platelets also increase, and the low hydration energy of adsorbed K is not sufficient to overcome the electrostatic attraction forces and the clay platelets collapse (Chapter 3).

The effect of EPP on the HC of smectite-sand mixtures leached with DW is illustrated in Fig. 5.4. The numbers in parentheses next to the smectites names, give the EPP

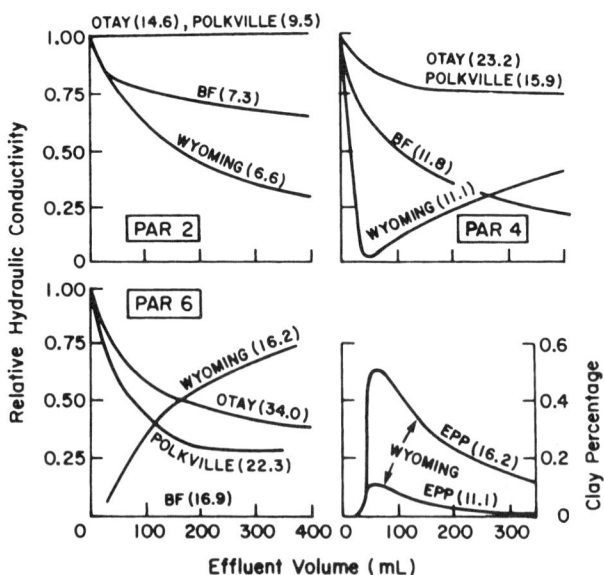

Figure 5.4 Relative hydraulic conductivity and effluent clay concentration of K–Ca smectite-sand mixtures leached with distilled water. Exchangeable potassium percentage (EPP) in equilibrium with given solution of potassium adsorption ratio (PAR in parentheses next to smectite name. (BF = Bell Fourche) [Shainberg *et al.*, *Clays Clay Min.* 35:305-310, (1987b)].

values of the clay in equilibrium with a given potassium adsorption ratio (PAR) solution. The decrease in HC in 0.01 M Cl⁻ solution was due to clay swelling. In equilibrium with a 0.01 M Cl⁻ solution with PAR of 4.0 (EPP varied between 11.1% for Wyoming and 23.2% for Otay), no decrease in HC was observed, and no swelling occurred for the four clays. Some swelling was observed for Wyoming smectite in equilibrium with the solution with PAR of 6.0 and none for the other smectites (even up to EPP = 34.0 for Otay smectite). When the 0.01 M Cl⁻ solutions with PAR of 2.0, 4.0, and 6.0 were replaced by DW, the relative HC of the various clay-sand mixtures varied markedly. The smectites with high charge densities were affected only slightly by EPP up to 34.3, whereas the HC of the smectites having low charge density (Wyoming and Belle Fourche) changed markedly. The dispersive effect of exchangeable K on low charge density smectites was similar to that of exchangeable Na. The low hydration energy of the K cations, coupled with the strong electrostatic attraction forces between platelets of smectites with high charge density, accounts for the "inefficiency" of K in dispersing these smectites. Hence, the dispersive effect of exchangeable K on smectites with high charge density was negligible. It can thus be concluded that the effect of exchangeable K on the HC of smectites depends strongly on the charge density of the mineral. The higher the charge density of the smectite the smaller the adverse effect of K.

Effect of Dilute Solutions (< 3 mmol꜀ L⁻¹) on the HC of Smectite-Sand Mixtures

The effects of displacing 0.01 M NaCl + CaCl₂ solutions having SAR of 10 and 20 with solutions of the same SAR, but with electrolyte concentrations of 0.003, 0.002, and 0.001 M Cl⁻, on HC are illustrated in Fig. 5.5. Displacing the 0.01 M solution (SAR = 10) with 0.001 M Cl⁻ resulted in a curve similar to that found for DW, i.e., a sharp drop followed by a sharp increase in relative HC (Shainberg *et al.*, 1987b). Clay dispersion in 0.001 M solutions and the resulting increase in relative HC were, however, appreciably less than for DW. Replacing the 0.01 M solution (SAR = 10) with 0.002 and 0.003 M solutions of the same SAR resulted in a sharp decrease in relative HC (Fig. 5.5) but no turbidity was observed in the leachate, indicating that solutions more concentrated than 0.002 M Cl⁻ prevented macroscopic movement of the clay. As the ESP of the Wyoming clay-sand mixtures increased to 21.6, clay dispersed and the relative HC increased for electrolyte concentrations of 0.001, 0.002, and 0.003 M. Only at an electrolyte concentration of 0.004 M was the CFC of the clay exceeded and no clay dispersion was observed.

Keren and Singer (1988) noted that when 0.003 M Cl⁻ solutions were replaced by DW, the HC continued to drop but clay did not become mobile. However, when the 0.01 M solution of the same SAR was replaced by DW, the HC decreased sharply and rapidly increased again with clay becoming mobile (Fig. 5.2). The difference in behavior

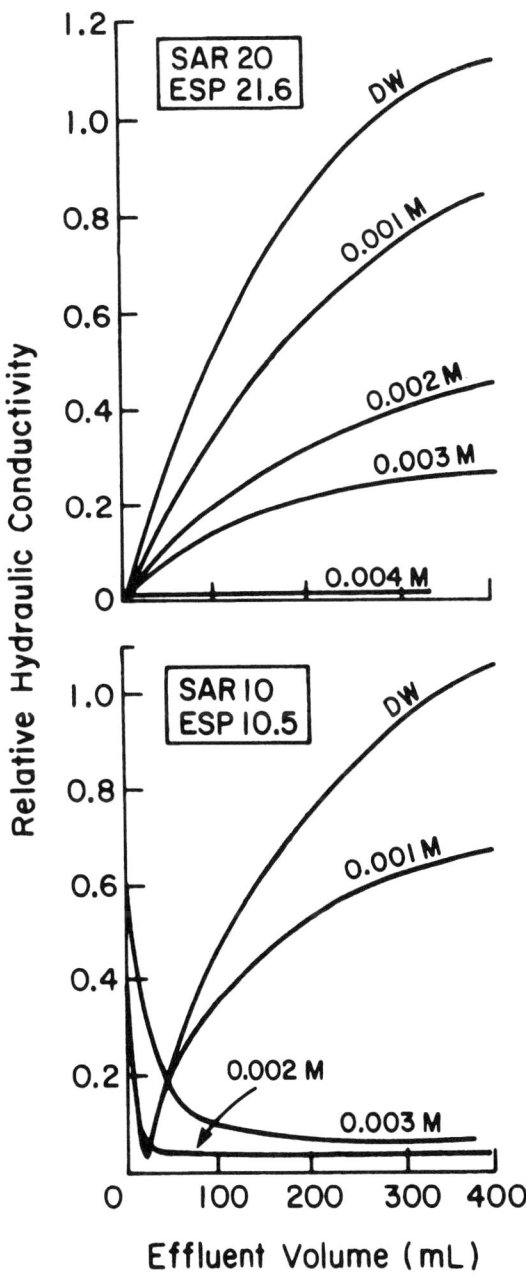

Figure 5.5 Relative hydraulic conductivity of Na–Ca Wyoming smectite-sand mixtures leached with dilute salt solutions, DW = distilled water, SAR = sodium adsorption ratio, ESP = exchangeable sodium percentage [Shainberg *et al.*, *Clays Clay Min.* 35:305-310, (1987b)].

between the two clay-sand systems (SAR = 10) was due to the degree of swelling of the clay before replacing the solution with electrolyte below the CFC. Clay swelling is a continuous process and increases gradually with decreases in solution concentration, but clay dispersion is only possible below the CFC. By leaching the mixture with a solution of electrolyte concentration below 0.01 *M* but above the CFC further swelling occurs, decreasing the conducting pore radii. The subsequent introduction of a solution with electrolyte concentration below the CFC further decreases pore radii, causing the dispersed particles to be trapped in the narrowed pores. As a result of this sieving effect, the HC further decreases, but no clay appears in the leachate.

HC of Illite-Sand and Kaolinite-Sand Mixtures

The effect of sodicity and electrolyte concentration on the HC of mixtures of illite (Fithian, Illinois) and sand were studied by Alperovitch *et al.* (1985). In the SAR range up to 30, as long as the electrolyte concentration in the leaching solutions exceeded 0.01 *M* Cl⁻, no changes in HC and clay dispersion were observed, presumably because illite is a nonexpandable clay. When the Na-Ca illite mixtures were leached with DW, the clay concentration in the effluent was high and exceeded that of Wyoming smectite, showing that illite is more dispersible than smectite. Oster *et al.* (1980) arrived at a similar conclusion.

The effects of sodicity and electrolyte concentration on the HC of kaolinite-sand mixtures were studied by Levy *et al.* (1991b). As expected, no swelling was observed in 0.01 *M* Cl⁻ solutions, but when the kaolinite-sand mixtures in equilibrium with 0.01 *M* Cl⁻ solutions of SAR 10 and 20 were replaced by DW, some reduction in HC was observed. The reduction in HC was related to increases in the pH of the effluent and partial dispersion of the clay. The pH of the 0.01 *M* effluent solution was 6.1 (due to CO_2 dissolution in the solution from the atmosphere) while the pH values of the DW effluents were 6.9, 7.1, and 7.6 for SAR values of 5, 10, and 20, respectively. The increase in pH with an increase in ESP was due to exchangeable Na hydrolysis (Bar On *et al.*, 1970), resulting in some clay dispersion and a small decrease in HC (Levy *et al.*, 1991b). Another possible explanation for the small reduction in HC upon increasing the ESP to 10 and 20 could be the presence of 2:1 clay impurities. Schofield and Samson (1954) recognized that only a pure kaolinite would remain flocculated when saturated with Na. In the presence of some micaceous and smectitic impurities, which neutralize the positively charged edges of the kaolinite, the kaolinite clay may disperse in the presence of exchangeable Na.

5.3 SODICITY AND HYDRAULIC CONDUCTIVITY OF DISTURBED SOIL SAMPLES

5.3.1 Introduction

The flocculation and dispersion behavior of soil clays differs significantly from that of reference clays, possibly because soil clays usually occur as mixtures and because of their intimate association with other minerals, oxides, and organic matter present in the soil. For instance, smectitic impurities in predominantly kaolinitic soils (Frenkel *et al.*, 1978) or dissolved fractions of organic matter (Durgin and Chaney, 1984; Frenkel *et al.*, 1992a) generally impart a high degree of dispersion to soil clay. Similarly, high soil pH, which, in turn, increases the net negative surface charge of the soil clay, also adversely affects the susceptibility of the clay to dispersion (e.g., Chiang *et al.*, 1987). Conversely, the presence of hydrous oxides (McNeal *et al.*, 1968), or sparingly soluble minerals such as $CaCO_3$ (Shainberg *et al.*, 1981b), reduces the severity of clay dispersion. It is, therefore, essential to validate and quantify the susceptibility of soil clays to sodic conditions.

To describe the relationship between HC and solution composition, Quirk and Schofield (1955) developed the concept of threshold electrolyte concentration (TEC), which is the concentration of the percolating solution that causes a 10–15% decrease in the soil permeability at a given ESP value. Soil permeability can be maintained, even at high ESP values, provided that the electrolyte concentration of the water is above a critical TEC level. For their particular soil, a noncalcareous silty loam, the TEC values were approximately 0.6, 2.3, 9.5, and 250 $mmol_c$ L^{-1} for soils with ESP values of 5, 8, 21, and 100, respectively. According to Quirk and Schofield (1955), even a Ca soil (ESP = 0) may show a reduction in HC, provided that the electrolyte concentration is below 0.6 $mmol_c$ L^{-1}. This has been verified by Emerson and Chi (1977), who observed dispersion of soil saturated with divalent cations at salt concentrations below 0.5–2 $mmol_c$ L^{-1}. When water of salinity below 1 $mmol_c$ L^{-1} (rain or snow) is applied to a soil, even Ca soils and soils of low ESP may disperse and their permeabilities may decrease.

The basic approach of Quirk and Schofield (1955) has been extended to a large number of additional soils. However, in most of these studies, the electrolyte concentrations of the percolating solutions were maintained above 3 $mmol_c$ L^{-1} (e.g., McNeal and Coleman, 1966; McNeal *et al.*, 1966a, 1968; Yaron and Thomas, 1968; Rhoades and Ingvalson, 1969) and, consequently, HC was controlled by swelling (Section 5.2). Typical effects of electrolyte concentration and ESP on soil HC are demonstrated in Fig. 5.6.

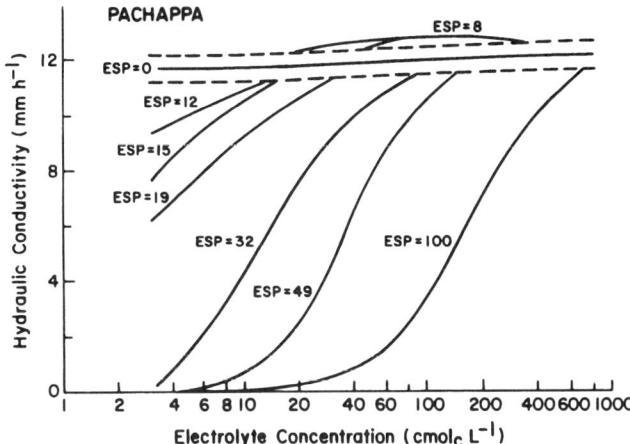

Figure 5.6 Hydraulic conductivity (HC) of Pachappa sandy loam as related to electrolyte concentration and exchangeable sodium percentage (ESP) [McNeal and Coleman, *Soil Sci. Soc. Am. Proc.* 32:190-193, (1966)].

5.3.2 Response of Sodic Soils to Leaching with Distilled Water

Even in arid regions where irrigation is essential for maintaining agricultural production, intermittent rain may lower the electrolyte concentration sufficiently to promote clay dispersion, depending on soil properties. For instance, in Australia under rainfed conditions, soils (especially clays) often become waterlogged during the winter rainy season because what little salt is present is leached in the first rains, causing the clay to disperse and the soil to seal (Fitzpatrick *et al.*, 1994). In Israel, where water having an SAR value as high as 26 with an EC of 4.6 dS m^{-1} is used for commercial irrigation, no permeability problems were experienced during the summer irrigation season even though ESP exceeded 26 (Frenkel and Shainberg, 1975). The electrolyte concentration in the irrigation water was sufficient to prevent the dispersive effect of Na. However, upon applying DW (to simulate rainwater in winter), the soil HC dropped to a small fraction of the initial 9–12 mm h^{-1} value (Frenkel and Shainberg, 1975). Similarly, in some areas in the Central Valley of California, permeability problems arise where waters of very low salinity (0.05–0.2 dS m^{-1}) are used for irrigation (Mohammed *et al.*, 1979).

Felhendler *et al.* (1974) found that for two smectitic soils (a sandy loam and a silt loam), HC was only slightly affected below SAR values of 20 as long as the concentration of the percolating solution exceeded 10 mmol$_c$ L^{-1} (Fig. 5.7). However, when the percolating solution was replaced by DW, simulating rainfall, the responses of the two soils

differed drastically. The HC of the calcareous silt loam dropped to 42 and 18% of the initial values for soils having ESP values of 10 and 20, respectively. On the other hand, the HC of the noncalcareous sandy loam soil dropped to 5 and 0% of the initial values under the same conditions, with mobile clay appearing in the leachate, whereas the leachate of the silt loam was clear. Felhendler *et al.* (1974) postulated that the HC response was associated with the potential of clay to disperse when the soil was leached with DW. The behavior of the two soils could be explained in the same way as the behavior of the clay-sand mixtures in dilute solutions (Section 5.2.2). When the electrolyte concentration of the soil solution exceeds the CFC, dispersion does not occur and soil HC decreases due to the swelling mechanism alone, which was the case for the calcareous silt loam. However, when the electrolyte concentration of the soil solution fell below the CFC, clay dispersed and clogged the conducting pores causing the soil to seal, which occurred in the noncalcareous sandy loam. This hypothesis will be verified in the Section 5.3.4.

5.3.3 Effect of Low Electrolyte Concentration on HC

Shainberg *et al.* (1981a,b) studied the effect of replacing 0.01 M Cl$^-$ solutions with SAR of 10 and 15 by DW or salt solutions containing 0.5, 1, 2, or 3 mmol$_c$ L^{-1} on the relative HC of, and the clay dispersion from, Fallbrook soil-sand mixture (1:1) (Fig. 5.8). When leached with DW, even an ESP value of 10 was sufficient to reduce the HC appreciably, but electrolyte concentrations of 2 and 3 mmol$_c$ L^{-1} in the percolating solution were adequate to prevent the

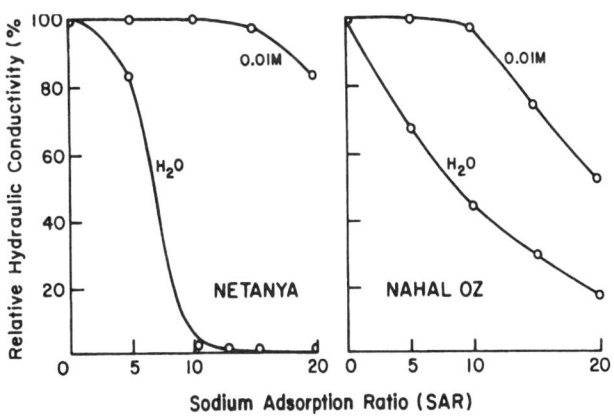

Figure 5.7 Hydraulic conductivity (HC) of sandy loam (Netanya) and silt loam (Nahal-Oz) soil as a function of the sodium adsorption ratio (SAR) and concentration of leaching solutions [Felhendler *et al.*, *Trans. 10th Int. Congr. Soil Sci.* 1:103-112, (1974)].

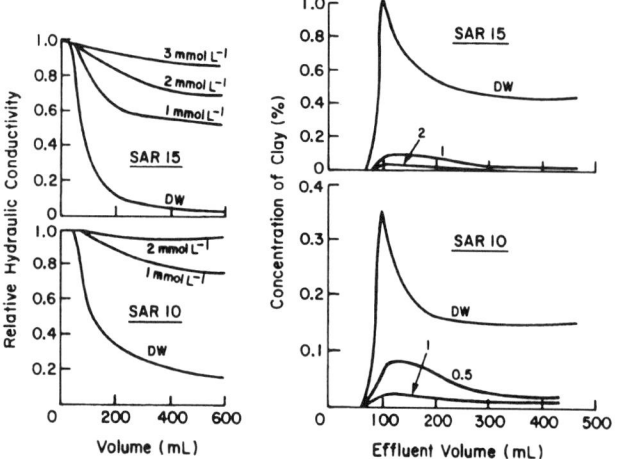

Figure 5.8 Relative hydraulic conductivity (HC) and concentration of clay in the effluent of Fallbrook soil-sand mixture (1:1), equilibrated with 0.01 M Cl⁻ solutions, SAR = 10 and 15 and leached with distilled water (DW) or salt solutions of 0.5, 1, 2, and 3 mmol$_c$ L⁻¹ [Shainberg *et al., Soil Sci. Soc. Am. J.* 45:273-277, (1981a)].

electrolyte concentrations above 3 mmol$_c$ L⁻¹ were used in determining soil HC (Chapter 1).

5.3.4 Soil Properties Affecting Response to Leaching with Distilled Water

Effect of Mineral Weathering

Arid land soils can release 3–5 mmol$_c$ L⁻¹ of Ca and Mg to solution as a result of the dissolution of plagioclase, feldspars, hornblende, and other sparingly soluble minerals (Rhoades *et al.*, 1968). The solution composition of a calcareous soil at a given ESP when brought in contact with DW can be calculated (Shainberg *et al.*, 1981b). As CaCO₃ dissolves, Ca replaces Na on exchange sites until the solution is in simultaneous equilibrium with exchange sites and with the CaCO₃ solid phase. Assuming that the apparent CaCO₃ solubility product is 10⁻⁸ (mol L⁻¹)² and the atmospheric CO₂ pressure is 30 Pa, the EC values of solutions in equilibrium with soils having ESP values of 5, 10, and 20 are 0.4, 0.6, and 1.2 dS m⁻¹, respectively (Shainberg *et al.*, 1981b). These concentrations are sufficient to counter the deleterious effects of exchangeable Na, even when the soil is leached with rainwater.

Unlike clay-sand mixtures, the sensitivity of sodic soils to solutions of low electrolyte concentration is subject to great variation. Shainberg *et al.* (1981a,b) hypothesized that a major factor causing differences among various sodic soils in their susceptibility to hydraulic failure when leached with low electrolyte concentrations was their rate of salt release from mineral dissolution, which determines the electrolyte concentrations of percolating solutions. They postulated

adverse effects of SAR values of 10 and 15, respectively, on the HC of this soil. Similarly, peak clay concentrations of 3.5 and 10 g L⁻¹ were obtained in the DW effluent at SAR values of 10 and 15, respectively, with the clay concentration decreasing markedly with an increase in the solution concentration (Fig. 5.8).

The response of the Fallbrook soil to various combinations of SAR and electrolyte concentration is illustrated in Fig. 5.9. When the concentration of the percolating solution was maintained at greater than 3 mmol$_c$ L⁻¹, no reduction in HC took place until the SAR exceeded 12. For solution concentrations of 2, 1, and 0.5 mmol$_c$ L⁻¹, threshold SAR values for a given decrease in the HC were 9, 6, and 4, respectively. When Fallbrook soil was leached with DW, the detrimental effect of exchangeable Na started at an ESP value of about 1, and once this was exceeded, the HC was always very sensitive to further increases in exchangeable Na. The more dilute the leaching solution, the more sensitive the soil became to incremental increases in ESP beyond the threshold value.

The effect of ESP on the HC of soils leached periodically with rainwater should be reemphasized; even a low ESP may cause loss of HC. At an ESP of 5, the HC of a soil like the Fallbrook was reduced to 20% of its initial value. The high sensitivity of soils to low levels of exchangeable Na when leached with "high-quality" (0.7 mmol$_c$ L⁻¹) water explains why an ESP value of 5 was accepted in Australia to separate sodic from normal soils (McIntyre, 1979). On the other hand, the USSL Staff (1954) suggested the use ESP greater than 15 to designate sodic soils because

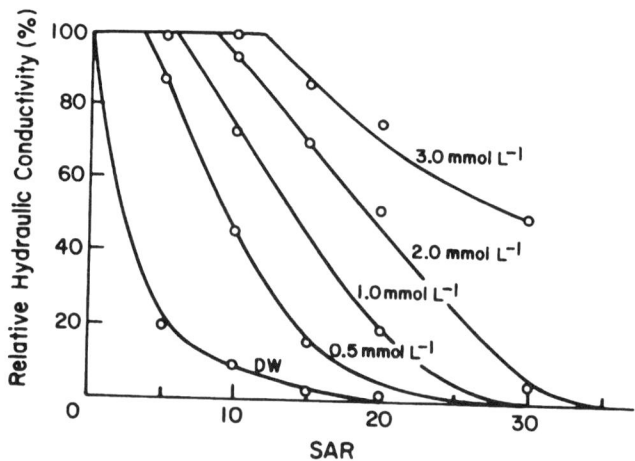

Figure 5.9 Relative hydraulic conductivity (HC) of the Fallbrook soil-sand mixture as a function of electrolyte concentration and sodium adsorption ratio [SAR (ESP)] of system [Shainberg *et al., Soil Sci. Soc. Am. J.* 45: 273-277, (1981a)].

that sodic soils containing minerals ($CaCO_3$ and a few primary minerals) that readily release soluble electrolytes will not readily disperse when leached with DW at moderate ESP values, because a sufficiently high electrolyte concentration (>3 $mmol_c$ L^{-1}) could be maintained to prevent clay dispersion. In addition, the ESP value will be reduced, because most of the cations released are Ca and Mg. Conversely, soil solution concentrations in soils lacking readily weatherable minerals are likely to be below the CFC, making the soils more susceptible to clay dispersion. Consequently, the lodgment of dispersed clay particles in the conducting pores may markedly decrease the HC of such soils.

The data presented in Fig. 5.7 for the two Israeli soils support the hypothesis of Shainberg et al. (1981a,b). The soil that was more physically stable with respect to chemical changes (Netanya sandy loam) was also more susceptible to sodic conditions when leached with DW. This hypothesis was further supported by measurements of clay dispersibility, HC, and the mineral weathering properties of three California soils. The Fallbrook soil, which contains few weatherable minerals, was most sensitive to clay dispersion and loss in HC from increasing ESP despite the presence of sesquioxides and kaolinite in its clay fraction. The Gila and Pachappa soils, which have higher mineral dissolution rates, were less affected by exchangeable Na.

Effect of Lime

Many sodic soils commonly contain salts and have pH values above 7.5. Under such conditions, where the solubility of $CaCO_3$ is very low, Ca from this source is not effective in displacing exchangeable Na and in the reclamation of these soils. sodic soils containing $CaCO_3$ are common in many semiarid and arid regions of the world. The presence of fine $CaCO_3$ particles in soils can improve the physical condition of sodic soils (USSL Staff, 1954). To explain this effect, the USSL Staff (1954) and Rimmer and Greenland (1976) suggested that lime acts as a cementing agent which stabilizes soil aggregates and prevents clay dispersion. Another mechanism which explains this beneficial effect is the potential of $CaCO_3$ to dissolve and to maintain the soil solution at concentrations above the CFC values of the soil clays, thus preventing their dispersion (Shainberg and Gal, 1982; Naidu et al., 1993b). According to the dissolution mechanism, exchange reclamation (replacement of Na by Ca) is still negligible because of the low concentration of Ca ions in the soil solution; however, clay dispersion is prevented mainly because of the electrolyte concentration effect (Shainberg et al., 1981a). The effect of $CaCO_3$ dissolution on the HC of sodic soils was investigated by Shainberg and Gal (1982) by mixing lime-free soils with low percentages of powdered lime (0.5 and 2.0%). Although the HC of the lime-free soils dropped sharply when 0.01 M solutions with SAR values of 20 were replaced by DW, mixing the soils with powdered lime prevented both HC decay and clay dispersion (Fig. 5.10). The increase in soil solution concentration due to $CaCO_3$ dissolution was proposed as the mechanism responsible for the beneficial effect observed (Shainberg and Gal, 1982). Similar results in a leaching column study with a Xeralf were obtained by Naidu et al. (1993b).

The relative importance of each of the two mechanisms described by which $CaCO_3$ stabilizes soil structure can be estimated from its effect on infiltration rate (IR) and crust formation under rainfed conditions (e.g., Agassi et al., 1981; Ben-Hur et al., 1985). If $CaCO_3$ acts as a cementing material, calcareous soils should be less sensitive to crust formation than noncalcareous soils. However, if the dissolution mechanism is dominant, the electrolyte concentration at the soil surface exposed to rain will be insufficient to prevent dispersion, and both types of soil will be sensitive to crusting. Results to verify that the latter mechanism dominates will be presented in Section 5.4.2.

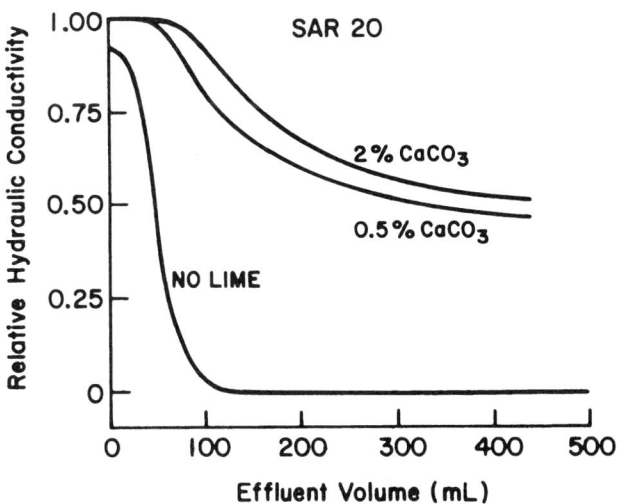

Figure 5.10 Effect of lime on hydraulic conductivity (HC) of noncalcareous soil equilibrated with 0.01 M Na (SAR = 20) and subsequently leached with distilled water (DW) [Shainberg and Gal, *J. Soil Sci.* 33:489-498, (1982)].

Effect of Soil Texture

Mobile clay in the leachate is usually only observed in sandy soils. When the clay content is high, the small size of the conducting pores usually ensures that dispersed clay moves only short distances before it clogs the pores, resulting in reduced HC. Thus in loams and clays, the dispersion mechanism still operates, but no macroscopic movement of the clay particles is observed. When a clay soil is mixed with sand, the dispersion mechanism and macroscopic clay movement become evident.

The importance of pore size distribution on HC was demonstrated by Keren and Singer (1988), who leached a low Na clay-sand mixture with dilute solutions($\sim$3 mmol$_c$ L^{-1}) for a long period of time. This permitted the clay in the mixture to swell and decrease the radii of conducting pores. A subsequent leaching with DW further decreased the pore radii, but also dispersed and mobilized the clay particles (Keren and Singer, 1988) which became trapped in the narrow pores, reducing the HC still further. Such behavior will also occur in loam and clay soils.

Effect of Exchangeable Magnesium

Although in many soils the exchangeable Ca-Mg molar ratio is high, the reverse is true in some soils. The effect of adsorbed Mg on soil hydraulic properties is a controversial issue. While the USSL Staff (1954) grouped Ca and Mg together as being similar in promoting and maintaining soil structure, van der Merwe and Burger (1969) found that Na-Mg saturated soil was structurally less stable than the Na-Ca counterpart, concluding that exchangeable Mg can have a deleterious effect on soil structure and permeability. McNeal *et al.* (1968) also showed that soils high in exchangeable Mg had a lower HC relative to soils high in exchangeable Ca. A distinction has been made between the direct effect of exchangeable Mg in causing decreases in HC, which was termed a *specific effect*, and the inability of Mg in irrigation waters to counter the accumulation of exchangeable Na in the soil (Chi *et al.*, 1977; Emerson and Chi, 1977; Rahman and Rowell, 1979). Curtin *et al.* (1994) compared the effects of both Mg and Ca on soil selectivity for Na, dispersibility, and HC of six Canadian oils. They observed that the adverse effects of Mg on clay dispersion and HC were greater than could be explained by the higher exchangeable Na level in the Mg system, concluding that the specific effect of Mg was dominant. The effect of sodicity on the HC of three Israeli soils, with Mg and Ca as complementary cations, was studied by Alperovitch *et al.* (1981), who found that in a calcareous soil (Kesufim) exchangeable Mg had no specific adverse effect on the HC, whereas in the noncalcareous soils (Hermon and Netanya) Mg caused a decrease in the HC of the soils beyond that of the Na-Ca system (Fig. 5.11). They suggested that the presence of Mg enhanced the dissolution of CaCO$_3$ in the calcareous soil, producing electrolyte which prevented clay dispersion and HC decay. The observed favorable effect of

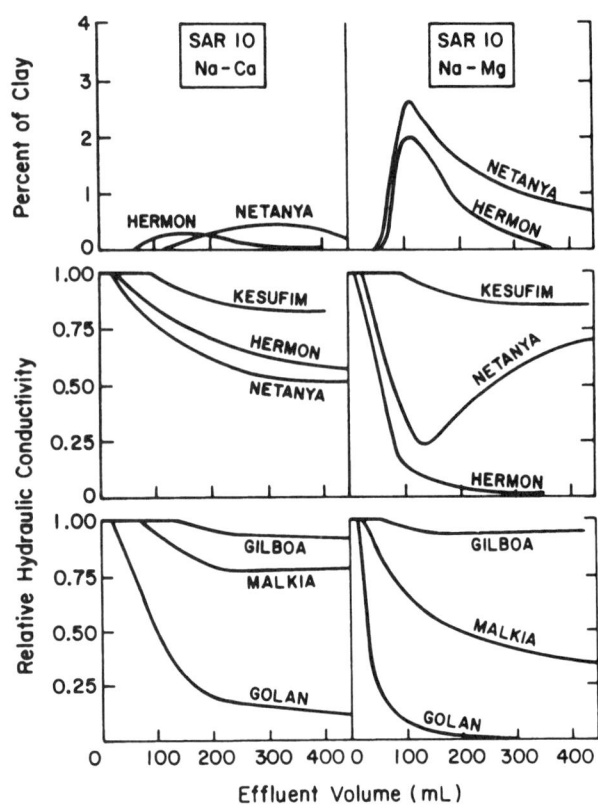

Figure 5.11 Relative hydraulic conductivity (HC) and concentration of clay in effluent of three Israeli soils equilibrated with 0.01 *M* solutions (SAR = 10) and leached with distilled water. Where no data for clay concentrations are given, no clay appeared in effluent [Alperovitch *et al., J. Soil Sci.* 32:543-554, (1981)].

Ca, compared to Mg, on the HC of the noncalcareous soils was explained by the higher soil solution concentration in the Ca relative to the Mg system in the soils exhibiting little mineral dissolution (Alperovitch *et al.*, 1981). Furthermore, Shainberg *et al.* (1988) concluded that the lower HC values of the Na-Mg system relative to the Na-Ca system are related to the effect of Mg on the hydrolysis of these clays. The presence of high concentrations of Mg at the clay surface slowed down the release of octahedral Mg from the clay lattice (Kreit *et al.*, 1982) and lowered the EC of the Na-Mg clay (Shainberg *et al.*, 1988). Conversely, in kaolinitic soils, the EC$_e$ and the resulting HC of Na-Mg systems were comparable to those of Na-Ca systems (Levy and van der Watt, 1988). Apparently, the aforementioned adverse effect of Mg on the electrolyte composition does not occur in kaolinitic soils because of the limited isomorphous substitution in kaolinites, which results in very small amounts of octahedral Mg in the crystal (Levy *et al.*, 1989).

5.4 INFILTRATION RATE OF SODIC SOILS

5.4.1 Mechanisms Controlling Infiltration Rate

Soil infiltration rate (IR) is defined as the volume flux of water flowing into the profile per unit of soil surface area under any set of circumstances. The mechanisms controlling the change in the IR depend on the mode of water application to the soil. Under conditions where water is supplied to the soil without energy, the IR depends on the HC of the soil matrix and its variation with time, which depends on initial water content and suction, soil texture and structure, profile uniformity (Hillel, 1980), and level of salinity and sodicity in the applied water and soil solution (Section 5.3). When water is supplied with energy, the IR decreases from its initial high rate due to the formation of a thin layer (< 2 mm) at the soil surface, termed the seal. This seal is characterized by greater density, high shear strength, finer pores, and lower saturated HC than the underlying soil (McIntyre, 1958; Bradford *et al.*, 1987). A structural seal, usually caused by the impact energy of waterdrops, should be distinguished from a depositional seal formed by translocation and deposition of fine soil particles at a certain distance from their original location (Arshad and Mermut, 1988; Chen *et al.*, 1980). The discussion to follow focuses on the former.

Structural seal formation is due to two complementary mechanisms (McIntyre, 1958; Agassi *et al.*, 1981): (a) the physical disintegration of soil aggregates on wetting, and

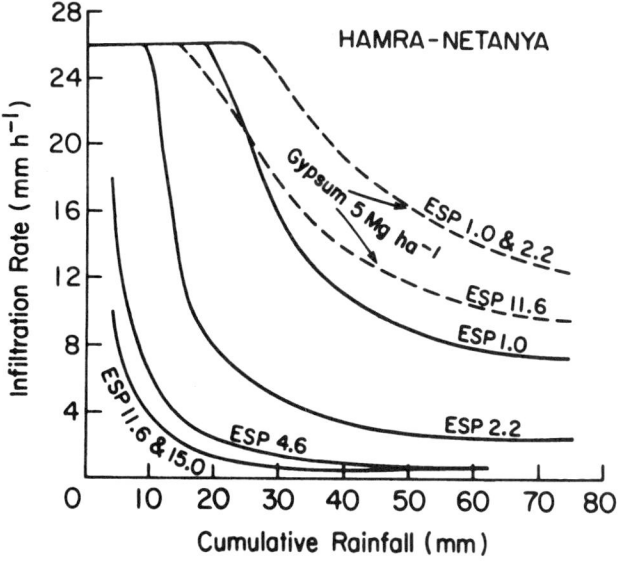

Figure 5.12 Effect of soil exchangeable sodium percentage (ESP) and phosphogypsum on infiltration rate of Netanya soil under simulated rainfall using distilled water [Kazman *et al., Soil Sci.* 35:184-192, (1983)].

compaction caused by the impact of the raindrops, and (b) the chemical dispersion and movement of clay particles into a region at 0.1–0.5 mm depth, where they lodge and clog the conducting pores. Such a seal forms during rain (Duley, 1939; Epstein and Grant, 1973; Morin and Benyamini, 1977) or sprinkler irrigation (Aarstad and Miller, 1973) and is responsible for the decrease in IR.

5.4.2 Effect of Sodicity

Na-Ca System

The seal formation and IR of four smectitic soils were highly sensitive at low ESP levels when exposed to distilled water (Kazman *et al.*, 1983) (Fig. 5.12). Even at the lowest sodicity (ESP = 1.0), a seal was formed and the IR dropped from an initial value above 100 to a final infiltration rate (FIR) of 7.0 mm h^{-1}. An ESP value of 2.2 was sufficient to cause a further drop in the FIR of the sandy loam to 2.4 mm h^{-1}. The amount of rain required to approach the FIR (i.e., the rate of seal formation) was also affected by the ESP (Fig. 5.12). Since the raindrop impact energy was the same in all the experiments, the differences in IR curves for the various treatments were the result of chemical dispersion of the soil clay caused by sodicity. The high sensitivity of the soil surface to low ESP values is explained by three factors (Oster and Schroer, 1979; Kazman *et al.*, 1983): (a) the mechanical impact of the raindrops, which enhances chemical dispersion, (b) the absence of a surrounding soil matrix (sand particles), which when present slows clay dispersion and movement, and (c) the almost total absence of electrolytes in the applied water. With respect to the first factor, it is very difficult to separate the mechanical and the chemical mechanisms as they are complementary. Agassi *et al.* (1985) noted that in the absence of the former, the chemical mechanism does not come into effect at ESP levels below 5. Evidently, in the lower ESP range, the chemical mechanism needs some activation energy to start operating at the soil surface, which, in the case of rain, is provided by the impact energy of the raindrops.

Na-Mg System

Elevated exchangeable Mg levels can cause deterioration in soil structure, resulting in the development of a "magnesium Solonetz" (Ellis and Caldwell, 1935). In addition, Mg enhances dispersion in montmorillonitic and illitic clays compared to Ca (Bakker and Emerson, 1973; Chapter 3). With respect to IR, the effect of exchangeable Mg was found to depend on the kinetic energy of the waterdrops.

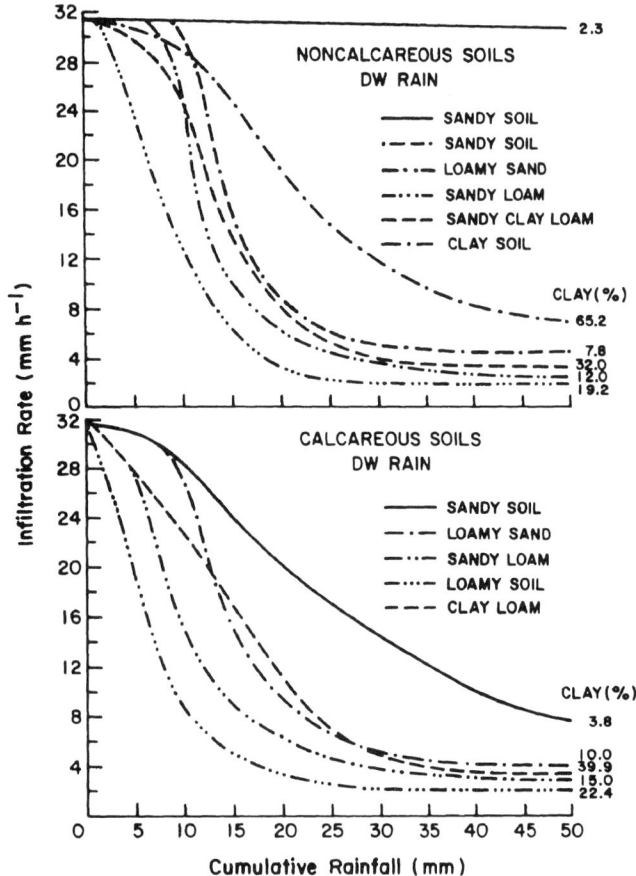

Figure 5.13 Infiltration rate (IR) of various soils as a function of clay percentage [Ben-Hur *et al.*, *Irrig. Sci.* 6:281-296, (1985)].

When high-energy rain (22.9 kJ m^{-3}) was used, exchangeable Mg had an effect similar to that of Ca (Levy *et al.*, 1988), but with low to medium energy (8.0–12.5 kJ m^{-3}), the IR in Mg-Na treated soil was lower than that in the Ca-Na soil (Keren, 1990). These findings suggest that the adverse effect of Mg on clay dispersion and hence on IR is pronounced only under conditions where chemical dispersion is dominant in controlling IR (i.e., raindrops with low to medium kinetic energy).

5.4.3 Effect of Clay Content and Mineralogy

Soils with ESP less than 5 have been reported to respond differently to seal formation and IR reduction when exposed to rain. Studies have indicated that a number of soil properties and especially clay content and mineralogy significantly affect the IR of soils. The effect of clay content on the IR of soils with ESP lower than 2.5 was studied by Ben-Hur *et al.* (1985) using soil samples with clay contents between 3 and 60%. Soils with 10–30% clay were the most susceptible to seal formation and had the lowest IR (Fig. 5.13). With increasing clay content, the soil

structure was more stable and seal formation was diminished. In soils with lower clay contents (< 10%), the amount of clay available to disperse and clog soil pores was limited, and as a result, a poorly developed seal was formed (Ben-Hur *et al.*, 1985).

Most of the studies on seal formation and IR were conducted on soils in which the dominant clay minerals were smectites, which are known to be dispersive in both HC and IR determinations. The effect of exchangeable Na on the HC of kaolinitic soils is quite small (Frenkel *et al.*, 1978; Chiang *et al.*, 1987). However, a number of kaolinitic and illitic soils from the southeastern United States (Miller, 1987; Miller and Scifres, 1988) and South Africa (Levy and van der Watt, 1988; Stern *et al.*, 1991a) were found to be susceptible to seal formation. Evidently, upon mechanical stirring by the beating action of raindrops, aggregates from kaolinitic soils may disperse (according to the level of exchangeable Na) and, consequently, form a seal with low IR. In addition, Stern *et al.* (1991a) observed that the

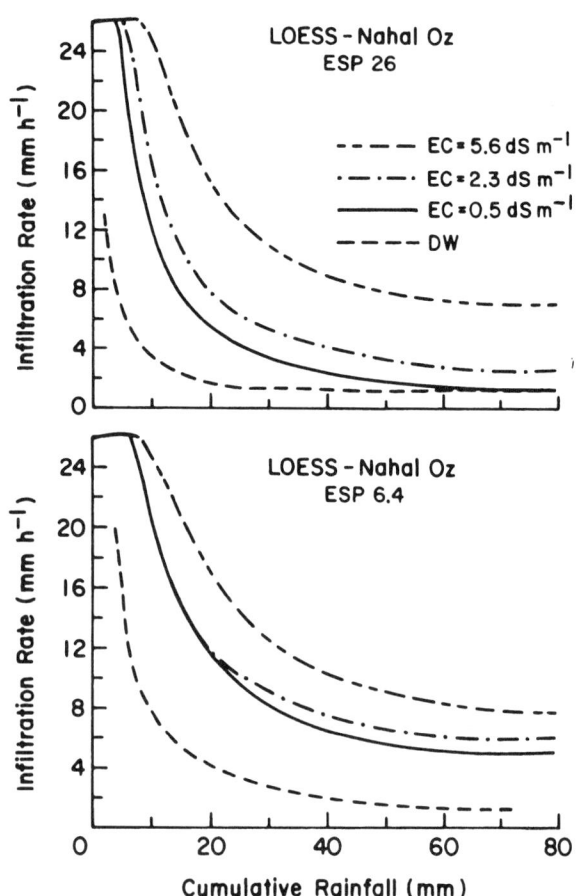

Figure 5.14 Effect of electrical conductivity (EC) of applied water and soil exchangeable sodium percentage (ESP) in rain simulation experiments on infiltration rate of loess Nahal Oz soil [Agassi *et al.*, *Soil Sci. Soc. Am. J.* 45:848-851, (1981)].

susceptibility of the kaolinitic soils to seal formation was positively correlated with smectitic impurities. They concluded that kaolinitic soils with small amounts of smectite are dispersive and susceptible to seal formation. Conversely, kaolinitic soils which contain no smectite impurities are less susceptible to sealing. The stability of kaolinitic soils can also be ascribed to (a) low pH values, which result in an increased amount of positive charge on the edges of the clay particles, and (b) the presence of hydroxy-Al and -Fe polymers, which act as cementing agents (El-Swaify and Swindale, 1969).

5.4.4 Effect of Electrolyte Concentration

Clay dispersion is very sensitive to the chemistry (electrolyte concentration and cationic composition) of the applied water (Shainberg and Letey, 1984, and references cited therein). This is particularly true in the case where the soil surface is exposed to the mechanical action of falling water droplets, which enhance clay susceptibility to chemical dispersion. Agassi et al. (1981) noted that the lower the concentration of the electrolyte solution, the faster was the rate at which the IR decreased (Fig. 5.14), and that, for the same electrolyte concentration, increasing the soil ESP resulted in a sharper decrease in and a lower FIR. Similarly, Oster and Schroer (1979) found that electrolyte concentration greatly affected IR, even at low SAR values. They observed an increase in FIR from 2 to 28 mm h⁻¹ as electrolyte concentration in the applied water, having SAR levels between 2 and 4.6, increased from 5 to 28 mmol$_c$ L⁻¹. These results, as well as others, indicate that the IR is far more sensitive than the HC to the electrolyte concentration of the applied water (Shainberg and Letey, 1984).

5.4.5 Soil Amendments

Traditional strategies to control sodicity hazards resulting from sealing and low IR were based mainly on agrotechnical measures. Modification of some soil properties responsible for seal formation, such as clay dispersion and aggregate stability at the soil surface, may serve as an alternative strategy and can be effected by applying chemical amendments to the soil. The realization that seal formation and the resulting low IR are surface phenomena suggests that only the soil surface needs to be treated, thus reducing the quantities of amendments required.

By-product Gypsum
Chemical dispersion is prevented when electrolytes are present in the soil solution at levels above the CFC. Under natural rain conditions, a readily available electrolyte source,

which releases salt over an extended period, should be applied on the soil surface to compensate for the lack of electrolytes in the rainwater. Phosphogypsum (PG) and flue gas desulfurization gypsum (FDG), which are by-products of the phosphate and power generation industries and readily available, meet these requirements. Spreading PG at rates of 5–10 Mg ha⁻¹ on the surface of smectitic (Kazman et al., 1983), and kaolinitic soils (Miller, 1987; Miller and Scifres,1988) was effective in slowing down the rate of IR decline and maintaining a higher FIR compared with the control even in soils with ESP values of 1.0 (Fig. 5.12). This indicates that some chemical dispersion takes place even at very low ESP values, and possibly even in Ca-saturated soils. Agassi et al. (1986) suggested that the favorable effect of PG on IR should be attributed not only to its effect on the EC of the percolating water but also to (a) the physical interference with the continuity of the seal, and (b) the partial mulching of the soil surface, which protects the soil from the beating action of the raindrops.

Organic Polymers
The use of organic polymers, mainly polysaccharides (PSD) and polyacrylamides (PAM), for improving aggregate stability, maintaining high IR, and reducing seal formation has recently been studied extensively (Shainberg and Levy, 1994, and references cited therein). Addition of small amounts of polymers (10–20 kg ha⁻¹), either sprayed directly onto the soil surface or added to the applied water,

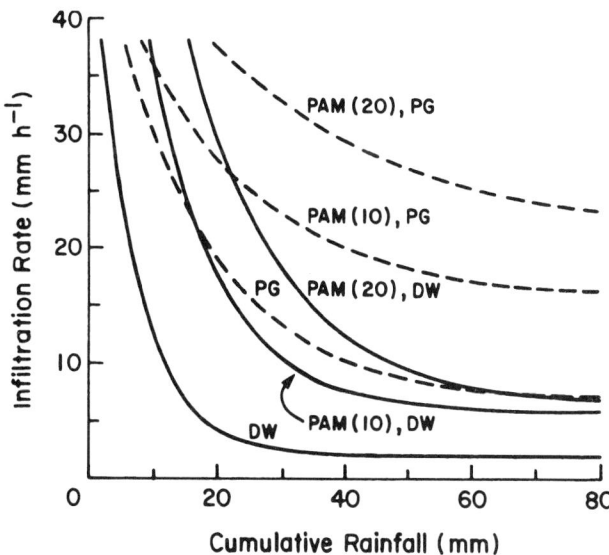

Figure 5.15 Infiltration rate (IR) of loess as a function of cumulative rainfall for different levels of polyacrylamide (PAM) (10 and 20 kg ha⁻¹) and phosphogypsum (PG) (5 Mg ha⁻¹) treatments under simulated rainfall distilled water (DW) [Shainberg *et al.,* *Soil Sci.* 149:301-307, (1990)].

was effective in stabilizing and cementing aggregates together at the soil surface and, hence, maintaining high IR values in soils with ESP below 5, under both laboratory (Helalia and Letey, 1988a; Levy *et al.*, 1992; Shainberg *et al.*, 1990) and field conditions (Levy *et al.*, 1991a; Stern *et al.*, 1991b). The efficacy of anionic polymers in preventing seal formation is enhanced when the soil clay is maintained in a flocculated state (Shainberg *et al.*, 1990), and the resulting FIR could be tenfold higher than the control (Fig. 5.15). Of the polymers currently available and under study, anionic PAM which has the longest residual effect, has been found to be the most effective in controlling seal formation and maintaining high IR (Levy *et al.*, 1992). On the other hand, when added to soils with ESP above 20, PAM was ineffective in controlling seal formation and maintaining high IR values. In soils with moderate ESP (10–15), the effect of PAM was inconsistent and depended on water quality and the amount of PAM added (Levy *et al.*, 1995).

5.5 SODICITY AND AGGREGATE STABILITY

5.5.1 Aggregate Stability and Its Importance

Aggregation in soils is the arrangement of primary particles (sand, silt, and clay) into secondary structural units, typically on a millimeter scale in size. Soil aggregates thus consist of clay particles, in quasicrystals or domains (Quirk and Aylmore, 1971), interspersed with larger quartzitic primary particles and held together by physicochemical forces that are a product of the chemical and microbiological environment of the particular soil. The stability of aggregates is typically assessed relative to wetting and/or physical disruption, while its importance in soil use and management relates largely to the voids created between aggregates. Aggregation can indeed be defined in terms of porosity, as discussed by Quirk (1978). The larger pores (often described as *macropores* or water transmission pores, greater than 50 *μm* in size) conduct water into the soil profile from the surface during rainfall, and their stability is crucial in facilitating water entry and return of aerobic conditions once rainfall has ended (Greenland, 1979). The breakdown of unstable aggregates results in pore collapse, which slows infiltration greatly, resulting in runoff and erosion from the soil surface. Subsequent gas diffusion into the root zone is likewise slowed by the presence of smaller, predominately water-filled pores, and root growth may suffer as a result.

The traditional view of soil structure, particularly the water-stable aggregates of surface soils, is that (a) humified organic materials are the binding agents that hold

microaggregates together, and (b) increasing soil organic matter levels implies improved structure (Allison, 1973). Humic materials are conceived of as long-chain molecules capable of "bridging" between soil particles, thereby physically forming an aggregate more or less resistant to breakdown by wetting and abrasive forces, as in the model of Edwards and Bremner (1967).

Quirk (1978), however, recalled Bradfield's early statement that "granulation is flocculation plus," and emphasized that a necessary condition for the formation of water-stable granules is the flocculation of the clay particles, plus further stabilization by organic and/or inorganic cementing agents. The principles of flocculation and dispersion discussed previously are central in the formation of clay quasicrystals and floccules ($<20\,\mu m$) that will become the building blocks of aggregates. Such floccules, are further aggregated by inorganic (clay particles or metal oxides) or organic materials (humified or residual microbial products) into microaggregates, ranging from 20 to 250 μm in size. Macroaggregates, typically defined as larger than 250 μm, are agglomerates of microaggregates and primary sand and silt particles, held together largely by root hairs, fungal hyphae, and large biomolecules (Tisdall and Oades, 1982). The picture in Fig. 5.16, modified from the diagram of Greenland (1979), illustrates these constituents and associations. A detailed discussion of the role of organic matter in structure development is given in Chapter 4.

Aggregate stability is typically measured using a wet sieving technique, in which rapid wetting and minimal energy input are used to disrupt the macroaggregates, and further energy input (severe agitation or ultrasonic energy) leads to the breakdown of the microaggregates. Stable macro- or microaggregates remaining on the sieve (typically >250 μm or >50 μm, respectively) after mild or severe agitation in water are a measure of soil resistance to "slaking", i.e., a measure of the ability of long-range binding of organic materials to hold smaller aggregate units together. The term "slaking" is used to refer to the breakdown of soil aggregates into particles >20 μm in size as opposed to dispersion, which is the further breakdown of fine soil particles and release of <2 μm clay (Chapter 3). Abu-Sharar *et al.* (1987) postulated that slaking of aggregates results from a random breakdown of the aggregates at planes of weakness which occurs prior to clay dispersion. Dispersion of individual clay platelets may follow slaking if the hydration and osmotic forces within microaggregates are sufficient to overcome the attractive forces operating between particles within the aggregate. Concerning clay dispersion in aggregates, Rengasamy *et al.* (1984b) made a distinction between spontaneous dispersion, where the aggregates in the suspension were not mixed, and mechanical dispersion, where the aggregates were shaken in water. High ESP (> 10) values and low electrolyte concentration were needed for spontaneous dispersion. Under mechanical stress (e.g., impact of raindrops, shear force of flowing water, mechanical energy of cultivation machinery) aggregate breakdown and clay dispersion were already observed at low ESP (< 10) levels (e.g., Shainberg *et al.*, 1992a). However, in the presence of Al and Fe oxides in the soil, the effect of high ESP and low electrolyte concentration was negligible in enhancing aggregate breakdown and clay dispersion (McNeal *et al.*, 1968; El-Swaify, 1973). The cementing action of the Al and Fe oxides prevented clay swelling and dispersion with Al oxides being more effective than Fe oxides in impeding the adverse effects of sodic conditions (Alperovitch *et al.*, 1985; Keren and Singer, 1988). Alperovitch *et al.* (1985) proposed that the higher efficacy of Al oxides in controlling clay swelling and dispersion was due to the higher charge density and planar

structure, compared with the spherical shape and low charge density of Fe oxides.

5.5.2 Sodicity Effects on Aggregation

The effect of exchangeable Na in inhibiting macroaggregation may occur by weakening the covalent associations between organic materials and soil minerals, and/or by increasing the osmotic and hydration forces during wetting that cause particle repulsion (Chapter 3). A limited number of studies in the literature have provided evidence that partial Na saturation does affect slaking of either macro- or microaggregates. Using aggregates smaller than 2 mm of three California soils, Abu-Sharar *et al.* (1987) showed that increasing the SAR of equilibrating solutions from 0 to 10 caused a significant increase in slaked material (< 20 μm) with mild shaking in water. Only small increases in clay dispersion (< 2 μm particles) were ob-

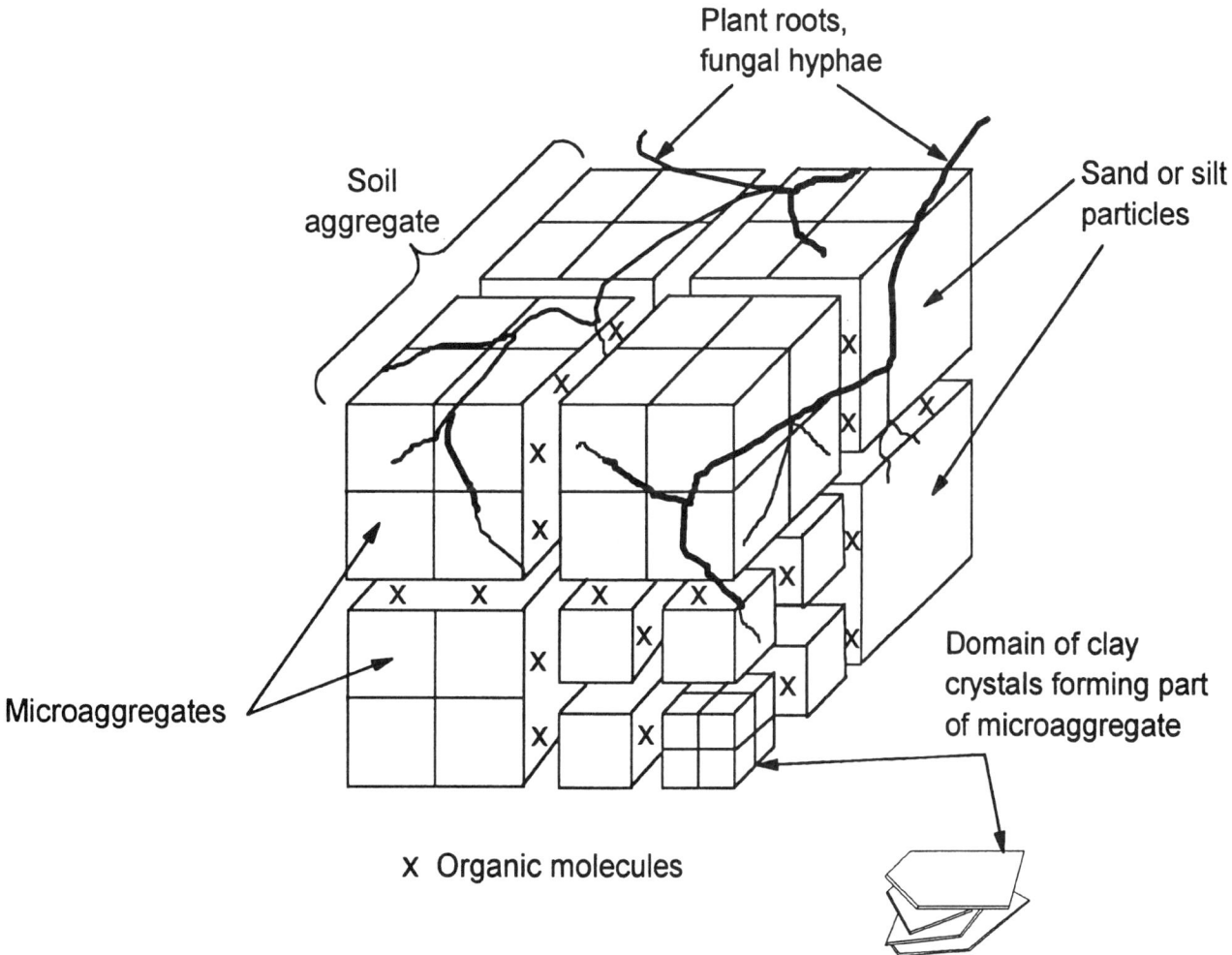

Figure 5.16 Heirarchical structure of aggregates (Adapted from Greenland, *Soil Physical Properties and Crop Production in the Tropics*, pp. 47-56, John Wiley, New York, (1979)].

served over the same range of SAR. There was little further increase in slaking at an SAR of 20, although clay dispersion did increase, leading Abu-Sharar *et al.* (1987) to conclude that aggregate slaking occurred independently of dispersion. These investigators also found that slaking of aggregates increased with a decrease in the electrolyte concentration. Other studies, however, have suggested that macroaggregate slaking was affected by fractional Na coverage of exchange sites only in the ESP range above 5. Using 2–4.75 mm aggregates on a 0.8 mm sieve under a rainfall simulator, Shainberg *et al.* (1992a) found no difference in aggregate stability at ESP levels below 5, but appreciable decreases in the aggregation of two soils when the ESP was increased above 10. Aly and Letey (1990) found a similar trend in a Vertisol, where the stability of 2–6 mm aggregates decreased from about 30% at an SAR of 1 and 5 to 8% at an SAR of 15. Other studies (Coughlan and Loch, 1984; Goldberg *et al.*, 1988) reported no correlation between ESP and aggregate slaking in the low ESP range. Unlike the findings of Abu-Sharar *et al.* (1987), Shainberg *et al.* (1992a) demonstrated that slaking of aggregates into microaggregates was rapid and independent of electrolyte concentration. However, further breakdown of the microaggregates into clay floccules (< 20 µm) was controlled by the ESP of the soil and by the electrolyte concentration of the soil solution. These findings of Shainberg *et al.* (1992a) suggest that part of the aggregate breakdown observed by Abu-Sharar *et al.* (1987), which depended on the electrolyte concentration, was, in fact, deflocculation of clay floccules and not slaking of microaggregates.

The stability of 105 µm microaggregates was studied by Levy *et al.* (1993a) using ultrasonic energy on six kaolinitic and illitic South African soils adjusted to varying ESP levels. Levy *et al.* (1993a) found that in four other soils (two Aridisols, a Mollisol, and an Oxisol), increasing the ESP from 1 to 25 had little effect on microaggregate stability. Conversely, two Alfisols showed a decrease in coarse (50–100 µm) aggregates and an increase in both silt- (2–20 µm) and clay-sized (< 2 µm) particles when the ESP was increased from 1 to 25, indicating a similar effect of Na on both aggregate breakdown and clay dispersion. In these soils, a greater input of ultrasonic energy resulted in a linear decrease in coarser aggregate sizes and a simultaneous increase in clay dispersion at an ESP of 1. At higher ESP values, however, dispersion increased dramatically with energy input, resulting in the loss of coarser (50–100 and 20–50 µm) aggregates. Thus it might be suggested that in soils where microaggregate stability is affected by changes in ESP, dispersion is the driving force controlling the breakdown of microaggregates at high ESP, resulting in immediate production of dispersed clay.

The influence of Na on aggregate stability should vary with soil type, amount of binding agents, and magnitude of the attractive forces between particles. Rengasamy and Olsson (1991) have noted that organic matter additions do not tend to improve aggregation of highly sodic soils and conversely, that Na additions have little adverse effect on highly organic soils. In each case, the overriding effect of one variable negates the other. Most soils, however, are intermediate with respect to these extremes, and increases in ESP above 10 are likely to have adverse effects on aggregation as well as promoting dispersion.

5.5.3 Aggregation, Soil Sealing and Erodibility

The intrinsic tendency of a soil to seal and erode under given rainfall conditions has commonly been related to soil structural stability. Most indices of soil erodibility use aggregate stability and soil organic matter levels as predictors of the erosive potential of soils (Wischmeier and Mannering, 1969). The mechanisms invoked are twofold: (a) stable aggregates prevent surface sealing and enhance rain infiltration, resulting in less transport capacity in runoff, and (b) by resisting breakdown, fewer fine particles are available for detachment and transport by overland flow.

Clearly, soil susceptibility to sealing depends on aggregate stability. However, only a few studies evaluated the relative importance of aggregate stability in the process of seal formation. Shainberg *et al.* (1992a) observed that aggregate slaking took place much faster than seal formation. Only 9 mm of rain were needed to disintegrate aggregates compared with more than 40 mm of rain needed for seal formation. They showed that soil ESP enhanced aggregate slaking in the upper ESP range, whereas the effect on seal formation occurred in the lower ESP range. Shainberg *et al.* (1992a) also showed that electrolyte concentration in the applied water had a negligible effect on aggregate slaking but a significant effect on the rate of seal formation and FIR. They suggested that aggregate slaking was the first and rapid step in seal formation, which is followed by the kinetically slow processes of clay dispersion, movement, and clogging of the conducting pores. Recently, Reichert and Norton (1994) reported an apparent inconsistency between aggregate stability (determined by wet sieving) and soil susceptibility to sealing in 12 clayey soils from the United States due to differences in both energy inputs and aggregate sizes in the two different tests.

Furthermore, despite the clear relationships between sodicity and aggregation, only a few studies have correlated sodicity directly with soil erodibility. This fact probably relates to the historical focus of erosion research on humid region soils where the relatively low Na levels have not been considered problematic, and also to the experimental difficulties in separating covarying chemical and physical soil effects on erosion losses. Early qualitative studies of erodibility emphasized dispersion (Middleton, 1930), but later research has focused more on aggregation.

Notable exceptions are investigations of California and Israeli soils exposed to simulated rainfall in small pans (Singer et al.. 1982; Levy et al., 1994). In California, the effect of ESP in the range of 0–100 on soil loss varied from being minimal on stable soils to a three-fold increase on unstable soils. Interestingly, in two of the soils, runoff amounts were not affected by Na treatments. Hence, the effect of Na on increasing soil loss was related to increased detachment, presumably due to aggregate breakdown and dispersion (Singer et al., 1982). Levy et al. (1994) studied three soil types (Typic Chromoxerert, Calcic Haploxeralf, Typic Rhodoxeralf) at various ESP levels exposed to simulated rain using three different waters, with EC values of 0.001, 0.8, and 5.0 dS m^{-1}. Seal formation characterized by the FIR was enhanced with an increase in soil sodicity and a decrease in EC for each soil type. Nearly 70% of the variation in FIR (and seal formation) was explained equally by EC and ESP. On the other hand, most of the variation in soil loss was explained by clay content (37.6%), supplemented by ESP (14.8%) and EC (9.7%).

Levy et al. (1994) also obtained a moderate inverse correlation between FIR and soil loss, suggesting that it is incorrect to draw conclusions about soil susceptibility to interrill soil erosion from FIR data. Except when the soil had an ESP below 5 and was rained on with DW, the degree of surface roughness and soil loss increased linearly with an increase in clay content. Surface roughness can serve as a qualitative estimate of the strength and degree of development of the seal formed. The smoother the surface, the stronger and more developed the seal. It is thus concluded that, under conditions where soil loss is determined by the strength of the seal formed, clay content is a better indicator than the FIR of the potential for interrill erosion from soils susceptible to sealing.

Shainberg et al. (1992b) compared rill and interrill erosion on a sandy Xeralf in small runoff pans. They measured soil loss at adjusted ESP levels of 2, 8, and 19 under high- and low-energy rainfall and on shallow or steep slopes. Erosion rates at 5% slope (interrill-type erosion) under high-energy rainfall were low and not affected by ESP level. When the slope was increased to 35%, however, soil loss increased dramatically due to rilling and was three times higher at an ESP of 19 compared to an ESP of 2. Low-energy rainfall applied on steep slopes still caused rill formation, and higher ESP again produced a significant increase in soil loss. When tap water (EC = 1 dS m^{-1}) was

applied as rainfall, erosion decreased dramatically due to the flocculating effect of the salts present. The authors concluded that rill initiation is especially sensitive to soil ESP, although interrill sediments produced under high energy and low EC rain contributed substantially to the total soil loss at higher slopes.

Other anecdotal evidence has linked erodibility with sodicity and dispersion. Applications of sodium nitrate fertilizers have been observed to increase dispersion and erosion rates (Hall, 1904). Application of 100 kg N ha^{-1} as NaNO$_3$ increased the ESP of the top 2.5 cm of an Ultisol to 60 and resulted in a sixfold increase in soil loss at 9% slope under simulated rainfall (Miller and Scifres, 1988). More generally, measures of soil dispersion (water-dispersible clay) commonly can be correlated with erodibility in studies that span a wide range of soils (Meyer and Harmon, 1984; Wischmeier and Mannering, 1969; Miller and Baharuddin, 1987a,b; Fitzpatrick et al., 1994). Such relationships link dispersion with aggregate instability and erosion mechanisms, and can be related back to fundamental soil properties such as colloid charge, mineralogy and cation type, which ultimately determine soil behavior (Chapter 3). The complexities of such relationships, however, remain to be elucidated in a form that will allow a prediction of erodibility from more basic soil properties.

5.6 CONCLUSIONS

Low (ESP < 5) and moderate (5 < ESP < 10) sodicity levels have a marked, and often irreversible, effect on the physical and hydraulic properties of most soils under conditions of low levels of electrolyte concentration in the soil solution. At these sodicity levels, the predominant mechanism controlling changes in soil hydraulic properties is clay dispersion. Clay swelling, which becomes effective only at ESP values above 10, plays a more minor role. Studies of clay-sand mixtures demonstrate the importance of the dispersion mechanism and the effect of low levels ($\leq 0.003 M$) of electrolyte in preventing clay dispersion and the consequent decrease in HC. Properties such as mineralogy, charge density, specific surface area, etc., are important in determining clay susceptibility in sand-clay mixtures to hydraulic failure when leached with dilute electrolyte solutions. The HC of soils of low sodicity was also similarly very susceptible to low EC. When leached with DW, the ability of the soils to maintain electrolyte levels in the soil solution above the CFC of the soil clay determines their sensitivity to sodic conditions. The HC of soils containing sparingly soluble minerals, such as CaCO$_3$, will be significantly less affected by low levels of sodicity than that of more highly weathered soils which do not release electrolytes into the soil solution.

The IR of soils is markedly more sensitive to low levels of sodicity than is HC. The higher vulnerability of the soil surface to lowsodicity is caused by the mechanical impact of the raindrops which disintegrate the aggregates at the soil surface and enhance clay dispersion. Macroaggregates (> 250 μm) slaking into microaggregates (20–250 μm), which takes place rapidly, is not affected by low levels of ESP (< 5). Conversely, the breakdown of microaggregates strongly depends on dispersion forces, soil ESP, and the electrolyte concentration of the soil solution. Hence, observed differences in FIR among soils having low ESP values arise from differences in microaggregate susceptibility to dispersion.

In soils having low ESP values, the electrolyte concentration of the applied water determines the physical and hydraulic properties of the soils. Recognition of the following situations is therefore of importance. In such soils, irrigation with waters containing some electrolytes (> 3 $mmol_c$ L^{-1}) will cause little or no damage to the soil due to sodicity because the EC of the irrigation water is high enough to counter the dispersive effects induced by sodicity. On the other hand, during storms, rainwater dilutes the soil solution, resulting in very low electrolyte concentrations, which thus enhance clay dispersion with subsequent deterioration in the physical and hydraulic properties of soils.

6

Root Zone Sodicity

R. J. Shaw, K. J. Coughlan, and L. C. Bell

6.1 INTRODUCTION

Root zone soil properties have direct and indirect effects on plant growth. Direct effects arise from soil physicochemical properties that determine soil structure and the stability, density, and strength of the soil as a medium for root development, while indirect effects are those where soil water relations and aeration are impacted by soil properties and behavior of the root zone. The ease of subsoil management and its sensitivity to degradation resulting from soil compaction, erosion, or tunneling can be directly related to soil sodicity and other basic soil properties. While many of the processes affecting soil structure and stability in surface layers are applicable to subsoils (Chapter 5), there are major differences resulting from the much lower energy of wetting and lower rates of subsoil hydraulic conductivity (HC), as well as the impact of overburden on soil behavior. For subsoils, the creation of porosity and the replacement of sodium (Na) by calcium (Ca) to stabilize the soil structure are important issues to be evaluated in assessing the viability and sustainability of amelioration strategies.

Water relations of soils are important to agriculture and environmental issues associated with land use. Three aspects of the hydraulic behavior of soils are particularly important: (a) water entry, and the consequences of low infiltration rates (IR) resulting in runoff and/or waterlogging, (b) root zone water storage and productive use of the water by vegetation through evapotranspiration, and (c) drainage below the root zone and groundwater recharge. In climates with periods of significant rainfall excess, key issues are surface and profile drainage of excess water to minimize waterlogging. In semiarid and arid climates, maximizing water entry and storage within the root zone for vegetative productivity are the priorities. In semiarid areas, agricultural development involving forest clearing and replacement by pasture or cropping systems can reduce water use by vegetation and increase deep drainage. Irrigation can also significantly increase deep drainage below the root zone and, hence, groundwater recharge, which may lead to the development of shallow water tables and surface salting

through evaporation. In these instances, the hydraulic behavior of subsoils is important in determining the environmental consequences (Chapter 9).

This chapter focuses on the hydraulic behavior of the root zone of sodic soils, where important aspects are soil wetting, available water storage, salt accumulation, and deep drainage below the root zone. Soil processes that determine subsoil behavior based on soil particle packing and threshold electrolyte concentration–exchangeable sodium percentage (TEC-ESP) relationships are discussed for soil dispersion and HC. These processes provide a framework to interpret subsoil behavior and identify viable soil management and reclamation options.

6.2 BACKGROUND

Sodic soils have been very widely associated with degraded soil structure and poor soil-water-air relations (USSL Staff, 1954; Rengasamy and Olsson, 1991; Jayawardane and Chan, 1994). The implication of Na in this process has long been known, as indicated by Warington (1900), "This evil admits of cure by treatment with gypsum which converts the sodium carbonate to the less injurious sodium sulphate" and de' Sigmond (1926), "In alkali (sodic) soils, a considerable part of the exchangeable cations is represented by sodium and this combined sodium may be responsible for the bad physical properties."

Eaton (1940), who had earlier proposed the use of sodium percentage of irrigation water as a measure for assessing the effect of irrigation on soil behavior, stated: "The effect of adsorbed sodium is related to the amount and character of the clay of the soil, to the extent to which other ions are replaced by sodium and to the extent to which flocculating electrolytes are removed." This latter approach has been the emphasis of much of the research since that time, as reviewed by Shainberg and Letey (1984) and Sumner (1993a), and forms a basis for the evaluation of the management implications for sodic soils.

Soils with sodic subsoils are characterized by moderate to high exchangeable Na. Various schemes have been developed to provide criteria for the determination of what constitutes a sodic soil (USSL Staff, 1954; Northcote and Skene, 1972) and will be evaluated in Section 6.3. Soils with sodic subsoils often show strong texture contrast between the A and B horizons with higher clay contents in the upper B horizon (Stace *et al.* 1968; Northcote, 1979; Hubble *et al.*, 1983), except if soils have high montmorillonite clay contents and swell and shrink significantly. In texture contrast soils with subsoils of low permeability, a bleached A^2 horizon of variable thickness may be present immediately above the B horizon or a hardpan. This feature is usually accepted as clear evidence of periodic waterlogging consistent with the presence of a sodic subsoil (Hubble *et al.*, 1983).

In general, soils which have the highest potential for agriculture usually have low levels of exchangeable Na throughout the profile and, therefore, are the easiest to manage. In more marginal agricultural regions, soils with some sodic properties in the root zone are often used, particularly under irrigation or where reliable rainfall occurs in the cropping season, which minimizes restraints due to low water availability. In strongly sodic soils, physical properties such as low plant available water capacity (PAWC), low HC, increased swelling, high bulk density ρ_b, and uneven soil wetting can restrict plant growth. An example of such a soil with columnar structure in the upper B horizon and restricted subsoil wetting is illustrated in Fig. 6.1. Holmes and Stace (1968) suggested that this columnar structure was formed by surface soil falling into shrinkage cracks and subsequent swelling of the B horizon into a smaller volume on wetting, thus causing the doming of columns. However, many of these soils currently show very limited swelling and shrinkage and, as a result, are hostile environments for root development. The widespread occurrence of soils with strongly differentiated profiles consisting of sandy to loamy A horizons over sodic clay B horizons is essentially unique to Australia and Africa (Hubble *et al.*, 1983; Soil Classification Working Group, 1991), although they do occur elsewhere.

Soils with sodic surface properties are amenable to ameliorative treatments such as gypsum and organic matter additions, or minimum tillage practices. However, amelioration of sodic subsoils as in Fig. 6.1 is very expensive and of variable effectiveness. Several of these management approaches have been reviewed by Jayawardane and Chan (1994) (Chapter 8). Optimizing cropping system management is the only viable low-cost technique for management.

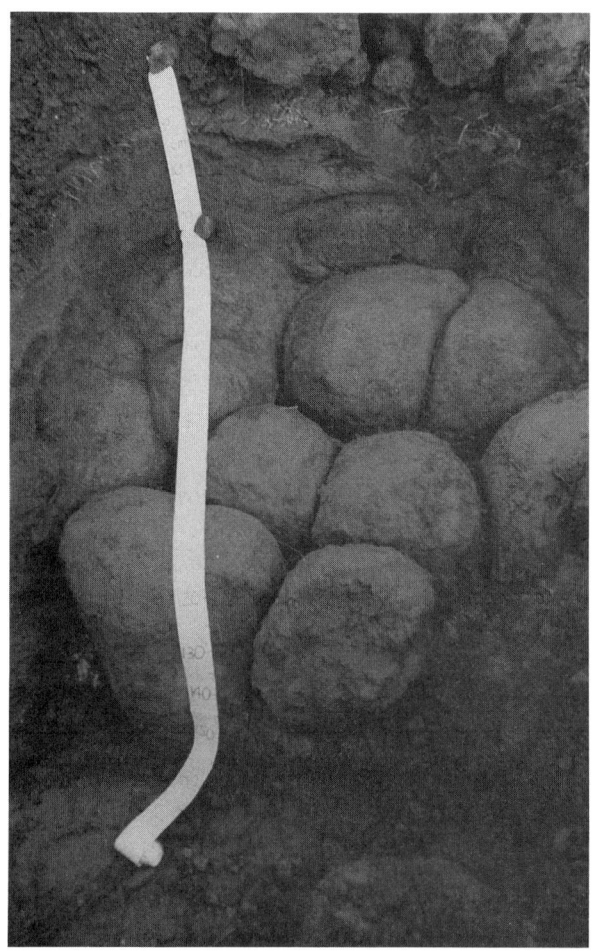

Figure 6.1 Columnar structure in clay B horizon of sodic soil from central Queensland.

6.3 PROPERTIES OF SODIC SUBSOILS

6.3.1 Plant Available Water Capacity

The PAWC which is defined as the size of the soil water storage capacity available for plant growth, is calculated as the difference between the upper and lower water storage limits summed over the root zone depth (Gardner *et al.*, 1984). McCown *et al.* (1976), who evaluated the PAWC of a range of mainly sodic soils in north Queensland from measurements of soil water content after ponding, found that the depth of wetting was related to electrical conductivity (EC) and ESP profiles. There was a significant negative relationship between the volume of water added during ponding and the ESP at the top of the B horizon.

Field studies of the physical behavior of soil in 5 m by 5 m plots with galvanized iron walls inserted to a depth of 1 m were used to assess the suitability for irrigation of a range of soils in Queensland (Shaw and Yule, 1978; Gardner and Coughlan, 1982). Water entry and root zone water availability were assessed using irrigation waters with EC values between 0.17 and 0.25 dS m^{-1}, sodium adsorption ratio (SAR) values between 0.5 and 1.5, and residual sodium carbonate (RSC) values between 0.3 and 0.7 mmol$_c$ L^{-1}. Average soil profile clay contents to 0.9 m depth ranged from 23 to 83% and ESP in the root zone from 0.5 to 47. The strong relationship between PAWC and subsoil sodicity (at 0.6 m depth) is illustrated in Fig. 6.2, which shows a two-stage response to ESP. For soils which are nonsodic, soil properties such as texture, structure and soil depth dominate PAWC. As the ESP increases, there is an ESP effect on PAWC which dominates the effect of other soil properties. Although increasing ESP results in reduced PAWC, consistent with the results of McCown et al. (1976), no threshold ESP value above which the reduction in PAWC begins is obvious.

Gardner et al. (1984), who examined the behavior of sodic subsoils, found a distinctive wetting deficit in the top of the sodic B horizon (0.3 m), as shown in Fig. 6.3. The predicted maximum field water content was based on validated empirical predictive relationships with field measured maximum wet soil profiles and -1500 kPa soil water content. The authors attributed the wetting deficit to the "throttle" effect on water movement caused by the poor physical conditions promoted by the low EC$_e$ (2.9 and 11.3 dS m^{-1} at 0.3 and 0.6 m) and high ESP (23.5 at 0.3 m and 46.5 at 0.6 m). McIntyre et al. (1982a) offered a similar explanation for the slow wetting of sodic clay subsoils under

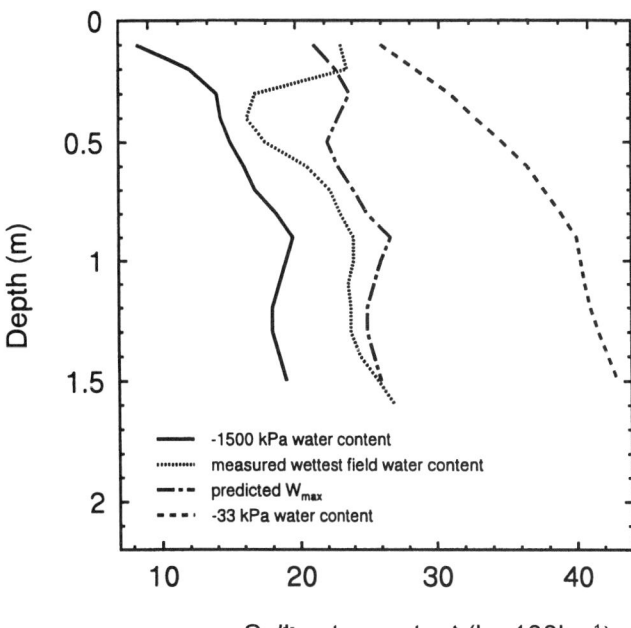

Figure 6.3 Gravimetric water content profiles for sodic texture contrast soil illustrating wetting deficit in upper B horizon from Gardner and Caughland [Agricultural Chemistry Branch Technical Rep. 20, Queensland Department of Primary Industries, Indooroopilly, Queensland, Australia (1982)]. The W$_{max}$ profile is the predicted maximum water content in the field 2 to 3 days after irrigation [Shaw et al., Aust. J. Soil Res. 32:143-172, (1994)].

prolonged ponding. In their ponded field plots, the limitation (throttle) to soil wetting occurred between 0.25 and 0.55 m in the profile due to a combination of high ESP, clay, and exchangeable magnesium (Mg) contents, and low EC. The potential errors incurred in using the traditional -33 kPa laboratory technique to estimate the "field capacity" of subsoils based on ground and dried samples (<2 mm) are illustrated in Fig. 6.3. The grossly inflated -33 kPa water content values in the subsoil can be attributed to the additional porosity induced by grinding soils, the enhanced swelling due to lack of confinement, a high ESP, and the faster rate of saturation wetting compared to the slower rates of wetting in subsoils restrained by overburden. Aspects of wetting rates and their consequences on soil water measurements, particularly of swelling clay soils, are discussed in detail by Thorburn et al. (1989).

6.3.2 Hydraulic Conductivity

Because HC is determined by the pore size distribution of the soil, factors that influence the size and stability of larger pores will affect saturated HC. Many studies have shown that HC is determined by the combined effects of clay mineralogy, clay content, sodicity, and EC (McNeal

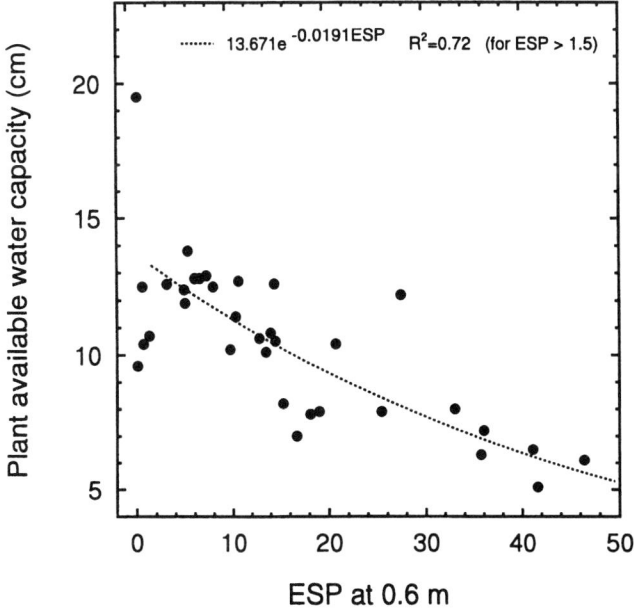

Figure 6.2 Relationship between plant available water capacity (PAWC) and exchangeable sodium percentage (ESP) at 0.6 m depth measured in 5 m by 5 m plots [Shaw, et al., Aust. J. Soil Res. 32:143-172, (1994)].

and Coleman, 1966; Yaron and Thomas, 1968; Cass and Sumner, 1982b; Shainberg and Letey, 1984). Frenkel *et al.* (1978) identified two processes contributing to a decrease in HC: (a) clay dispersion, resulting in clogging of soil pores, and (b) increased soil swelling, resulting in pore size restrictions, with the relative magnitudes of these effects varying with clay content, mineralogy, and ρ_b. Much of this work has been conducted on disturbed samples in columns in the laboratory, which results in the inability to extrapolate directly to field conditions. For example, Rowell *et al.* (1969) found that the level of mechanical stress caused by soil drying and clay extraction changed the electrolyte concentration at which clay dispersion occurred by about 100-fold, while Hamblin (1985) showed that soil dispersion occurs only at high water-soil ratios. Thus it is likely that dispersion will be less important in subsoils than in surface soils, except in the vicinity of saturated macropores. Consequently, techniques developed to assess the HC of surface soils are not necessarily applicable to subsoils, as will be illustrated in the next section.

Laboratory versus Field Hydraulic Conductivity

The HC of swelling clay soils (0–0.1 m) in 0.12 m diameter permeameters was quite different from that in 5 m by 5 m plots in the field. Laboratory HC values using distilled water taken between 3 and 5 h for four dried, ground (< 2 mm), and sieved (0.5–1.0 mm) clay soils were compared to the measured field infiltration under ponding of ten similar cracking clay soils (Shaw and Yule, 1978) in the same area (Fig. 6.4). Some of the corresponding soil properties are presented in Table 6.1. Laboratory HC was inversely related to clay content, or swelling percentage, with no significant relationship to ESP, which was low and varied little. This indicates that the saturated HC in laboratory columns is largely determined by the swelling properties of the matrix in a confined system, at least for the clay soils in Table 6.1. On the other hand, Sumner (1993a) on reviewing the literature suggested that the effect of swelling on HC is only important at SAR values above 10 (ESP ≈ 12), but these criteria are not appropriate for Vertisols as illustrated earlier.

Variations in ponded field IR are related to changes in clay content, with the trend being opposite to that of the laboratory method, as shown in Fig. 6.4. The introduction of subsoil ESP, which ranged from 0.3 to 19 at 0.5–0.6 m, improved the relationship ($r^2 = 0.827$ to 0.94), depending on the shape of the function adopted for ESP over the narrow range of the data. The difference between using the soil properties of the 0–0.1 and the 0.5–0.6 m depths in evaluating the field relationships was small only because of the dominant effect of the clay content in determining the

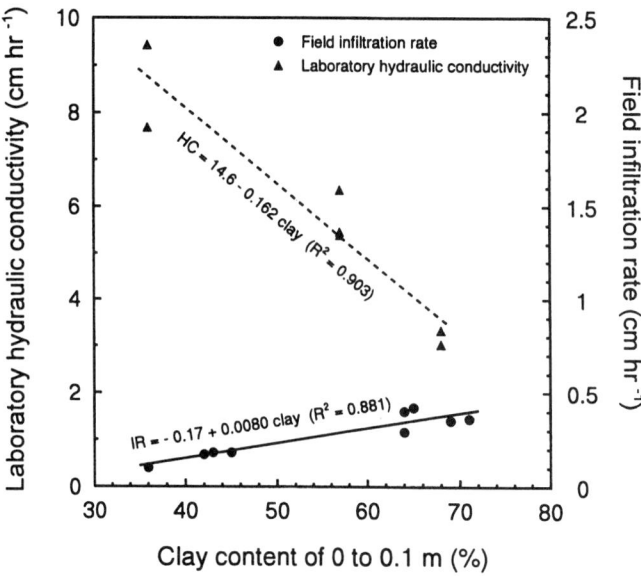

Figure 6.4 Comparison of laboratory measured hydraulic conductivity (HC) (3-5 h) of 0–0.1 m depth with field measured infiltration rate (IR) (3–5 hours) in 25 m² plots under ponding on clay soils.

water movement through the matrix of large and small cracks.

The essentially opposite effects of the clay content on laboratory and field measurements highlight the problems of evaluating swelling behavior in clay soils using disturbed laboratory samples. Field soils can swell partially unconfined into the shrinkage void space whereas in rigid laboratory columns of ground soil, swelling occurs into the available pore space, resulting in more dramatic effects. Field assessments of the suitability of these soils for irrigation were opposite to those based on the laboratory evaluation of HC at the soil surface (Shaw and Yule 1978), which supports the views of Ritchie *et al.* (1972), who stated that field determined HC values are the only appropriate parameters for characterizing swelling clay soils.

6.3.3 Subsoil Salt Content

Soil salinity can be a significant limitation to agricultural productivity where soils contain high salt concentrations in the root zone, or at the soil surface due to capillary rise of saline groundwater. Soil salinity is also important in sodic subsoils with low to very low permeabilities and mixed mineralogies (clay 35–55%) under annual rainfalls of between 280 and 2200 mm from northeastern Australia, where EC_e increases with ESP, as shown in Fig. 6.5. There is a good logarithmic relationship between EC_e and ESP for the range of rainfalls, with the possible exception of those less than 300 mm per year which tend to have slightly higher

TABLE 6.1 Measured laboratory hydraulic conductivity (HC), soil swelling and related soil properties for cracking clay soils near Emerald, central Queensland

Site	Laboratory saturated HC mm h[-1]	Swelling mm mm[-1]	Bulk density (ρ_b) kg m[-3]	CEC cmol$_c$ kg[-1]	Clay %	CCR[a] mol$_c$ kg[-1] clay	ESP %
B3A	30.3	0.16	1034	67.8	68	0.9	1.1
B3B	33.3	0.14	1015	67.8	68	0.9	1.1
TB3 cultivated	53.9	0.11	984	65.9	57	1.04	0.2
TB3 non-cultivated	63.6	0.13	950	65.9	57	1.04	0.2
TS4A	94.3	0.02	1085	21.8	36	0.58	1.9
TS4B	7.68	0.02	1085	21.8	36	0.58	1.9
AR6 cultivated	54.6	0.10	1021	41.6	57	0.68	1.6
AR6 non-cultivated	20.6[b]	0.11	1025	41.6	57	0.68	1.6

[a] CCR is the CEC of the clay fraction ($< 2\mu$m), expressed as mol$_c$ kg[-1]

[b] This value was atypical and was excluded from the regression.

salt concentrations. The CEC-to-clay ratio (CCR) has been used as an index of clay mineralogy on the basis of its correlation with the charge density of the clay mineral (Coughlan and Loch, 1984; Shaw and Thorburn, 1985; Brubaker *et al.*, 1992). The relationship between CCR and constituent clay minerals based on a detailed analysis of the mineralogical data for Australian soils presented in Stace *et al.* (1968), is as follows:

<0.20	kaolinite
0.20–0.35	illite and kaolinite
0.35–0.55	mixed clay mineralogies
0.55–0.75	mixed clay mineralogies with a higher proportion of montmorillonite
0.75–0.95	dominantly montmorillonite with the possibility of feldspars
>0.95	montmorillonite, plus feldspars or CEC from other than the clay fraction.

Subsoil salt contents in soils not affected by a shallow water table result from the selective uptake of water from rainfall by vegetation, with salt consequently accumulating in the soil. Over time, a balance is reached between the leaching effects of rainfall and the concentrating effects of evapotranspiration (ET), resulting in an equilibrium EC_e. This can be deduced by combining the steady-state salt and water mass balance equations, assuming negligible runoff (Bernstein, 1967; Shaw, 1988). The ratio of the electrolyte concentration of the rainfall to that at the bottom of the root zone [leaching fraction (LF)] (USSL Staff, 1954) is related to the quantity of rainfall and ET,

$$C_i / C_o = 1 - \frac{ET}{Q_i} \quad [6.1]$$

where

C_i = salt concentration in rainfall

C_o = salt concentration at bottom of root zone (assumed to be in equilibrium with water draining below root zone)

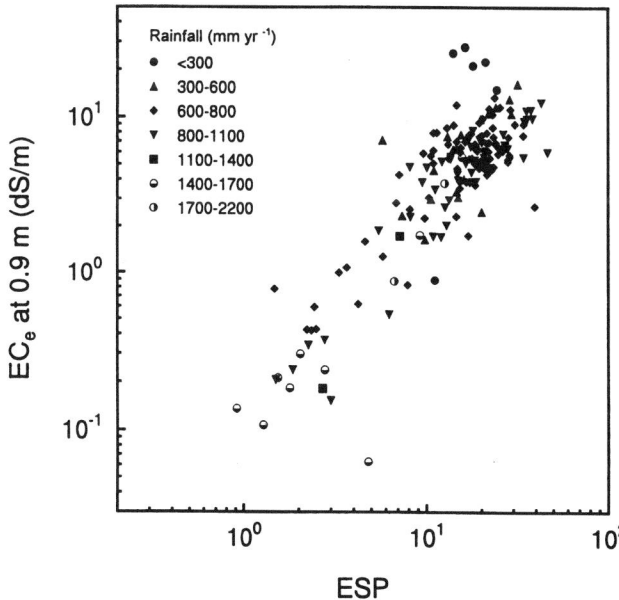

Figure 6.5 Relationship between electrical conductivity (EC_e) and exchangeable sodium percentage at 0.9 m depth for soils with annual rainfalls between 280 and 2200 mm y[-1] and clay contents of 35–55% clay and cation exchange capacity to clay ratio (CCR) values of 0.55–0.95.

ET=quantity of evapotranspiration
Q_i =quantity of rainfall

The lower EC_e values at higher rainfalls are due to the extra water available for leaching compared to the more arid environments (Fig. 6.5).

Sodic soils with high subsoil salinity have two effects on plant growth: (a) a direct limitation on the available water due to the osmotic effect of the soil solution, and (b) the limited soil wetting and water availability resulting from soil degradation due to the effects of high sodicity (Fig. 6.2). When water uptake weighting functions are used to integrate salinity effects on plants over the root zone depth (Hoffman and van Genuchten, 1983), salinity at deeper depths makes a smaller contribution than averaging over the root zone. As a result, in situations of water stress under dryland cropping, soils with lower salinity (and sodicity) at the bottom of the root zone will be more productive.

Clay content and mineralogy, expressed as CCR, play important roles in addition to sodicity in determining subsoil salinity levels (Shaw and Gordon, 1994). The proportion of the soils in given clay content and CCR ranges that would exceed the salt tolerance level for cotton at 0.9 m depth (EC_e = 7.7 dS m^{-1}) (Maas and Hoffman, 1977) are presented in Table 6.2. There is a strong tendency for soils with intermediate clay contents and mixed mineralogies to show the highest frequency of saline subsoils. Processes causing this distribution of subsoil salinity are discussed in Section 6.4.

6.3.4 Subsoil Stability

Mechanisms responsible for the structural stability of soils such as dispersion and swelling in surface soils have been well described in the literature (Emerson, 1954; Schofield and Samson, 1954; Shainberg and Letey, 1984; Rengasamy and Olsson, 1991). Soil stability has been commonly assessed by laboratory HC (Quirk and Schofield, 1955; McNeal and Coleman, 1966) or dispersion tests (Emerson, 1967; Rengasamy *et al.*, 1984b; Cook, 1988). The general consensus is that soil stability is a function of TEC-ESP relationships, which affect attractive and repulsive forces between clay particles during wetting and inputs of energy (Rengasamy and Olsson, 1991). Soils with high sesquioxide and/or organic contents are more stable (Chapter 3). Quirk and Schofield (1955), who were the first to introduce this concept to determine the boundary between stable and unstable permeability chose an arbitrary 15% reduction in HC relative to that at an initially high electrolyte concentration as the TEC value. While McNeal and

Table 6.2 The percentage of soils in each category within a rainfall range of 400 to 1000 mm (from northeastern Australia) that have a soil salinity at 0.9 m depth that would exceed the salt tolerance for cotton (EC_e 7.7 dS m^{-1}) under dry land conditions. There were 660 soils within this rainfall range.

Clay content (%)	Percentage of soils where subsoil Ec_e exceeds 7.7 dS m^{-1}				
	CEC/clay ratio (CCR) (mol$_c$ kg^{-1})				
	< 0.35	0.35 to 0.55	0.55 to 0.75	0.75 to 0.95	> 0.95
15-25	0	0	0	0	0
25-35	0	0	8	38	10
35-45	0	33	21	24	0
45-55	0	33	29	42	0
55-65	0	31	16	18	0
65-75	0	0	7	14	0
75-85	0	0	0	0	0

Coleman (1966) and Shainberg and Letey (1984) suggested that 25 and 50% reductions be used as the sensitivity of laboratory measurements was known to be higher than that in the field, Cass and Sumner (1982a,b,c) developed a Na stability model that was based on the slope of the linearized response curves for a series of HC reductions. This overcame the problem of selecting an arbitrary HC reduction by providing a single relationship from which any nominated reduction in HC could be chosen.

Jayawardane (1979) proposed the Equivalent Salt Solution Series (ESSS) concept (Chapter 8) to predict changes in HC based on the need for a change in the SAR of the percolating water to be matched by a corresponding change in EC to maintain the same level of porosity, which requires knowledge of how porosity changes with EC. The advantage of this method is that HC reduction can also be predicted across the full range of values apart from the threshold level chosen. However, the need to obtain porosity or swelling data over a range of sodicity and EC requires additional experimentation to normalize the response. McNeal (1968) and Jayawardane and Blackwell (1991) have used swelling of ground samples to predict the HC response of soils, but the validity of this approach is questionable.

The stability of subsoils is commonly assessed from the response of soils to wetting or HC measurements in the laboratory. Two processes which affect stability are soil swelling and clay dispersion (Frenkel et al., 1978). Soil swelling has a marked effect on reducing HC, and the enhanced swelling of sodic soils results in reduced subsoil wetting, and hence, reduced available water storage. Pore clogging has been attributed to dispersion of clays, which also leads to reduced HC and subsoil wetting, as illustrated in Fig. 6.2. These processes can be related to clay mineralogy. Quirk and Schofield (1955) and others have found different critical flocculation concentrations (CFC) for different clay minerals. Ion demixing, which occurs with smectite clays can result in smectitic clays being more stable to changes in ESP at low ESP values. Thus there is some divergence of opinion in the literature over the relative magnitude of these factors in determining soil stability. This can partially be explained by the effects of using different laboratory HC methods for soil stability assessment.

Laboratory techniques for the evaluation of TEC-ESP relationships using HC have been criticized in the literature. The initially concentrated electrolyte pretreatment commonly used (800–1000 $mmol_c$ L^{-1}) is well above that of all but severely salt-affected soils and has been shown to alter clay dispersion behavior and subsequent HC irreversibly (Reeve and Tamaddoni, 1965; Waldron and Constantin, 1970; Ferreiro and Helmy, 1974). The rate of change of electrolyte concentration from the high initial to subsequent lower values in saturated HC experiments had a large effect on the dispersed clay concentration in the effluent (Keren and Singer, 1988), while for swelling soils, confined swelling into the available pore space resulted in unrepresentatively low HC values (Merrill and Doering, 1985). Where swelling rather than particle dispersion, movement, and pore clogging is the dominant process, results from unconfined experiments will find limited application to subsoils. On the other hand, Lal et al. (1970) found that saturated wetting of a high smectite content clay in laboratory columns resulted in the swelling and slaking of the surface layers, creating a limitation (throttle) to water movement with the remainder of the column remaining unsaturated. Thus the type of clay material is important since, while dispersion can occur with all clays, slaking, swelling, and consequent aggregate breakdown, or swelling into pore space, will only be significant with swelling clays.

Few field studies of HC on ponded plots or irrigation furrows which have evaluated the effect of electrolyte or sodicity have been reported in the literature (Lyle et al., 1986; Evans et al., 1990; Singh et al., 1992). However, the difficulty with field studies is the time required for soil equilibration with the applied water, which is of the order of years to decades (Thorburn et al., 1990), and the integrated response of soil behavior with soil depth on the measured HC. It is not always the soil surface horizons that present the largest problems to soil hydrologic behavior, as was seen in Fig. 6.2.

Similar problems can arise in laboratory techniques for the determination of soil stability. Collis-George and Smiles (1963) used a range of dispersion and stability techniques to distinguish between the properties of scalded and normal soil surfaces. No laboratory technique was adequate to account for the observed differences in field behavior. Slow controlled wetting because of its low-energy input was too sensitive to small changes in structural stability caused by soil drying in the field. On the other hand, fast wetting, which disrupted structural stability, was unable to distinguish between soils, while the soil solution cation exchange approach ignored the other significant factors contributing to soil stability, such as soil drying, organic matter, and sesquioxide contents.

Field Subsoil Salt-Sodicity Equilibrium

Because of the problems with laboratory studies, an alternative approach of defining a soil response to sodicity and electrolyte concentration by considering the field EC_e-ESP equilibrium at the bottom of the root zone in 900 nonirrigated soils was adopted. The approach was based on the salt mass balance in the soil profile and accounts for clay content, mineralogy (expressed as CCR), and rainfall. The soils under rainfalls of 280–2200 mm per year were separated into clay content-CCR groups because of the nonlinear response of EC_e to these properties. The derived EC_e-ESP equilibrium relationships for the 45–55% clay, 0.75–0.95 CCR soil groups are presented in Fig. 6.6. The consistency of the data from a wide range of soils indicates that natural soils have come to an EC-ESP equilibrium which is assumed to be a reflection of the EC-ESP threshold for a given rainfall. These results for three annual rainfalls are compared to the sodium stability model analysis of Cass and Sumner (1982b) (Fig. 6.6) after converting SAR to ESP using the relationship of USSL Staff (1954) and averaging the threshold slope value b_t for the same soil group.

A comparison of the two families of curves indicates that while the general response is similar, the laboratory HC method gives results which are different at low and high ESP values. At an ESP value of 30 or greater, the divergence is related to the greater sensitivity of the laboratory technique to high ESP, probably through enhanced swelling at high ESP levels, and partly to the uncertainties in the relationship between SAR and ESP, as discussed by Sposito and Mattigod (1977). At low ESP levels, there is a greater sensitivity in the laboratory technique, probably due to the response of ground soil samples to saturation wetting compared to natural structured subsoils where the energy of wetting is less than that at saturation. In neither case is there a definitive threshold between stable and unstable conditions, which confirms the findings of McIntyre (1979). While he selected a soil ESP of 5 rather than 15, as proposed by USSL Staff (1954), no definitive stepwise change in HC occurred at this ESP (Chapter 1).

In terms of field data, differing rainfall has a similar effect on the threshold values for the selected reductions in laboratory HC. Under field conditions, the variation with rainfall can be considered as the energy (quantity of rainfall) required to move water through a restriction at the bottom of the root zone (ESP and soil properties). While various boundary conditions of EC-ESP thresholds have been set for classification and other reasons, the evidence from both the field equilibrium and the laboratory threshold curves in the literature indicates strongly that the response is a continuum determined by the degree of HC reduction considered or, in the case of the equilibrium curves, the quantity of rainfall available for leaching. In any system

where there is a constraint to flow, an increase in pressure or energy will overcome the restriction (within practical limits). Thus increasing rainfall for a given ESP will result in greater leaching. By inference, the same net leaching could occur at a low ESP under low rainfall. By analogy with electricity ($i = V/r$ where i is current, V is voltage, and r is resistance), the root zone HC is $\propto$ rainfall/ESP, indicating that the stability or HC of a soil is not unique and depends on the degree of stress or energy under which the soil is placed.

Oster and Schroer (1979) measured the IR on cropped undisturbed soil cores (0.2 m diameter, 0.53 m long) previously treated with waters of varying water quality after 19 months. Drainage from the columns was classed as insignificant. A range of irrigation water and soil salinity and sodicity measurements were used and calculated. Only data for the surface layers (0–76 mm) are used here, as the shape of the ESP profiles indicated that the lower depths of the columns had not reached equilibrium. After converting their cation concentrations to EC based on the relationship of Marion and Babcock (1976), the desired subsoil EC-ESP equilibrium curves at rainfalls of 250, 500, 1000, and 2000 mm for the same clay-CCR group fit the isoinfiltration rate data well in Fig. 6.7. This means that the selection of an acceptable stability level has to be related to the management issues for the particular soil. In winter rainfall situations, where waterlogging is a major problem, higher soil stability is necessary to maintain HC and soil deep drainage in order to minimize the duration of waterlogging. For a semiarid climate with a summer rainfall pattern, subsoil stability is of less importance and soil water storage is the property of primary importance. Thus different situations may require different threshold values. In situations where subsoils are exposed, rainfall or flowing water can both leach salt out of the material and increase energy levels well above those occurring naturally so that higher energy stability tests are required.

6.4 PROCESSES DETERMINING SODIC SUBSOIL BEHAVIOR

Soil structure and, hence, the hydraulic response of different soils result from the abundance and organization of primary particles of clay, silt, and sand (Panayiotopoulos, 1989). In Australia, Brewer (1979) concluded that the soil fabric could be "predicted with some degree of certainty from particle size analysis." Dalrymple and Jim (1984), who confirmed Brewer's observations in relating soil fabric to soil texture after considering the importance of wetting and drying cycles, identified three categories based on

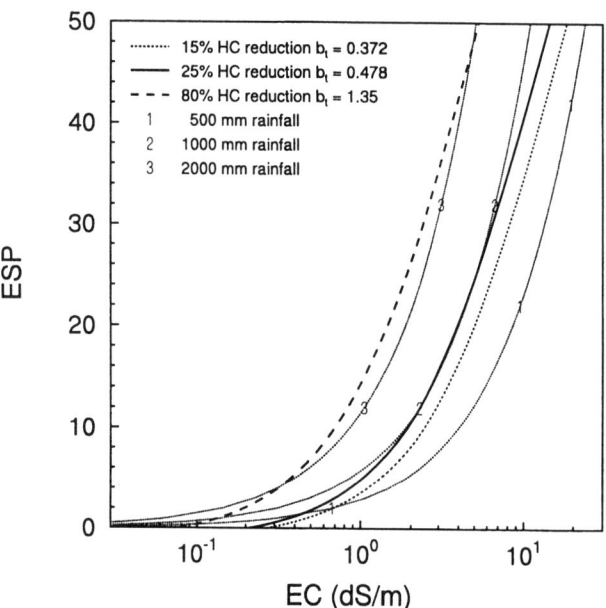

Figure 6.6 Threshold electrolyte concentration (TEC)–exchangeable sodium percentage curves based on equilibrium values in field subsoils compared to saturated laboratory hydraulic conductivity (HC) threshold relationships. Soils with 45–55% clay and CCR of 0.75–0.95 are used as the example. b_t values are average slopes of sodium stability model [Cass and Sumner, *Soil Sci. Soc. Am. J.* 46:507-512, (1982b)].

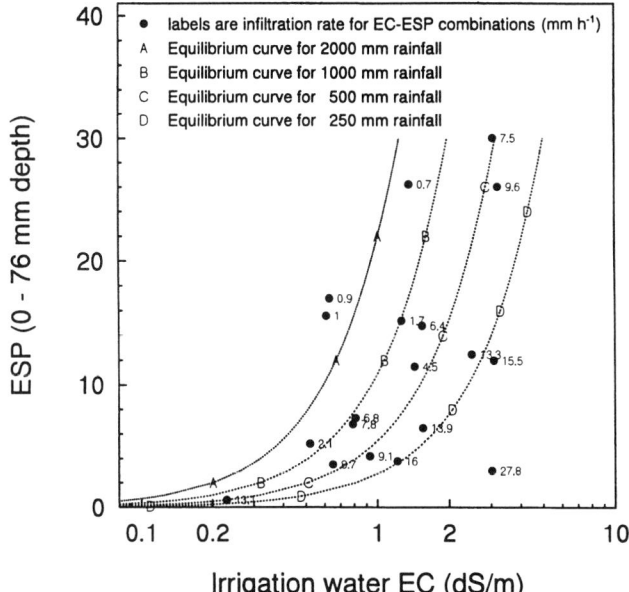

Figure 6.7 Infiltration rates (IR) (0–76 mm) of cores and equilibrium field subsoil EC-ESP curves for soil properties of cores at four annual rainfalls [Oster and Schroer, *Soil Sci. Soc. Am. J.* 43:444-447, (1979)].

clay content: (a) microfabrics resulting from the packing of skeletal grains between 0 and 10% clay, (b) a coarse to fine matrix, which is influenced by a sand to silt ratio of between 20 and 40% clay (if the clay-to-silt ratio is low, a coarse matrix is favored), and (c) a fine matrix above 60% clay, regardless of the sand-to-silt ratio. The activity of the clay had some influence on the degree of organization of the microfabric. Sodium and Ca-Na montmorillonite developed more fabric organization due to greater swelling than Ca montmorillonite. Brewer (1979) was of the opinion that the extent of microfabric development was a useful extension of soil structural assessment and thus a better estimate of soil behavior, which makes it feasible to use the clay content as a surrogate measure of soil fabric and, hence, an estimate of soil hydrologic behavior.

Bodman and Constantin (1965), who investigated the role of particle size distribution in the compaction of binary soil component mixtures, found that the greater the difference in particle diameter, the greater was the reduction in bulk volume on mixing. Minimum porosity occurred at close to 25% clay content under compaction. However, Smith *et al.* (1978) and Coughlan *et al.* (1978) found that a minimum porosity for soil aggregates occurred at about 50% clay content rather than at the lower clay content derived theoretically and experimentally by Bodman and Constantin (1965). Because in soils a continuous distribution of particle sizes exists, the theory of particle

packing is more difficult to apply and does not necessarily hold. Although HC is a function of soil porosity and water content (Marshall, 1958), the relationships between HC and total porosity have not been very satisfactory. Shaw (1988) used the binary packing theory of Bodman and Constantin (1965) and the work of Smith *et al.* (1978) and Coughlan *et al.* (1978) to show that the broad concepts could be applied to the salt content of subsoils and, by implication, the porosity of a soil based on its clay content. Soils with a clay content of about 40–50%, corresponding to the transition from a coarse to a fine matrix, had the highest salt content. Similar relationships for natural soils and sand-clay mixtures have been obtained by Chretien and Bisdom (1983), while Ben-Hur *et al.* (1985) showed that IR through soil crusts decreased sharply with increasing clay content up to about 20% clay, above which it rose slightly. In this case, the minimum IR at about 20% clay is presumably due to the energy of rainfall causing particle sorting.

Panayiotopoulos (1989) identified some of the problems that precluded definitive predictions of HC for natural soils based on particle size distribution. These included particle shape (clays being platelike and sand being spherical), friction between particles, cohesion and bridging, and differences in the density of different particle sizes. Coughlan *et al.* (1978) found that, at clay contents of less than 10%, the clay formed films around the sand particles without filling the void space, which actually increased the porosity of the coarse matrix.

The particle packing concepts are confirmed by the EC_e values at 0.9 m for the field soils under essentially constant rainfall (600 to 800 mm) in relation to clay content, CCR, and ESP (Fig. 6.8). Soils with 40–60% clay have the highest EC_e values, with the effect becoming more pronounced with increasing ESP. The effect of ESP becomes obvious around an ESP value of 3–5. The higher salt contents are attributed to reduced porosity and lower soil permeability. The traditional view that smectite-rich soils are the most sensitive to increasing ESP is not confirmed in Fig. 6.8, probably due to the greater resilience of smectitic soils in restructuring on wetting and drying. The change in EC_e in response to CCR (mineralogy) is very similar to the change in CFC for the different mixtures of kaolinite and montmorillonite clays presented by Arora and Coleman (1979). Emerson (1954) showed that a lower electrolyte concentration was required to flocculate acid compared to alkaline clays. At high pH values in equilibrium with $CaCO_3$, some Ca dissolution on wetting resulted in sufficient Ca available in solution to flocculate the soil samples. On the other hand, in some subsoils, the presence of $CaCO_3$ in

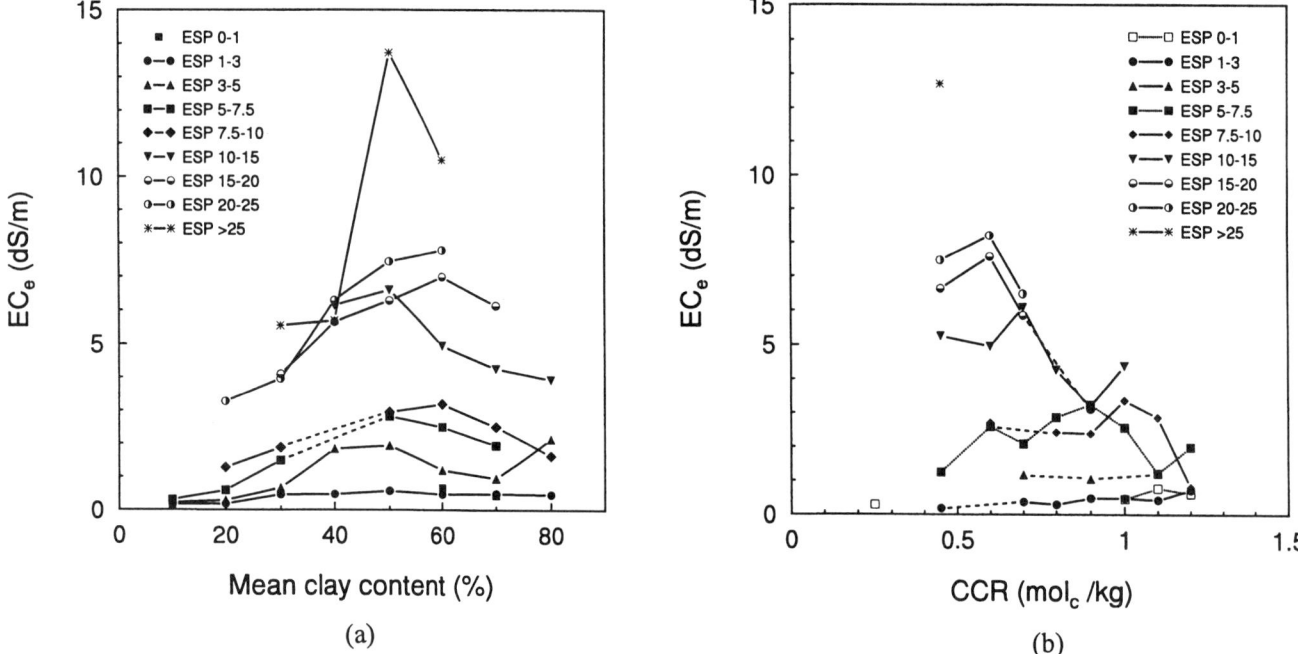

Figure 6.8 Relationship between averaged electrical conductivity (EC$_e$) (0.9 m) at different exchangeable sodium percentages (ESP) for field soils in a rainfall range of 600–800 mm and (a) mean profile clay content, and (b) CEC/clay ratio (CCR) for soils with 55–80% clay. Solid lines are drawn between data points that represent the average of two or more values and dashed lines between points with missing intermediate values.

alkaline soils can act as a sink for Ca, resulting in enhanced levels of ESP because CaCO$_3$ precipitates on drying. This situation leads to an increased potential for clay dispersion.

The pattern of behavior shown in Fig. 6.8 can be attributed to the movement of dispersed clays into the soil matrix, filling available pore spaces. In soils with limited swelling and shrinkage, a dense subsoil will result with limited HC. Vinten and Nye (1985) considered that most water movement in saturated soils occurred in pores with a diameter greater than 50 μm, and thus dispersed clay platelets could move readily. Dispersed clay will be filtered out of suspension where pores become restricted or by adsorption on pore faces. They demonstrated that complete blockage of pores by dispersed clay did not occur and that dispersed clay can travel long distances in soil pores. In soils with shrink-swell properties, wetting and drying processes will align clays to form orientated aggregates. Soil dispersion will occur in structured soils adjacent to macropores, and in regions where electrolyte concentration may be lower due to preferential flow. As a result, detached clay will move with the water to other parts of the root zone, where accumulation in the void spaces can form a soil matrix that is less permeable to water.

Harris (1958) investigated the processes leading to the formation of gilgaied and badly structured subsoils in Iraq. He considered that poorly structured subsoils were strongly

associated with wetting and drying cycles and reorientation of clay particles under overburden stresses. Olsen and Mesri (1970) examined two models for swelling and compression of pure clay systems. In the first, a mechanical model based on the interaction of particle shape, particle friction, and geometric arrangement of particles, swelling was not very responsive to changes in electrolyte concentration. In the second, a physicochemical model, electrolyte concentration and ESP cause major changes in clay behavior. Their results confirm the concepts presented here, namely, that in clay suspensions the physicochemical model dominates, but once a clay structure has been established, the response depends on the clay mineral present. Olsen and Mesri (1971) found that kaolinite fitted the mechanical model closely since the crystals were large and not particularly responsive to electrolyte; on the other hand, montmorillonite was responsive to the physicochemical nature of the medium, with illite being intermediate. In addition, montmorillonite exhibited long-range double-layer forces, which contributed to stability on swelling.

Thus the differences between laboratory and field responses in HC as an index of subsoil stability are largely due to soil swelling. In field soils, swelling and shrinkage can restructure subsoils and increase porosity, and as a consequence, result in a lower EC for a given ESP level. In disturbed soil cores under saturated wetting in the laboratory, these processes do not occur, and swelling virtually becomes an artifact of the measurement technique, thereby downgrading smectitic soils.

These processes can be described in terms of threshold or equilibrium curves, which have been idealized into three regions (Fig. 6.9) in a plot similar to the conventional EC-ESP stability relationships. Line 1 is almost parallel to the ESP axis, indicating that increasing ESP has a very limited effect on EC. Such soils do not respond to physicochemical effects but to the mechanical effects of bonding and cementation of clays in soils with low clay contents. Kaolinite associated with large amounts of iron oxides, or acid clay soils with aluminum (Al) as the dominant exchangeable cation, are typical of this response. In the intermediate region of true chemicophysical response (Line 2), an increase in ESP results in a corresponding increase in EC, which is typical of most soils. Kaolinite is least sensitive to the physicochemical response, while illite and mixed mineralogy soils are most sensitive to Na. Because montmorillonite-rich soils can restructure on wetting and drying, they are intermediate in sensitivity to Na as assessed by EC. In the third region (Line 3), a large increase in EC results in only a very small increase in ESP. This is typical of soils with no CEC and no effective leaching below the root zone, which is a highly improbable situation for normal field soils.

A unified soil property-based model can be proposed that provides an explanation of the effect of sodicity on the processes and behavior of field subsoils with respect to

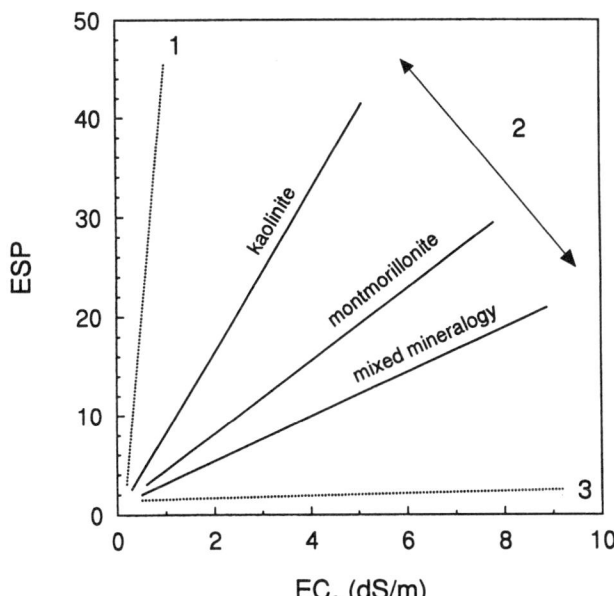

Figure 6.9 Conceptual framework for soil response to exchangeable sodium percentage (ESP) and electrical conductivity (EC) based on threshold soil stability and clay mineral type.

EC and ESP (Shaw, 1996). The conceptual basis for the model and the relationships between salt content, rainfall, and key soil properties are illustrated in Fig. 6.10. Subsoil salt content is used as a tracer for water movement. Water movement depends on the quantity of water available for leaching (as rainfall) and the properties of the soil matrix determined by clay content, clay mineralogy, and ESP. The illustrative relationships are derived from the regressions and represent soils with a CCR of 0.55–0.75 and rainfall of 500 mm for Fig. 6.10b, a clay content around 30% and rainfall of 1000 mm for Fig. 6.10c and a clay content of 45–55% and a CCR of 0.35 for Fig. 6.10d.

Soils with low ESP have low EC_e values due to adequate leaching, even under relatively low rainfall (Fig. 6.10a). The influence of particle packing and clay mineralogy on salt accumulation, which is strongly influenced by ESP, is illustrated in Fig. 6.10a and b. Soils with high clay content dominated by smectite mineralogy show lower salt contents than lower clay content soils with mixed mineralogy, particularly at ESP values greater than 5. The effect of rainfall on the threshold or equilibrium relationships between ESP and EC is presented in Fig. 6.10d. The range in threshold values is related to the energy available to move water through the restriction of the soil matrix.

Such a model, which links particle packing and its impact on soil porosity together with clay mineralogy and the response of the clay minerals to physicochemical factors, can be used to identify (a) subsoils that are amenable to improvement by the addition of gypsum as a Ca source and electrolyte, and (b) whether the subsoil is capable of restructuring through wetting and drying. It can also be extended to provide a framework for soil surface behavior, for example, in the development of the soil shown in Fig. 6.1. The soil surface layer will come to equilibrium with exposure to the energy of rainfall. If the clay on the soil surface is sodic, it will disperse and migrate leaving silt and sand fractions behind. The dispersed clay that moves below the A horizon may form the upper B horizon and reduce its permeability until the soils come into a quasi equilibrium. The depth of the A horizon would reflect the degree of sodicity of the B horizon and, hence, its limited permeability during soil formation. If the soil swells and shrinks, restructuring of the soil surface will constantly occur (Coughlan, 1984) and pedoturbation will bring some subsurface clay to the surface. Such soils are unlikely to show the strong texture contrast of the nonswelling sodic soils.

6.5 CONCLUSIONS

The hydraulic behavior of the root zone of sodic soils has a major effect on soil wetting, available water storage, salt accumulation, and deep drainage below the root zone.

Thus waterlogging, plant growth, and leaching to groundwater are affected by subsoil sodicity.

Commonly used laboratory methods of assessing subsoil sodicity are based on HC and dispersion techniques. Both these techniques have been shown to have problems in correctly assessing the behavior of subsoils *in situ* due to the increased energy of wetting, subsoil structure disruption, confinement of soils, and possible artifacts of pretreatment.

Using the subsoil salt content as an index of soil hydraulic behavior and EC-ESP stability relationships, a unified soil property model has been presented. This model incorporates clay content as a determinant of particle packing and hence porosity, clay mineralogy as an index of the sensitivity of a soil to ESP and the resilience of a soil to restructuring due to wetting and drying cycles, and ESP as a measure of the degree of soil sodicity. These properties behave in nonlinear ways to provide a unified soil property model to interpret the response of soils to ESP and to integrate the published results in the literature.

Acknowledgments

The laboratory evaluation of the hydraulic conductivity of clay soils in Section 6.4.1 was jointly conducted with Don Yule.

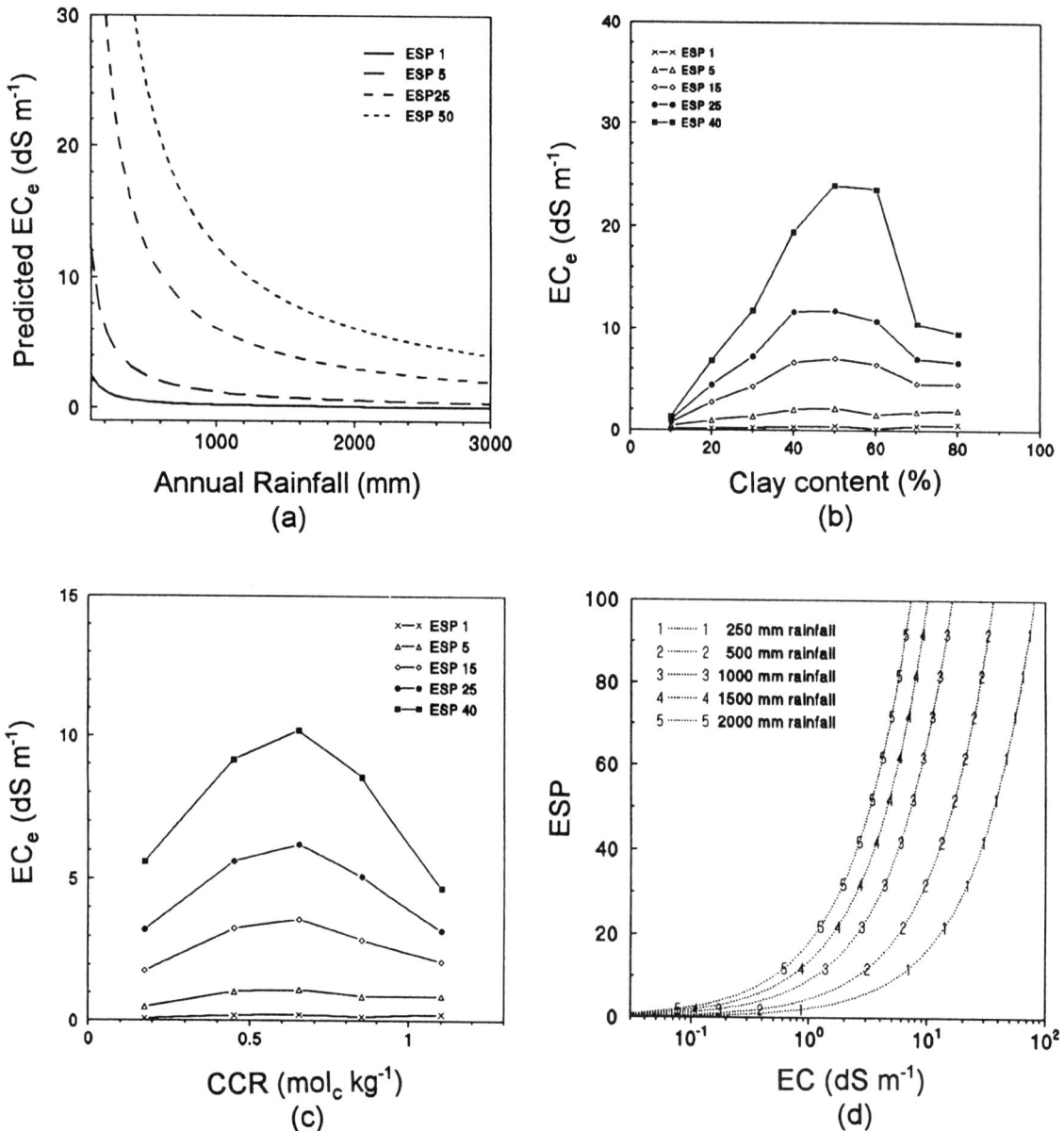

Figure 6.10 Conceptual framework for unified soil property model showing contribution of rainfall and major soil properties of clay content, mineralogy (as CCR), and ESP to subsoil salt content (as EC_e): (a) Relationship between EC_e, ESP, and rainfall, (b) Relationship between EC_e, ESP, and clay content, (c) Relationship between EC_e, ESP, and mineralogy (CCR), (d) Equilibrium lines for EC_e, ESP, and rainfall.

7

Fertility Constraints to Plant Production

D. Curtin and R. Naidu

7.1 CONSTRAINTS TO PRODUCTIVITY

Nutritional constraints (deficiencies and toxicities) in sodic soils are caused by (a) altered composition of the soil solution induced through changes in soil redox potential E_h, pH, dissolved organic carbon (DOC), or excess sodium (Na) concentrations, and/or (b) restricted root development and function caused by physical impediments to growth, poor aeration, and increased root diseases. These, either individually or collectively, reduce the ability of plants to acquire nutrients from the soil. The chemical behavior and acquisition of essential plant nutrients in sodic soils is complex. This chapter considers the interactions between sodicity and macro- and micronutrients as well as some of the effects of amelioration practices on plant nutrient availability.

Sodic soils are notoriously difficult to manage and often have limited productivity. The physical properties of these soils are considered to be the primary limitation to their productivity (Cairns et al., 1962). Nutritional constraints are site-specific and may vary with the plant species grown and the farming system used. The following characteristics have been reported to limit sodic soil productivity in Alberta, Canada (Peters, 1978), and in Australia (Northcote and Skene, 1972):

1. Hard, compact B horizon limits water, air, and root penetration. Rainwater usually remains on the surface and is lost by evaporation. This means that roots concentrate on the surface, and as a result, crops cannot withstand long periods of drought.

2. The B and C horizons and often the Ap horizon are slowly permeable to water. Because of restricted infiltration, there is less storage of water for future use by crops during drought periods.

3. Because of low organic matter content, surface crusting occurs soon after the soils are brought into cultivation. This can result in uneven and patchy seedling emergence and low crop yields.

4. Chemical conditions (e.g., high salt concentrations) may adversely affect the uptake of nutrients by plant roots.

5. Low pH of the Ap horizon of anthropogenically acidified soils may adversely affect the growth of acid-sensitive crops.

6. The extreme variability of sodic soils over short distances makes management difficult.

Many other fertility problems have been encountered in sodic soils. These include deficiencies of micronutrients such as zinc (Zn), iron (Fe), and manganese (Mn), especially in alkaline sodic soils (Gupta and Abrol, 1990a), ion toxicities such as boron (B) and aluminum (Al), and plant stress caused by periodic subsurface waterlogging and anoxic or suboxic conditions (Ford et al., 1993; Naidu and Rengasamy, 1993).

7.2 BEHAVIOR OF PLANT NUTRIENTS IN SODIC SOILS

There is limited information on plant nutrient relations in sodic soils. Nutrient availability and uptake by plants are related to (a) the activity of the nutrient ion in the soil solution, which depends on pH, E_h, and the concentration and composition of solutes in solution, (b) the concentration and ratios of accompanying elements that influence the uptake of the nutrient by roots, and (c) environmental and soil factors (Grattan and Grieve, 1992). Poor growth in sodic soils has been mainly attributed to physical impediments within the soil profile that restrict root development, disrupt water movement, and cause poor aeration.

Developing ways of ameliorating poor physical conditions has been a preoccupation of researchers, and relatively little attention has been focused on fertility problems in sodic soils. Field fertility studies have often produced inconclusive results because of the extreme spatial variability of sodic soils (Toogood, 1978), which is well illustrated in Table 7.1 where the response of barley to N and P fertilizer on a Natriboroll in Alberta, Canada, was examined. Although some of the differences were large, statistical analyses showed none of the yield increases or decreases to be significant. Thus no conclusion could be drawn as to a suitable combination of N and P. According to Toogood (1978), "dozens of tests (in Alberta), of every type and description, have yielded equally discouraging results." Consequently, much of the information on nutrient behavior presented in the following sections has, of necessity, been drawn from greenhouse and laboratory studies.

7.2.1 Cation Nutrition

Increases in exchangeable Na, accompanied by decreases in exchangeable calcium (Ca), can result in serious imbalances in cation nutrition. Sodic soils characteristically have elevated concentrations of Na and low Ca/Na ratios in the soil solution. Plant growth is influenced by the ionic balance of the soil solution rather than by the actual Na concentration. Apart from its contribution to osmotic stress, Na *per se* is not harmful to most annual crops (Läuchli and Epstein, 1990), but some tree crops that are specifically sensitive to Na may accumulate sufficient Na to cause burning and shedding of leaves (Bernstein, 1975).

One of the main nutritional disorders associated with sodic soils is impaired uptake of Ca caused mainly by high Na concentrations in the soil solution. Calcium, which plays a vital nutritional and physiological role in plant metabolism, is essential in processes that preserve the structural and functional integrity of plant membranes, stabilize cell wall structures, regulate ion transport, and control ion exchange behav-

ior, as well as cell wall enzyme activities (Grattan and Grieve, 1992). Because Ca can be displaced from its extracellular binding sites by other cations, these functions may be seriously impaired when the ratio of Ca to other solution cations falls below certain limits.

Selectivity coefficients, estimated from the relationship between activities of Ca^{2+} and Na^+ in culture solutions and the equivalent fraction of Ca and Na in plant tissue, show that the cation uptake process is strongly selective for Ca over Na (Suarez and Grieve, 1988; Curtin *et al.*, 1993a). Physiological disorders related to Ca deficiency occur when the ratio of Ca to total cation concentration (TCC) in the soil solution falls below a certain threshold. The Ca/TCC ratio is a better indicator of Ca availability to plants than is the actual Ca concentration in the soil solution (Adams, 1974; Carter *et al.*, 1979a; Carter and Webster, 1990). The literature is replete with reports of Ca deficiency in plants grown on sodic soils and in solution culture where the Ca/TCC ratio is low (Carter *et al.*, 1979a; Maas and Grieve, 1987; Grieve and Maas, 1988; Carter and Pearen, 1988; Suhayda *et al.*, 1992; Curtin *et al.*, 1993a; Naidu *et al.*, 1995c). Soil survey data also suggest that sodicity-induced Ca deficiency is widespread. For example, in a study of sodic soils (Aridic and Udic Natriborolls) in Alberta, Carter and Webster (1990) found that almost one-third of the A horizons and over one-half of the B horizons had a Ca/TCC ratio of less than 0.1 in saturated paste extracts.

The critical Ca/TCC value can vary widely from crop to crop. According to Grieve and Maas (1988), cereals are especially susceptible to Ca deficiency at low Ca/TCC ratios. Growth of barley (*Hordeum vulgare*) may be restricted when the Ca/TCC ratio is less than 0.10–0.15 (Fig. 7.1). While cereals show large intergeneric differences in their tolerance to low Ca (Grattan and Grieve, 1992), genotypes within a given cereal species may also vary in their susceptibility to Ca disorders at low Ca/TCC ratios (Grieve and Maas, 1988). These workers suggested that differences in tolerance of sorghum (*Sorghum bicolor*) genotypes to low Ca were sufficiently large to offer prospects for developing a variety suited to sodic soils. Halophytes often have a strong ability to selectively absorb Ca from root media low in Ca. Kochia (*Kochia scoparia*), a facultative halophyte that colonizes saline soils on the Canadian prairies, can take up ample Ca when the Ca/TCC ratio is as low 0.03 (Curtin *et al.*, 1993a). However, the ability to tolerate low Ca is not universal among halophytes (Hayward and Wadleigh, 1949), and the physiological basis for tolerance of low Ca/TCC values is not well established. Tolerance of low Ca in certain sorghum genotypes has been attributed to the ability of the plant to transport adequate Ca to the shoot meristem (Grieve and Maas, 1988).

Reported critical Ca/TCC values may vary depending on whether they are estimated using ion concentrations or ion

TABLE 7.1 Yield increases of irrigated barley from N and P applications to a Natriboroil in Alberta, Canada [Toogood, *Solonetzic Soils Technology and Management in Alberta*, University of Alberta, Edmonton, (1978)]

P rate (kg ha⁻¹)	Yield increases[a] over check plot average of 1990 kg ha⁻¹			
	N rate (kg ha⁻¹)			
	0	22	45	90
0	--	+ 269	+ 484	+ 323
5	− 430	+ 538	− 107	+ 430
10	− 430	− 215	+ 54	− 54
19	+ 753	0	+ 430	− 107

[a] No significant yield responses were obtained

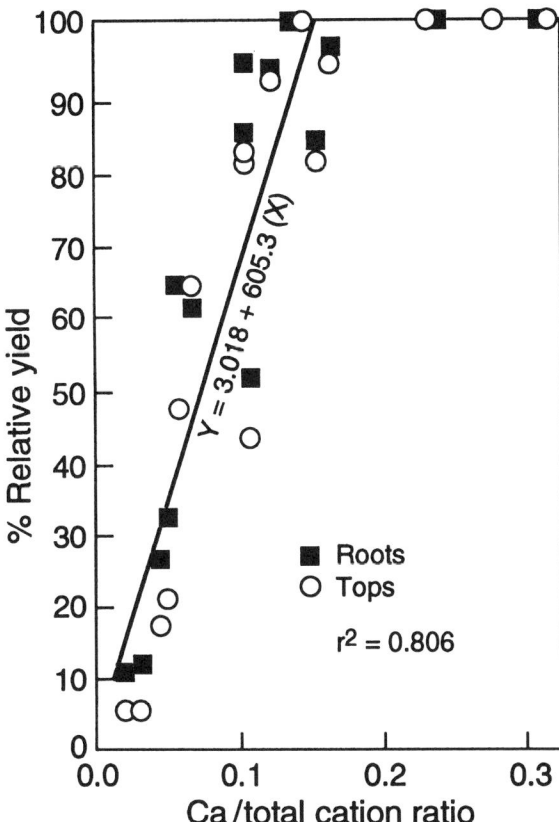

Figure 7.1 Relationship between Ca/total cation concentration (TCC) ratio of root medium and yield of barley [Carter, *et al., J. Soil Sci.* 60:161-174, (1979a)].

activities and on whether they are derived from data for solution culture, displaced soil solutions, or aqueous extracts of soils (Carter *et al.*, 1979b; Carter and Webster, 1990; Grattan and Grieve, 1992). The Ca/TCC ratio is usually based on ion concentrations, even though activities are more closely related to plant responses (Adams, 1974). Especially in areas where the soil solution is dominated by sulfate (SO_4) salts, Ca/TCC values may be much lower when estimated from ion activities than concentrations. Carter *et al.* (1979b) reported that the Ca/TCC ratio of the soil solution changed little as the soil water potential was varied from saturation to -1.5 MPa. They suggested that saturation extract data gave reasonable approximations of the Ca/TCC ratio for the *in situ* soil solution. Later, Carter and Webster (1990) confirmed that saturated paste Ca/TCC values accounted for 75% of the variation in barley tissue Ca in pot culture and 58% of the variation in field-grown barley Ca content. Critical plant tissue levels of 0.25% Ca in barley and wheat (*Triticum aestivum*) and 1% Ca in alfalfa (*Medicago sativa*) corresponded to a Ca/TCC ratio of 0.1 in the paste extract (Carter and Webster, 1990). Although the Ca/TCC ratio has proven to be a useful indicator of Ca availability, identifying specific ion competitions such as Ca/Na and Ca/magnesium (Mg) may be

preferable to reliance on the Ca/TCC ratio (Grattan and Grieve, 1992).

Because Mg is strongly competitive with Ca in the cation uptake process (Grattan and Grieve, 1992), high Mg/Ca ratios in the soil solution may induce Ca deficiency. The work of Carter *et al.* (1979a) suggests that Mg-induced Ca deficiency occurs in barley when the solution Mg/Ca ratio exceeds about 1 (Fig. 7.2). In nonsaline soils, the solution Mg/Ca ratio is largely determined by the composition of the soil exchange complex. Assuming that soils have about equal affinity for Ca and Mg (USSL Staff, 1954), a solution Mg/Ca ratio of 1 should not occur unless Mg represents about 50% of the exchangeable divalent cations, a situation that is uncommon in agricultural soils. However, sodic soils have been reported to have relatively high levels of Mg saturation (Bowser *et al.*, 1962; Williams and Colwell, 1977), and Mg/Ca ratios in excess of 3 have been recorded in soil solution displaced from sodic soils (Carter *et al.*, 1979a). In many Australian subsoils, the Mg/Ca ratio often exceeds 1, and this is one reason why Mg deficiency is not evident in Australian soils (Williams and Raupach, 1983).

Potassium (K), which is the metallic element required in largest amounts by plants, has not presented major concerns

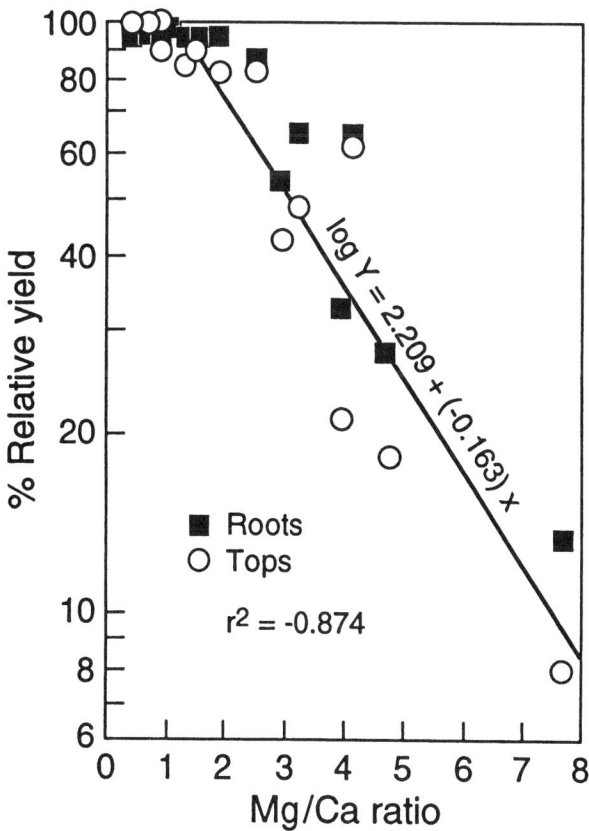

Figure 7.2 Relationship between Mg/Ca ratio of root medium and yield of barley [Carter, *et al., J. Soil Sci.* 60:161-174, (1979a)].

in sodic soils (Chhabra, 1985). According to Gupta and Abrol (1990a), the preponderance of micaceous minerals in sodic soils combined with Na-K exchange reactions ensures the release of ample amounts of available K. Highly weathered sodic soils of Australia appear to be an exception to this generalization in that K is among a suite of nutrients that are growth-limiting (Ford *et al.*, 1993).

In the determination of soil K fertility status, various categories (soil solution, exchangeable and fixed or nonexchangeable K) are of primary importance. While K in the soil solution is the form in which it is transported to the plant, the amount of K in solution is not necessarily a good indicator of availability, which is best predicted by the potential for K to be released to the soil solution (Evangelou *et al.*, 1994). Because K competes more effectively with Na than with divalent cations for exchange sites, the distribution of K between the solution and exchanger phases may be strongly influenced by the exchangeable Na level (Fig. 7.3).

At the practical level, the impact of sodicity on the acquisition of K by plants has not been fully explored. Under sodic conditions, where the concentration of Na in solution may be orders of magnitude higher than that of K, the K/Na ratio in the soil solution is likely to be a better indicator of plant K uptake than is the actual K concentration (Devitt *et al.*, 1981). Although the plant uptake process exhibits a strong affinity for K over Na (Grattan and Grieve, 1992), K uptake by plants can be markedly suppressed as sodicity increases (Bower and Wadleigh, 1948). Low tissue K in crops grown on Natric Cryoborolls that apparently contained adequate levels of

available K has been attributed to suppression of K uptake by Na (Carter, 1983). Sodium may partially substitute for K in some plants without affecting growth (Marschner, 1986) and indeed may be essential for some plants.

Selectivity for K over Na varies among plant species as well as among cultivars within a species (Grattan and Grieve, 1992). The degree to which different plants can tolerate soil sodicity may be partly related to their ability to selectively absorb K over Na. Sharma (1986) concluded that the ability to maintain high tissue K/Na ratios is a critical feature of sodicity tolerance in rice (*Oryza sativa*). Similar conclusions were reached by Joshi *et al.* (1980), who found that differential tolerance of sodicity among wheat varieties was correlated with the tissue K/Na ratio.

The Ca status of the plant root has an important bearing on its K-Na selectivity. When Ca supply is limited, root membrane permeability increases and K-Na selectivity is lost (Grattan and Grieve, 1992). The K/Na ratio of plant tissue has been shown to increase as the Ca supply in the root medium increases (Suhayda *et al.*, 1992). Low Ca availability in sodic soils may impair the K-Na selectivity of the root membrane, resulting in passive accumulation of Na in roots and shoots (Alam, 1994).

7.2.2 Nitrogen

Nitrogen (N) is an important growth-limiting element in many sodic soils (Cairns *et al.*, 1962, 1967; Cairns, 1963, 1968; Gupta and Abrol, 1990a,b; Malhi *et al.*, 1992a,b). Because the processes controlling the availability of N to plants (Fig. 7.4) have been reviewed extensively (Campbell, 1978; Haynes, 1986), the emphasis here will be on the effect of sodicity on soil N transformations and their impact on N use efficiency in sodic soils. Factors which may result in impaired N availability in sodic soils are low rates of N mineralization (Cairns *et al.*, 1962; Cairns, 1963), denitrification when aeration is restricted due to waterlogging (McGarity and Myers, 1968; Myers and McGarity, 1971), volatilization of ammonia, especially under alkaline conditions (Rao and Batra, 1983; Rao and Ghai, 1986; Rao, 1987), and inhibition of nitrate (NO_3^-) uptake by chloride (Cl^-) and sulfate (SO_4^{2-}) ions which are often present in high concentrations in sodic soils (Aslam *et al.*, 1984).

The poor productivity of Natric Cryoborolls in Alberta has been partly attributed to low N mineralization. Cairns (1963) found that a Natric Cryoboroll that was incapable of sustaining growth of barley released only 13 mg NO_3-N kg^{-1} of soil during a four week incubation, whereas a nearby productive soil mineralized over three times as much N. A lack of readily decomposable organic N was apparently the factor limiting N mineralization in the Natric Cryoboroll (Cairns, 1963). Laura

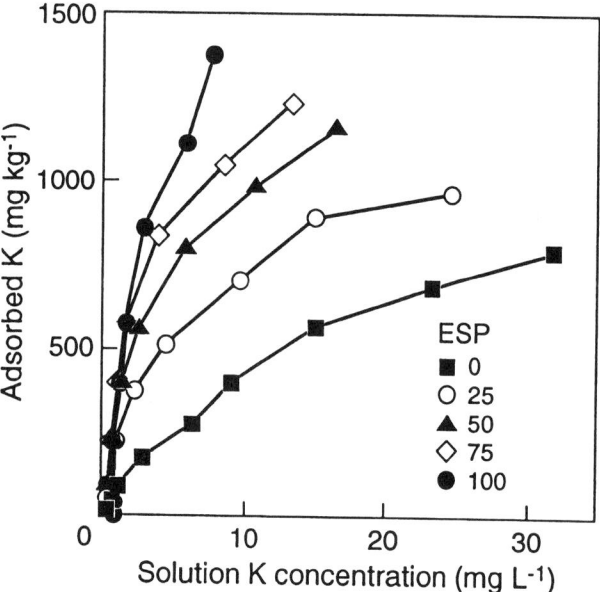

Figure 7.3 Adsorption of K by soil as influenced by exchangeable sodium percentage (ESP) [unpublished data of Curtin and Syers].

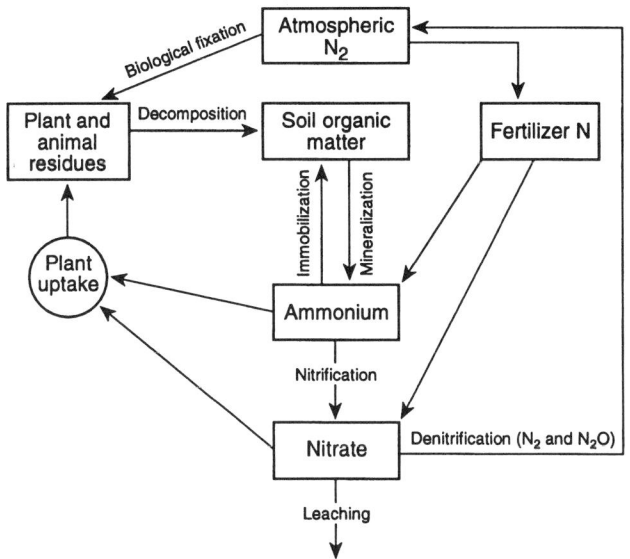

Figure 7.4 Soil nitrogen cycle.

(1976) examined the effect of sodicity, induced by treating a sandy loam with different amounts of sodium bicarbonate (NaHCO₃), on the mineralization of organic N (and C) in a soil amended with gulmohur (*Delonix regia*) leaves. The total amount of C mineralized during six months of incubation increased systematically as the soil ESP increased (Table 7.2). Sodicity was thought to have increased the availability of organic C to microbes, possibly due to organic matter solubilization. In contrast to C mineralization, the amount of N mineralized was constant between an ESP of 2 and 49, but increased between an ESP of 70 and 92. At the highest ESP (92), nitrification was completely inhibited and only ammonium (NH_4^+) accumulated in incubating soils. These results suggest

TABLE 7.2 Soil pH, EC_e, and total amounts of C and N mineralized (during 6 months of incubation) in a sandy loam treated with sodium bicarbonate to give exchangeable Na percentages (ESP) of 2 to 92 [adapted from Laura, *Plant Soil* 44:587-596, (1976)]

Soil ESP	pH[a]	EC_e[a] units	Mineralized N	Mineralized C
			(mg kg⁻¹)	
2	6.9	0.8	188	1720
23	7.6	2.3	184	2160
49	8.1	2.4	183	2540
70	8.8	2.5	221	2690
92	9.9	4.9	321	3060

[a] pH and electrical conductivity (EC) were measured in saturated paste extracts

that, at the ESP levels commonly found in the field, sodicity should have little effect on N mineralization. However, caution is required in interpreting these data because both pH and salt concentration increased concomitantly with ESP (Table 7.2), resulting in a confounding of the sodicity effects. Further work is needed to clarify the role of sodicity in relation to N mineralization.

Leaching losses of NO_3^- are probably small because sodic soils usually occur in dry climates and have poor water transmission characteristics. Gaseous losses of N due to denitrification and ammonia (NH_3) volatilization may, however, be significant. Work in Australia has shown that, in contrast to nonsodic soils where denitrification activity is largely confined to topsoils (Cho *et al.*, 1979), sodic B horizons exhibit substantial denitrification activity (McGarity and Myers, 1968; Myers and McGarity, 1972). Waterlogging arising from their impervious nature, coupled with a supply of available organic matter leached from the topsoil, makes sodic B horizons conducive to denitrification (Myers and McGarity, 1971). As yet, quantitative information on the magnitude of denitrification losses of N from sodic soils has not been forthcoming.

In sodic soils with high pH, NH_3 volatilization is the major pathway for N loss (Gupta and Abrol, 1990a,b), causing an increased (20–25%) fertilizer N requirement on alkaline sodic soils in India. Ammonia volatilization can be reduced substantially by incorporating N fertilizer into the soil. In a laboratory study in which urea was applied to an alkaline (pH 10.6) soil at a rate of 60 mg N kg⁻¹, NH_3 volatilization decreased by 19% for each 2.5 mm of soil covering the fertilizer (Rao and Batra, 1983). Other strategies which may curb volatilization losses of NH_3, are split application of N fertilizer (i.e., apply N in three to four small doses rather than a single large dose) or substitution of green manures for part of the inorganic fertilizer (Rao and Batra, 1983; Gupta and Abrol, 1990b). Although NH_3 volatilization is most acute at high pH, losses can also be serious in acid sodic soils when urea is broadcast on the surface. Low recovery of surface-applied urea-N by bromegrass (*Bromus inermis*) on a Natric Cryoboroll has been attributed to loss by volatilization (Cairns, 1968).

Sodic soils may occasionally be acid enough to restrict symbiotic N fixation. Alfalfa (*Medicago sativa*) growth can be poor on acid (pH < 6) sodic soils (Natriborolls and Natric Cryoborolls) of the Canadian prairies due to N deficiency. This has been attributed to low N fixation due to low numbers of rhizobia failing to nodulate plants adequately below pH 6 (Rice, 1978). The use of seed inoculated with rhizobia and coated with pulverized lime has proven to be successful in establishing legume crops under acidic conditions (Rice, 1978).

Nitrogen Fertilizers and Quality of Sodic Soils

Soil acidification from N fertilizers is a concern where acidity is potentially growth limiting, as in the sodic soils of the Canadian prairies. Large reductions in pH have been documented in field studies in Alberta, where high rates of N were applied for several years (Cairns, 1968; Carter, 1986; McAndrew and Malhi, 1992), resulting in decreased crop responses to fertilizer N over time because of increased acidity (Toogood, 1978). Microbial activity and N mineralization were not impaired when soil pH was reduced from 6.1 to 4.8 by applying ammonium nitrate (NH_4NO_3) (Carter, 1986), but reducing the pH much below 5 was detrimental to microbial activities and N fertility (Table 7.3). Nitrogen fertilization, by decreasing the sodium adsorption ratio (SAR), may have a slight ameliorating effect on sodic soil (Table 7.3).

7.2.3 Phosphorus

Phosphorus (P), which plays a vital role in energy transformations and is essential for the growth and development of plants and animals, does not appear to be a major fertility constraint in sodic soils. Numerous studies have indicated that soil P tends to be more readily available to plants in sodic than in comparable nonsodic soils (Pratt and Thorne, 1948; Chhabra, 1985; Gupta *et al.*, 1990; Curtin *et al.*, 1992), and crop responses to fertilizer P in sodic soils are rare. Chhabra (1985), for example, concluded that sodic soils of the Indo-Gangetic plain of India contained a sufficiently large amount of available P that fertilization was unnecessary for at least six years following soil reclamation. In terms of P fertility, the weathered sodic soils of Australia are exceptional in that acute

P deficiency is common (Naidu and Rengasamy, 1993; Naidu *et al.*, 1993b), but responses to P on the infertile Natrustalfs of eastern Australia tend to be long-lasting, suggesting that fertilizer P has a high residual value in sodic soils (Russell *et al.*, 1988).

Because P reactions in soils are well known (White, 1980; Barrow, 1985; Wild, 1988), the focus here will be on modifications of P behavior caused by sodic soil conditions. Regardless of the nature of the soil, P is largely absorbed by plant roots as orthophosphate ions ($H_2PO_4^-$ and HPO_4^{2-}) from the soil solution. The solution P concentration required for maximum growth differs between species, growth stage and growth rate, but is very low (0.03–0.3 mg P L^{-1}) (Wild, 1988). In the simplified soil P cycle (Fig. 7.5), soil solution P is the focal point, with several transfer processes operating to control the concentration of P in solution. With the possible exception of recently fertilized soils, adsorption-desorption is the dominant inorganic process influencing soil solution P concentration (Syers and Curtin, 1989). Adsorption of P by soil is strongly influenced by the nature and concentration of cations in the system, with the direction of cation effects being dependent on the soil surface charge. The discussion to follow applies to soils that have a net negative surface charge forming a large proportion of the world's soils. Negatively charged soils, when saturated with Na rather than with divalent cations, adsorb less P (Curtin *et al.*, 1987; Smillie *et al.*, 1987; Sharpley *et al.*, 1988). Decreasing P adsorption as exchangeable Na increases (Fig. 7.6) appears to be due to a decrease in surface electrostatic potential (i.e., the surface potential becomes more negative), making the surface less attractive to P ions (Barrow, 1985). This shift in the adsorption-desorption equilibrium tends to increase P concentration in solution.

Electrolyte concentration and pH modify the effect of sodicity on P solubility (Gupta *et al.*, 1990; Curtin *et al.*, 1992). In a laboratory leaching experiment with Canadian prairie soils, Curtin *et al.* (1992) found that the SAR of the leaching solution had relatively little effect on the P concentration in the leachate when EC was approximately 10 dS m^{-1}, but the effect of SAR became more marked as EC decreased (Fig. 7.7). The adsorption of P by sodic soils has been shown to decrease sharply as the pH increases (Barrow, 1984; Curtin *et al.*, 1993b). In contrast, P adsorption by nonsodic soils may show little pH dependence or may even increase when the pH is raised (Naidu *et al.*, 1990). As a result, the impact of sodicity on P solubility is greatest at elevated pH (Fig. 7.8). The importance of pH has been well documented by Gupta *et al.* (1990), who showed that the concentration of P in saturated paste extracts of highly sodic soils increased exponentially as pH increased from 8.5 to 10.5 (Fig. 7.9) with the solution P concentrations under strongly

TABLE 7.3 Effect of fertilization with amonium nitrate on chemical and biological properties of a Natric Cryoboroll at Vegreville, Alberta, Canada [adapted from McAndrew and Malhi, *Soil Biol. Biochem.* 24:619-623, (1992)]

N rate (kg ha^{-1})	pH[a]	SAR[a]	Soil organic		Biomass		Mineralizable	
			C	N	C	N	C	N
			g kg^{-1}				mg kg^{-1}	
0	5.7	10.5	23	3.3	465	41	392	286
76	5.6	12.8	29	4.2	454	49	410	264
152	4.7	7.0	33	4.6	332	30	316	280
305	4.2	7.2	32	4.6	285	5	199	171

[a] pH was measured in a 1:2 soil:10 mM CaCl$_2$ suspension; and SAR values were calculated from concentrations of Na, Ca and Mg in saturated paste extracts.

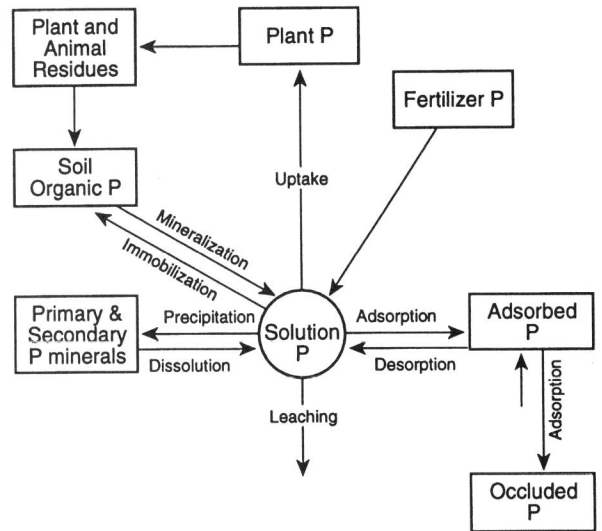

Figure 7.5 Soil phosphorus cycle.

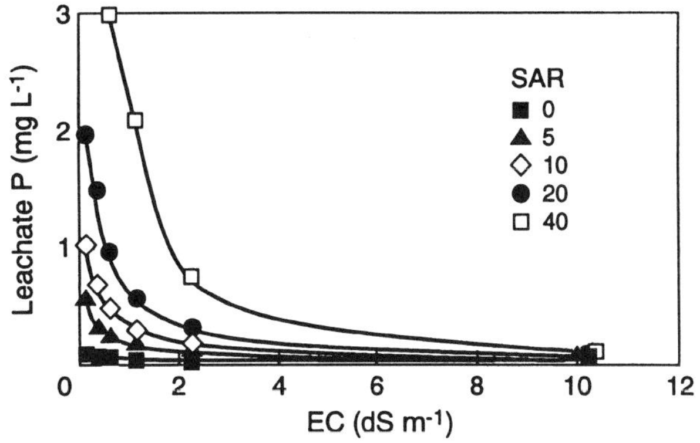

Figure 7.7 Influence of sodicity (SAR) and electrical conductivity (EC) of leaching solution on concentration of P in leachate from a Saskatchewan, Canada soil [Curtin, *et al., Soil Sci.* 153:409-416, (1992)].

alkaline, sodic conditions being far in excess of those needed for crop growth. The strong pH-dependence of the P adsorption-desorption equilibrium in sodic soils is due to a relatively large decrease in surface potential as pH is increased (Barrow, 1984).

Although the adsorption-desorption process is the one most susceptible to change induced by sodicity, the rate of dissolution of calcium phosphate compounds and the mineralization of organic P may also be affected to some extent (Sharpley *et al.*, 1988; Gupta *et al.*, 1990). Sharpley *et al.* (1988) induced dissolution of calcium phosphates by repeat-

edly washing Na-saturated soils with distilled water to maintain low concentrations of Ca and P. However, it is doubtful that, under field conditions, the solution Ca and P concentrations would be low enough to induce significant dissolution of calcium phosphate. In a laboratory study in which soils were rendered sodic by leaching with solutions of SAR ranging from 0 to 40, Curtin *et al.* (1992) found no evidence that sodic conditions resulted in a dissolution of calcium phosphates. Increasing sodicity tends to change the equilibrium between adsorbed (extractable in 0.5 M NaHCO$_3$) and solution P (water extractable) in favor of the latter (Table 7.4). Informa-

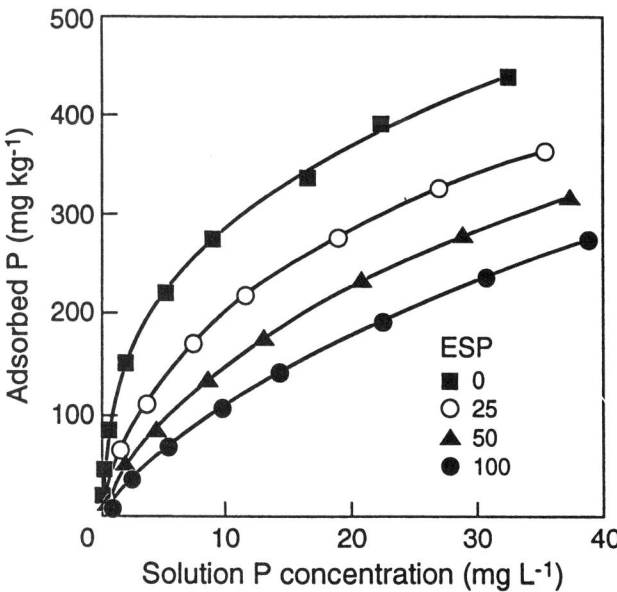

Figure 7.6 Phosphate adsorption by soil as influenced by exchangeable sodium percentage (ESP) [adapted from Curtin, *et al., Aust. J. Soil Res.* 31:137-149, (1993b)].

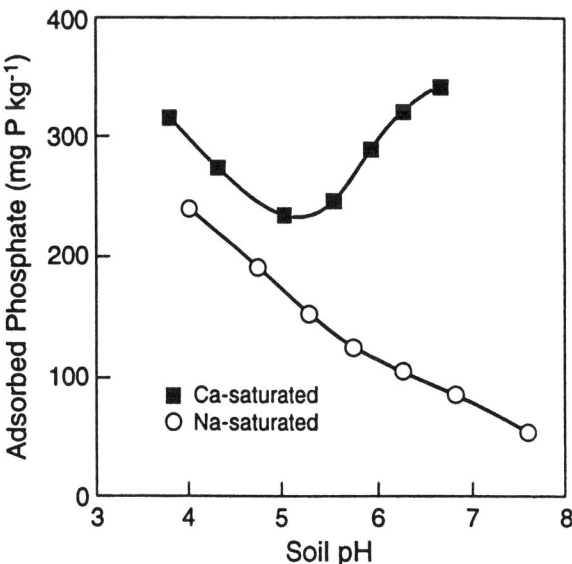

Figure 7.8 Adsorption of phosphate by Ca- and Na-saturated soil as a function of pH [adapted from Curtin, *et al., Aust. J. Soil Res.* 31:137-149, (1993b)].

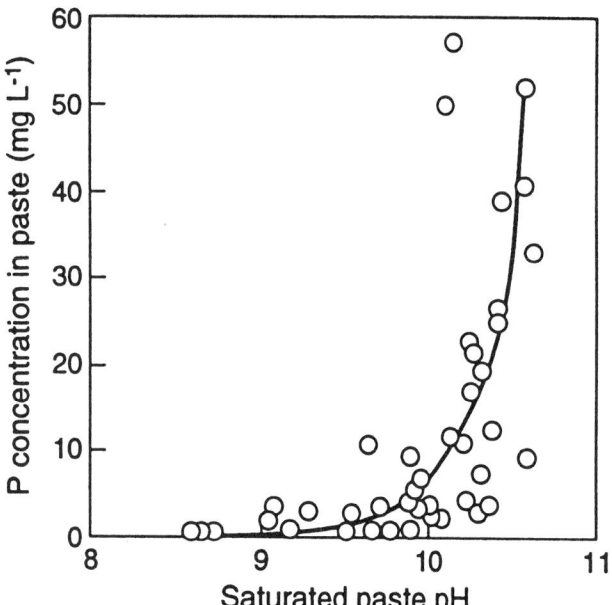

Figure 7.9 Phosphate concentration in saturated paste extracts of sodic soils from India as influenced by soil pH (redrawn from Gupta, *et al.*, *Soil Sci. Soc. Am. J.* 54:1254-1260, (1990)].

tion on the effects of sodicity on the mineralization and transformations of organic P is lacking.

The transport of P through the soil solution to the root surface occurs mainly by diffusion (Wild, 1988). Because the P concentration gradients between the root surface and the soil solution are normally small, the P diffusive flux is slow. Most of the P taken up by plant roots diffuses over only very short distances (millimeters) so that, to obtain an adequate supply of P, roots must proliferate throughout the soil. In sodic soils, poor physical conditions, high salt concentrations, and/or unfavorable moisture regimes are likely to hamper root growth, thereby limiting the ability of plants to acquire adequate P unless soil solution concentrations are high.

In summary, Na saturation of the exchange complex, high pH, and low electrolyte concentration shift the adsorbed-solution P equilibrium to give high P concentrations in soil solution. Conditions that maximize P solubility are, unfortunately, those that destroy soil structure. Thus maximum P solubility may coincide with low plant demand. However, some enhancement of P solubility may occur in soils with relatively low sodicity levels, even in the presence of salts and at low pH. Russell *et al.* (1988) found that, because adsorbed P was fairly soluble in strongly acid Natrustalfs, only small additions of P were needed to raise the solution P to 0.2 mg L^{-1}, the concentration considered to be optimum for growth. Curtin *et al.* (1992) demonstrated that relatively small amounts of exchangeable Na could have a significant

effect on P solubility. For example, soils with about 10% exchangeable Na had twice as much water-extractable P as those with no exchangeable Na (Table 7.4).

Indices of P Availability in Sodic Soils

Soil tests for available P developed for nonsodic soils may prove inadequate when used on Na-affected soils. The widely used Olsen test (Olsen *et al.*, 1954), which involves extraction with 0.5 *M* NaHCO$_3$, has been shown to be insensitive to the effect of sodicity on P solubility (Rimmer *et al.*, 1992; Curtin *et al.*, 1993b) which, to a large extent, is obscured when a concentrated Na solution is used as P extractant. Rimmer *et al.* (1992) state that

> using the Olsen test (on sodic soils) may lead to underestimates of P availability to plants and needless additions of P fertilizer; a test using a more dilute solution or even water, might be a better predictor of P availability in such soils.

The use of anion exchange resin membranes are likely to give a better estimate of available P in sodic soils than other methods because they act as sinks for P in a manner analogous to plant roots.

Further work is needed to develop tests that are sensitive to the factors controlling P solubility in sodic soils.

TABLE 7.4 Inorganic P fractions in a Typic Haploboroll from Saskatchewan (Canada) after leaching with solutions with sodium adsorption ratio (SAR) ranging from 0 to 40 [Curtin, *et al.*, *Soil Sci.* 153:409-416, (1992)]

	P fraction[a]			
	H$_2$O-P	NaHCO$_3$-P	NaOH-P	HCl-P
SAR (ESP)[b]	(mg P kg^{-1})			
0 (0)	6.5	14.8	38.3	159.4
2.5 (1.9)	8.7	14.2	38.9	167.6
5 (3.3)	11.5	14.5	38.3	164.6
10 (5.6)	13.9	10.7	37.1	162.1
20 (9.5)	15.5	6.6	35.6	157.9
40 (19.3)	16.0	3.3	35.3	161.5

[a]The soil was extracted sequentially with distilled water, 0.5 M NaHCO$_3$, 0.5 *M* NaOH, and 1 *M* HCl. These reagents extract soluble P (H$_2$O), adsorbed P fractions (NaHCO$_3$ and NaOH), and primary and secondary calcium phosphate compounds (HCl).
[b]Values in parentheses are soil exchangeable sodium percentages (ESP).

7.2.4 Sulfur

As a general rule, sodic soils are well supplied with SO_4^{2-}, the form in which S is taken up by plants. Because soils with pH > 6 adsorb very little SO_4^{2-} (Curtin and Syers, 1990), it is usually soluble and readily available to plants in sodic soils but tends to be concentrated in the subsoil. Field experiments in Alberta suggest that there may be some benefit to S fertilization in the early growth stages but little or no benefit after roots reach the subsoil (Nyborg, 1978). Sulfur-containing compounds such as gypsum ($CaSO_4 \cdot 2H_2O$), elemental S and iron pyrite (FeS_2) are commonly used as amendments to reclaim sodic soils, but there is little evidence that increased availability of S is a factor contributing to improved crop performance in soils treated with these amendments (Cairns and Beaton, 1976).

7.2.5 Micronutrients

Total Concentrations and Soil Processes Controlling Availability

Poor productivity of sodic soils has partly been attributed to micronutrient deficiencies due to the high pH of sodic soils (Naidu and Rengasamy, 1993). However, information on micronutrient dynamics in relation to soil-plant relations in sodic soils is limited. The concentration of trace elements in both solid and solution phases varies significantly depending on the soil type, the nature of parent material, and the intensity and duration of weathering processes. Typical ranges for micronutrient concentrations in soils from Australia, the United States, and the USSR are presented in Table 7.5 and illustrate the extreme variability in composition that can be expected from the interaction of the various factors of soil formation and soil parent material.

The concentrations of trace metals in soil solution are influenced by factors that control the adsorption-desorption and solubility processes: pH, organic matter content, redox potential, nature of constituent minerals, and composition of the soil solution (Mortvedt, 1972). Trace metal adsorption in soils can be specific or nonspecific (Loganathan et al., 1977; Benjamin and Leckie, 1981), but little detailed information is available on the proportion of cations adsorbed by either of these two mechanisms. In the case of Zn adsorption on Al and Fe oxides, Kalbasi et al. (1978) estimated these proportions to be 60–90% for the specific and 40–10% for the nonspecific mechanisms, respectively, but Tiller (1983) demonstrated that pH determined the relative proportions, with specific adsorption becoming more dominant with increasing pH. Changes in soil pH may influence adsorption by (a) ionization of surface groups (Stumm et al., 1980), (b) displacement of the equilibrium of the surface complexation groups, and (c) competition

TABLE 7.5 Total trace element concentration ranges in soils from different countries [Tiller *et al., Soils: An Australian Viewpoint*, Commonwealth Scientific and Industrial Research Organization Publications, Melbourne/Academic Press, London, (1983)]

Country	Concentrations (mg kg⁻¹)				
	Co	Cu	Mn	Mo	Zn
Worldwide	1–40	2–100	200–3000	0.2–5	10–300
USSR	0.1–15	1.6–200	100–12000	1–12	20–120
USA	<3–70	<1–300	<1–7000	<3–7	<24–2000
Australia	<2–170	<1–190	4–5100	<1–20	<2–180

with H^+ and other cations for negative sites and variation in metal species with pH.

Irrespective of the nature of the soil, the plant availability of many trace metals is largely determined by pH and redox potential (Section 7.3). Generally, the concentrations of monomeric species decrease with increasing pH. Hodgson et al. (1966), who speciated soil solution Zn^{2+} into ionic and complexed forms for a range of soils, found the amount of Zn complexed varied from 28 to 99% and increased with increasing organic matter content. McBride and Blasiak (1979) demonstrated that complexed Zn increased with increasing pH. While pH largely controls the availability of Zn and Cu, both pH and E_h directly influence the solubility of Mn and Fe. As pH increases above 3, trace metal adsorption increases rapidly (Cavallaro and McBride, 1984; Bar-Yosef et al., 1980; Adriano, 1986), due partly to increased adsorption by soil constituents and partly to other factors such as nutrient interactions and precipitation reactions at high pH. However, the role of precipitation-dissolution reactions on trace metal solubility in soils is unclear, although Lindsay (1979) provides data on soils that relate trace metal solubility in soils to the effect of pH on numerous soil minerals. The variation in adsorption with pH depends on both the nature and the concentration of the adsorbing species. Kinniburgh and Jackson (1982) reported that, at equilibrium Zn concentrations of about 0.1 mM, adsorption increased by roughly tenfold per unit increase in pH between pH 5.5 and 6.5. However, at ultralow Zn concentrations (μM), which are commonly found in soils, adsorption was more strongly pH dependent and increased 45-fold for each unit increase in pH.

Because the transfer of nutrients from the soil solution to the roots occurs mainly by diffusion processes, soil properties affecting mobility will also influence the availability of trace metals to plants. With the exception of molybdenum (Mo), increasing soil pH is likely to decrease plant availability of micronutrients, which has been observed in aerated soils for Mn, Zn, and Fe, but uptake of Cu is poorly related to pH at moderate levels of soil Cu (Martens, 1968; Sims, 1986). According to Loneragan (1975), the lack of Cu deficiency, particularly on calcareous soils, may be related to the formation of soluble Cu-organic complexes (Stevenson and Fitch, 1981), which is consistent with the observations of Jeffrey and Uren (1983), who found that the proportion of soluble complexed Cu increases with increasing pH in Australian soils.

Zinc, Copper, and Manganese

The concentrations of micronutrients in sodic soil solutions are often low and range from <0.01 to 10 μM (Naidu *et al.*, 1995e) in both surface and subsurface soils, indicating the highly nutrient-deficient nature of these soils. Singh *et al.* (1988), who investigated the effect of sodicity on growth, yield, and chemical composition of different oilseed crops as influenced by sodicity, found that increased sodicity resulted in lower plant Zn, Cu, and Mn and higher Na concentrations. Sharma *et al.* (1992) investigated the distribution of micronutrients in arid zone soils of the Punjab with pH values ranging from 8.1 to 8.4, concluding that widespread deficiencies of Zn and Fe occurred in surface soils. Zinc deficiency has frequently been observed in plants growing in sodic soils (Singh and Franklin, 1974; Mehrotra *et al.*, 1986).

Plant availablility of micronutrients in sodic soils is controlled by pH, electrolyte concentration, surface charge density and composition (cations and ligand ions) of the soil solution. However, no consistent correlations have been observed between soil properties and trace element retention. Singh *et al.* (1988) studied the distribution of trace metals among various chemical forms in calcareous soils of India using a sequential extraction scheme. They found that 17, 41, 11, and 7% of the total soil Cu, Fe, Mn, and Zn, respectively was associated with the crystalline iron oxide fraction while 12, 6, 9, and 5%, respectively, was present in the amorphous iron oxide fraction. No consistent correlations were observed between soil properties and metals associated with carbonate, organic matter, and Mn oxide fractions.

The extent to which a change in ionic strength and sodicity affects the availability of most micronutrients has not been well established. Although sodic soils cover a large proportion of the world's arable land, no studies have been designed to specifically address sodicity-related fertility issues. Where electrolyte concentration and sodicity effects have been studied, only nonsodic soils treated with solutions of different SAR have been used. Such investigations do not provide a true picture of the processes that control the micronutrient

solubility in sodic soils. Of the essential micronutrients commonly found to be deficient in sodic soils, Zn has received most attention because of the severe problems encountered in India where much of the research on the chemistry of Zn has been conducted.

Micronutrient chemistry of sodic soils is further complicated by the presence of carbonate minerals such as calcite and sodium carbonate (Na_2CO_3), which provide active sorption sites and nuclei for the precipitation of many trace metals. Hodgson *et al.* (1966) showed that in calcareous soils, $CaCO_3$ influenced the concentration of Zn, which was often low and did not exceed μM concentrations. The solubility of micronutrients in soils is complicated by the formation of insoluble carbonates. While, in many normal soils, a reduction in soil pH enhances micronutrient solubility, extractable Zn concentrations in sodic soils are often found to be higher than in reclaimed soils with lower pH (Singh *et al.*, 1983). This increased concentration of Zn in solution at higher pH values has been attributed to the mobilization of Zn by soluble or colloidal organic matter (Jeffrey and Uren, 1983). Chemical precipitation of Zn as a hydroxide or carbonate mineral may also control the solubility of Zn in soils (Lindsay, 1979, Mikkelson and Kuo, 1977; Saeed and Fox, 1977). Singh and Abrol (1985) investigated the solubility and adsorption of Zn in a sodic soil over a wide range of pH values. They found that precipitation of Zn as $Zn(OH)_2$ or $Zn(CO_3)_2$ controlled the solubility at pH values greater than 7.9, while in the pH range of 6.0–7.9, Zn solubility was highly pH dependent and controlled by chemisorption reactions. At pH values below 6.0, a reduction in adsorption was attributed to an increase in the concentration of competing cations such as Al, Mn, Fe, and Ca and to the partial dissolution of soil carbonates. These results show that Zn solubilty in sodic soils is controlled by both adsorption and precipitation reactions which are modified by the composition of the soil solution. Since micronutrient availability is reduced with increasing soil pH, micronutrient deficiencies can be a major factor limiting crop production in high pH sodic soils.

Ionic composition and interactions on the exchange complex can also influence micronutrient adsorption in soils. Nelson and Melsted (1955) noted that the relative order of adsorption of different cations on clay was H > Zn > Ca > Mg > K. Given that, in sodic soils, the ratio of Na to other cations can vary widely, it is likely that variations in the exchange complex composition may also influence micronutrient availability. Elrashid and O'Connor (1982) investigated the influence of solution composition (Cl^-, NO_3^-, and SO_4^{2-}) and ionic strength on the adsorption of Zn at anion concentrations of 0.005, 0.002, and 0.1 M in nine soils of varying physical and chemical properties, concluding that under the experimental conditions examined neither ionic strength nor anion complex formation significantly affected Zn adsorption. Shukla *et al.* (1980) studied Zn adsorption in some Indian soils as affected by exchangeable cations. They found that Zn adsorption in soils saturated with different cations increased in the order H < Ca < Mg < K < Na (Fig. 7.10). These results indicate that, in sodic soils, Zn adsorption is likely to be high. However, soil pH values ranged from less than 3 to greater than 9.3 with pH being the dominant factor in both the Al- and Na-saturated soils. In a similar study involving adsorption of Zn by Ca- and Na-saturated soils from India (pH>8), Mehta *et al.* (1984) reported that the values of standard free energy changes $G°$ for the Ca^{2+}-Zn^{2+} exchange reaction were negative while those for Na^+-Zn^{2+} were positive, anomalously suggesting a preference for Zn^{2+} over Ca^{2+} and a strong preference for Na^+ over Zn^{2+}. This contrasts with the observations of Shukla *et al.* (1980), who reported positive $G°$ values, indicating the nonspontaneous nature of the reaction. The greater preference for Zn on the exchange complex in the presence of Na indicates that less Zn will be available for plant growth in sodic soils.

Williams and Raupach (1983) reported widespread occurrences of micronutrient deficiencies in many Australian soils, concluding that such deficiencies reflect the prolonged or intense weathering and leaching to which so many Australian soils or their parent materials have been subjected. These investigators also report widespread deficiencies of Zn, Mn, and Cu in many calcareous and alkaline soils, which include many sodic soils such as Solonized Brown Earths (Xerochrepts),

Grey-Brown Earths (Xeralfs) and Black Earths (Xererts) of South Australia, Victoria, and Queensland. Quoting the work of Seeliger (1968), Naidu *et al.*, (1993b) report that sodic soils in South Australia, especially Natrixeralfs, Pelloxererts, and Rhodoxeralfs, are deficient in most micronutrients. While reviewing the literature on vineyard soils in Australia, Northcote (1988) reported that many of the alkaline soils with high subsurface ESP values had serious micronutrient deficiencies, which included Fe, Mn, Zn, Cu, and Co. Of these trace elements, Mn and Zn deficiencies have been recorded in a variety of plants, including cereals, legumes, and many fruits.

Widepread micronutrient deficiencies have also been reported for crops grown in sodic soils in India (Abrol *et al.*, 1988), where field studies have shown significant increases in crop yields with application of Zn fertilizer. For instance, quoting Singh *et al.* (1982), Abrol *et al.* (1988) reported that the application of 10 kg Zn ha^{-1} as $ZnSO_4$ was sufficient to overcome the deficiency in rice grown in an amended highly sodic soil. Sodicity-induced Zn deficiency has been reported by Mehrotra *et al.* (1986), who grew plants in pot culture on a loamy alluvial soil from the Lucknow district of India, treated to give ESP values ranging from 16 to 55. Increase in sodicity affected plant growth at both the pre- and postemergence stages. For instance, when ESP exceeded 24.5, the emergence of maize (*Zea mays*) seedlings was delayed (Table 7.6). At or above an ESP of 34, visible foliar symptoms of Zn deficiency appeared after 30 days growth and gradually became pronounced with increasing age, until at 60

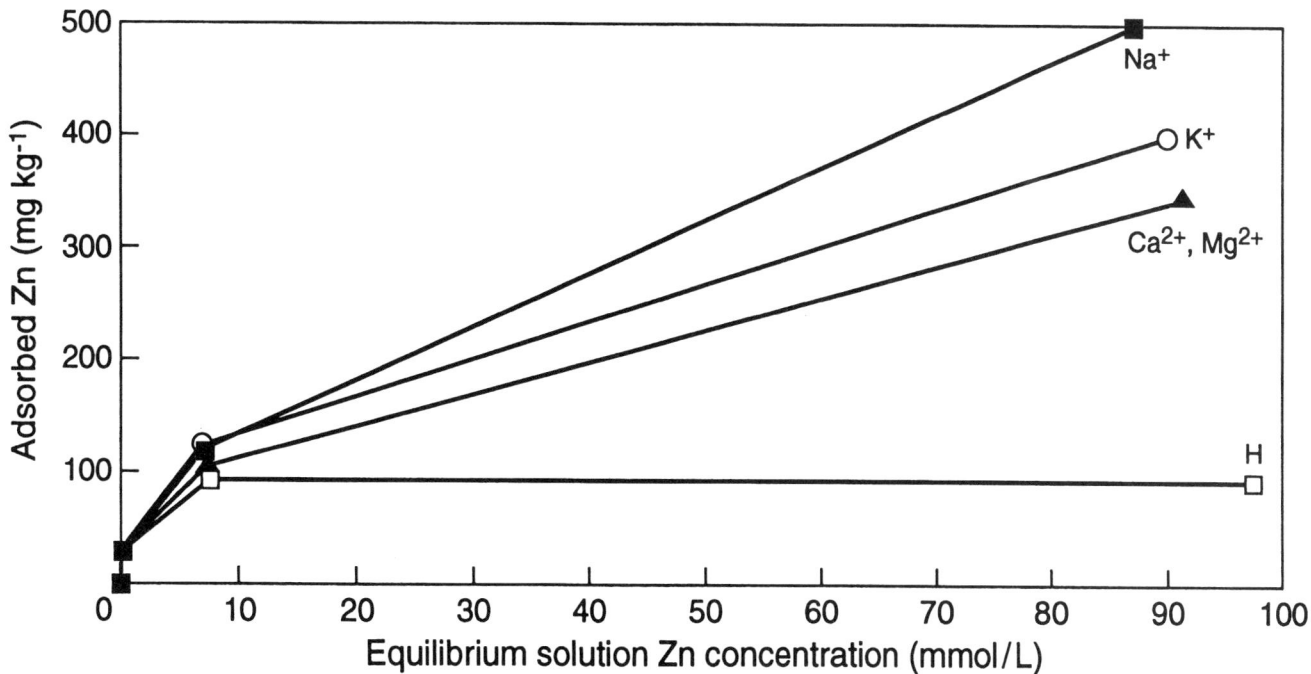

Figure 7.10 Adsorption of zinc by sodic soil from India as influenced by exchangeable cations [redrawn from Shukla *et al.*, *Soil Sci.* 129:67-371, (1980)].

days they even appeared at an ESP of 15.5 (Table 7.6). These observations supported their previous observations for maize plants grown under sodic conditions in the field. The severity of the deficiency symptoms increased with sodicity.

Iron
In addition to Zn, Fe is often found to be the limiting micronutrient in sodic soils due to high pH resulting from the presence of calcareous material (Singh *et al.*, 1982). Lime-induced iron chlorosis occurs in calcareous and solonized brown soils in Australia (Stephens and Donald, 1958), where surface layers also have an alkaline pH. Chen and Barak (1982) concluded that excess $CaCO_3$, bicarbonates, high soil pH, and excess water (poor drainage and root aeration) contribute directly or indirectly to Fe chlorosis. Similarly, Abrol *et al.* (1988) recorded increased yields in response to Zn, Fe, and Mn applications on alkaline sodic soils in India, but numerous researchers have found that the addition of Fe fertilizer was not effective unless the oxidation status of the soil was changed by prolonged submergence and addition of organic matter because the availability of Fe is controlled by the redox potential. Deficiency of Fe in many sodic soils in India has been eliminated through applications of FeS_2. Sinha and Jha (1984), who used FeS_2 as an ameliorant of sodicity and also as a source of Fe, S, and some other micronutrients, obtained fourfold increases in yield. Based on similar studies, Tiwari *et al.* (1984) also concluded that FeS_2 can be used as a multipurpose amendment in sodic soils.

Boron
Boron is normally present in soils as undissociated boric acid $[B(OH)_3]$ below pH 7 (Oertli and Grgurevic, 1975). Soil solution B concentration is controlled by both adsorption and precipitation reactions. Factors that influence adsorption reactions include pH (Hingston, 1964; Mezuman and Keren, 1981), the nature of exchangeable cations and ionic strength (Keren and Gast, 1981), and wetting and drying cycles (Biggar and Fireman, 1960; Keren and Gast, 1981). Increasing pH enhances B adsorption on clay minerals (montmorillonite, illite, and kaolinite), and hydroxy-Al compounds, which reaches a maximum in the alkaline pH range due to changes in the nature of B species in soil solution. As the soil pH increases to 9 and above, the concentration of the anion $B(OH)_4^-$ increases rapidly, leading to a marked increase in B adsorption. At higher pH values, increased concentrations of OH^- compete with B for adsorption sites, resulting in a rapid decrease in B adsorption.

With reference minerals, the extent of B adsorption depends upon the ambient conditions. For instance, Keren and Gast (1981) reported that the effect of index cations such as

Na^+ and Ca^{2+} on B adsorption by montmorillonite at pH values below 7 is minimal, but above pH 8 more B was adsorbed by Ca than by Na systems. Increasing the ionic strength to 0.07 or 0.36 M by the addition of NaCl increased the amount of B adsorbed on montmorillonite by 10 and 50%, respectively. Wetting and drying cycles and increased levels of K saturation may also increase B adsorption (Hadas and Hagin, 1972; Keren and Gast, 1981). The increased B concentrations at high pH values and the low sorption in the presence of Na ions have implications for B plant availability in alkaline sodic soils, resulting in the potential for B toxicity problems. Such toxicities have been widely observed in many cereal growing soils in South Australia, particularly on Calcic Natrixeralfs, Calciorthids, and Xerochrepts with free carbonate and high pH (Cartwright *et al.*, 1986) in which extractable B is correlated with ESP, CEC, and clay content. Holloway and Alston (1992) investigated the interaction between B and NaCl toxicities in alkaline sodic soils, concluding that excess B had a more deleterious effect on wheat yield than did salt at comparable concentrations of Na in the tissue. Soil amelioration with gypsum caused a marked reduction in the concentration of water-soluble B due to a change in the nature of the B species from water-soluble Na to relatively insoluble Ca metaborate (Gupta and Chabra, 1972). Gypsum applications which reduce ESP also lower B uptake by grasses. Because many Australian sodic subsoils have B concentrations exceeding 10 mg kg^{-1}, management to reduce B toxicity would presumably require subsurface applications of gypsum.

Molydenum
As with the other micronutrients, Mo availability is controlled by pH, the nature of clay minerals, and the soil solution composition, but little information is available on Mo levels in sodic soils. Because Mo availability increases dramatically with pH (Lindsay, 1979), crops grown on high pH sodic soils are likely to accumulate excessive levels of Mo, which may prove toxic to animals.

Ion Interactions and Micronutrient Availability
Because nutritional balance promotes growth (Sumner and Farina, 1986), nutrient interactions are important. The best known example is the interaction between P and Zn in relation to Zn availability, about which considerable controversy still exists concerning whether excess P causes Zn deficiency or low Zn concentrations in soil cause excessive P uptake by plants. Mehrotra *et al.* (1986) studied the effect of sodicity on concentrations and total contents of Zn, Na, P, Fe, Mn, and Cu and Na/Zn and P/Zn ratios on seedling emergence, dry matter yield, and tissue concentration in the middle (3rd–4th) leaves of maize. Sodicity reduced tissue Zn and increased Na and P

TABLE 7.6 Effect of soil sodicity on seedling emergence, dry matter yield and nutrient concentrations in the middle (3rd–4th) leaves of corn [data from Mehrotra *et al.*, *Plant Soil* 92:63-71, (1986)]

Soil ESP	Seedling emergence (%)	Yield (g pl^{-1})	Na	P	Zn	Fe	Mn	Cu	Na/Zn	P/Zn
			(%)			(mg kg^{-1})				
15.5	100	10.7	0.365	0.060	16.0	200	44.0	13.3	228	38
21.2	100	7.2	0.450	0.195	13.6	132	42.0	11.1	331	144
30.0	85	2.7	0.950	0.400	10.8	88	30.0	12.2	550	372
37.3	95	1.6	1.515	0.285	7.3	112	50.0	26.6	2075	392
43.0	75	1.1	1.640	0.270	6.4	157	53.0	15.8	2563	420
48.9	25	1.2	1.325	0.420	10.0	161	70.3	19.8	1525	429

(Table 7.6) which was reflected in the P/Zn ratio increasing with ESP up to 33.7. This highlights the effect that sodicity has on ion interactions, which can lead to deficiencies or excesses of nutrients. Many other examples of sodicity effects on tissue ion balance have been reported in sodic soils, for example, S and Zn (Kumar and Singh, 1980), Mg and Zn (Kumar *et al.*, 1981), Zn and P (Singh and Singh, 1980), and Zn, P, and Fe (Singh and Singh, 1983). These studies reveal that Zn and P and Zn, P, and Fe behave antagonistically toward each other. The role of Fe may be greater under submerged conditions which favor its solubility, permitting stronger interactions with Zn and P. Recent studies on pearl millet (*Pennisetum glaucum*) in pots (Kumar *et al.*, 1985) demonstrated a synergistic effect of N on tissue Zn concentrations that may translate into increased yield if these two nutrients are applied in appropriate amounts. On the other hand, heavy applications of Zn (>20 mg L^{-1}) to soil may decrease the content of N in crops.

7.3 INFLUENCE OF pH AND E_h ON NUTRIENT AVAILABILITY

Sodic soils undergo fluctuating redox conditions during wet winter months due to poor infiltration. Waterlogging of soils over long periods increases the concentrations of plant available micronutrients such as Fe and Mn. Sadana and Bajwa (1985) reported that submergence over a period of 10–12 weeks increased Fe^{2+} and Mn^{2+} concentrations in sodic soil solutions. However, in saline sodic soils, submegence led to only very small changes in Mn concentration. In a later study, Sadana and Takkar (1988) investigated, under greenhouse conditions, the effect of sodicity (ESP ranging from 3 to 60) and Zn (0 and 0.15 mmol kg^{-1} soil) on soil solution chemistry of Mn under submerged conditions. Soil submergence de-

creased pH and E_h and increased Mn concentrations in all treatments. Increasing sodicity levels increased soil solution pH and decreased the Mn concentration (Fig. 7.11). Thermodynamic calculations showed that the MnCO$_3$-Mn^{2+} system regulated the solubility of Mn^{2+}. The addition of Zn (ZnSO$_4$) did not have any appreciable effect on Mn concentrations.

Because organic matter promotes reducing conditions, increases in soluble Mn and Fe concentrations often occur when organic residues are incorporated (Swarup, 1980). The chelating effect of organic ligands released during decomposition may enhance the solubilization of trace metals such as Fe, Mn, and Cu. Although sodic soils are generally low in plant available Fe and Mn, the decrease in pH following amelioration and temporary ponding of water may reduce insoluble Fe and Mn oxides, increasing their availability (Ponnamperuma, 1972). Under such conditions, Fe and Mn are mobile and may be lost by leaching (Kovda and van den Berg, 1973; Sharma and Yadav, 1986).

Although Zn and Cu do not undergo valence changes under reducing conditions, their plant availability has been found to decrease under flooded conditions (Ponnamperuma, 1972; Iu *et al.*, 1981; Sajwan and Lindsay, 1986). This decrease may be related to the changes in surface chemical properties of oxyhydroxide minerals which provide active adsorption sites. In contrast to Zn and Cu, plant availability of Mn and Fe is controlled to a large extent by pH and E_h relations (Gotoh and Patrick, 1972). Depending on the original pH and organic matter content reducing conditions bring about changes in soil acidity, redox potential, and level of complexing agents. Turner and Patrick (1968) showed that with increasing time under flooded conditions, the concentrations of Fe^{2+} and Mn^{2+} increased while those of dissolved O$_2$, NO$_3^-$, Fe^{3+}, Mn^{4+}, and Mn^{3+} decreased. Working with rice, Sonar and Ghugare (1982) found that maintaining saturated conditions for 15 days prior to planting increased the availability of Mn. Earlier Jugsujinda and Patrick (1977) reported that under anaerobic soil conditions, Mn uptake by rice was apparently influenced by Fe uptake and soil pH.

7.4 EFFECT OF SOIL STRENGTH, WATERLOGGING, AND DISEASES ON ROOT DEVELOPMENT

The poor physical properties of sodic soils, in addition to directly limiting crop production through poor seedling emergence and root proliferation (Taylor, 1971), also indirectly have an effect on plant nutrition (Bernstein, 1975) by restricting water and/or nutrient uptake and gaseous exchange. Roots growing in hard soils are shorter and thicker and much more closely appressed to the soil particles than those in friable soils (Baligar *et al.*, 1975; Atwell, 1990). Because sodic soils are usually hard, particularly in the subsoil, restricted root proliferation reduces water and nutrient supply to leaves (Barraclough and Weir, 1988), causing growth to be stunted. This is particularly true under Australian conditions where subsurface sodic horizons are often nutritionally limited and dense.

Short periods of flooding can damage roots, resulting in reduced growth and necrosis (Jackson and Drew, 1984).

Although the effect of waterlogging on soil chemical and microbial changes have been extensively studied, limited information is available on the impact of waterlogging on root diseases in sodic soils. Increased incidence of root disease caused by *Pythium* spp. in wheat-growing soils of southern Australia has been reported by Rovira and Ridge (1983). Because of the adverse effects of anoxic and suboxic conditions in sodic soils on the physiological functions of plants (Drew and Sisworo, 1979), increased susceptibility to pathogens could result.

7.5 CROP TOLERANCE TO SODICITY

The tolerance of crops to sodicity varies widely, depending on their specific sensitivities to adverse nutritional and physical soil conditions (Pearson, 1960). Because the degree of sodicity-induced physical degradation can vary greatly from soil to soil, early research in the United States was designed to measure the tolerance of crops to sodicity in the absence of adverse physical conditions (Bower and Wadleigh, 1948; Chang and Dregne, 1955; Bernstein and Pearson, 1956; Pearson and Bernstein, 1958). This was achieved using conditioners such as vinyl acetate-maleic acid (VAMA) copolymer to maintain structural integrity (Bernstein and

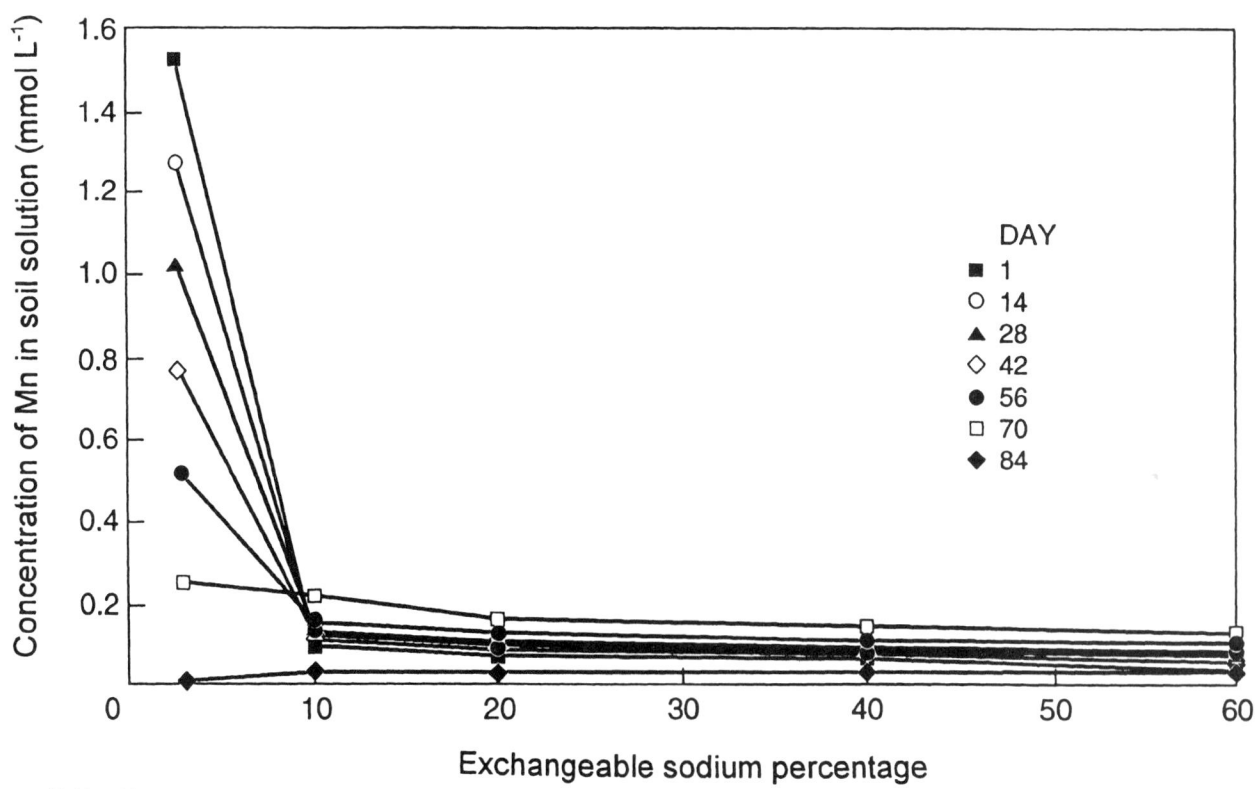

Figure 7.11 Effect of sodicity under submerged conditions on soil solution concentration of Mn in Indian soil [redrawn from Sadana and Takkar, *J. Agric. Sci. Camb.* 111:51-55, (1988)].

Pearson, 1956; Pearson and Bernstein, 1958) or using exchange resins as a substitute for soil (Bower and Wadleigh, 1948). Crops are listed in Table 7.7 approximately in order of increasing tolerance to sodicity based solely on nutritional factors. The growth of certain tree crops that are sensitive to Na may be retarded even in soils where the adverse physical effects of sodicity are scarcely noticeable (i.e., ESP 2–10) (Pearson, 1960). Most field crops are classed as moderately tolerant or tolerant of nutritional perturbations associated with sodicity (Table 7.7). Some grass species are apparently not affected nutritionally, even at very high ESP values (>60).

Differences in tolerance to sodicity-induced nutrient imbalances between plant species may be obscured by unfavorable growing conditions in the field (Bernstein and Pearson, 1956). In contrast to the U.S. studies discussed, crop sodicity tolerance work in India has generally been carried out under conditions where poor soil structure was a factor influencing plant responses (Abrol and Bhumbla, 1979; Chhabra *et al.*, 1979; Singh *et al.*, 1979). Results obtained in this manner may be site-specific and the findings are not necessarily applicable to other soil types with different physical conditions. For some of the sodicity tolerance information compiled by Gupta and Abrol (1990a), see Chapter 8 (Table 8.2).

The sodicity tolerance ranking of plants depends strongly on the conditions under which responses to sodicity are measured. For example, rice, which is insensitive to poor physical conditions, has been found to be highly tolerant of sodicity under field conditions (Abrol and Bhumbla, 1979). But when measurements are made under conditions where physical deterioration was prevented, rice ranked as only moderately tolerant (Table 7.7). There is probably no single critical sodicity level for a given crop. As stated by Carter and Peterson (1962):

> the ESP that can be tolerated by the plant probably varies with the physical conditions of the soil, the level and kinds of salts associated with the ESP level, the soil moisture, and other factors.

Some evidence of differential sodicity tolerance among genotypes within species such as rice and wheat has been presented (Joshi *et al.*, 1980; Mishra and Bhattacharya; 1980; Sharma, 1986). Varietal differences in sodicity tolerance appear to be related to nutritional factors. The ability to selectively absorb K from Na-dominated root media has been identified as the factor most closely associated with varietal differences (Joshi *et al.*, 1980; Sharma, 1986).

Growing sodicity-tolerant crops is a useful management option for coping with sodic soils. Although our knowledge of crop responses to sodic soil conditions is far from complete,

TABLE 7.7 Tolerance of various crops to exchangeable sodium percentage (ESP) [Pearson, U.S. Dept. of Agricul*ture* Inf. Bull. 216, (1960)]

Tolerance to ESP	Crop	Growth response under field conditions
Extremely sensitive (ESP 2–10)	Deciduous fruits Nuts Citrus Avocado	Sodium toxicity symptoms even at low ESP values
Sensitive (ESP 10–20)	Beans	Stunted growth at low ESP values even though the physical conditions of the soil may be good
Moderately tolerant (ESP 20–40)	Clover Oats Tall fescue Rice Dallis grass	Stunted growth due to both nutritional factors and adverse soil conditions
Tolerant (ESP 40–60)	Wheat Cotton Alfalfa Barley Tomatoes Beets	Stunted growth usually due to adverse physical condition of soil
Most tolerant (ESP > 60)	Crested and fairway wheatgrass Tall wheatgrass Rhodes grass	Stunted growth usually due to adverse physical condition of soil

the available data should enable reasonably informed decisions to be made as to the suitability of crops for a given soil.

7.6 EFFECTS OF AMENDMENTS ON NUTRIENT AVAILABILITY AND PLANT GROWTH

7.6.1 Gypsum and Lime

Productivity of sodic soils (Barley and Hutton, 1956; Loveday and Scotter, 1966; Noble and Kleinig, 1971) has generally been reported to improve following the application of gypsum (Overstreet *et al.* 1951; Carter *et al.*, 1986; Webster and Nyborg, 1986; Malhi *et al.*, 1992b; McAndrew *et al.*, 1992) due to improved physical properties (Loveday and Scotter, 1966; Grierson, 1978; So *et al.*, 1978; White and Robson, 1989), reductions in exchangeable Na (Loveday and Scotter, 1966; Grierson, 1978), and strength (modulus of rupture) of the soil (Grierson, 1978; So *et al.*, 1978; White and Robson, 1989), and better (less cloddy) seedbed conditions (Webster and Nyborg, 1986) allowing improved seedling emergence

(McKenzie *et al.*, 1993). Gypsum treatment invariably decreases sodicity and improves Ca availability (i.e., it increases the Ca/TCC ratio of the soil solution) (Malhi *et al.*, 1992b). Spectacular responses are sometimes achieved by deep ripping, which enables roots to penetrate the subsoil, and thus greatly increasing their effective supply of water and nutrients (Hamblin, 1985). Jayawardane and Blackwell (1985, 1986) found that deep ripping with gypsum slotting to ameliorate subsoil sodicity improved drainage, soil water content, porosity, and root development leading to improved yield, but the effect of gypsum was transient (Chapter 8).

Effects of gypsum on the biological properties of soil and on the availability of nutrients other than Ca have largely been ignored. Carter (1986) reported evidence that gypsum may have an adverse effect on N fertility under some circumstances. In a long-term field study, gypsum treatment reduced microbial and mineralizable N by 10–43 and 10–54%, respectively, possibly due to the decline in pH from over 6 in control soils to about 5 in gypsum-treated soil. Normally, sodic soils have moderately high pH values and gypsum application will not lower the pH to the point where acidity affects biological processes. The soil pH is reduced when Ca replaces exchangeable Na (Oster, 1982), but the effect is usually not large. Use of phosphogypsum (PG), which contains some phosphoric acid (1-2%), will cause additional acidification. A reduction in pH may increase the availability of trace metals (Oster, 1982).

Work in India (Chhabra *et al.*, 1981) suggests that when sodic soils are reclaimed by surface application of gypsum followed by leaching, losses of P may occur, increasing the likelihood of P responses to fertilization (Gupta *et al.*, 1985). These losses can be prevented by mixing gypsum into the soil rather than broadcasting it on the surface (Chhabra *et al.*, 1981), which is consistent with the work of Curtin *et al.* (1995b) who showed that water extractable P is greatly decreased when sodic soil is treated with gypsum (or $CaCl_2$).

Recently, Malhi *et al.* (1992b) reported that gypsum (20 Mg ha^{-1}) improved the K and Zn nutrition of barley on a Typic Natriboroll due to enhanced root growth rather than to any change in the availability of these nutrients. The low solubility of lime limits its usefulness as an amendment, but it can be of some benefit where acidity restricts growth on sodic soils (Webster and Nyborg, 1986; Wetter *et al.*, 1987).

Mixing Ca minerals (gypsum and lime) present in the C horizon into sodic A and B horizons by means of deep plowing has been widely studied in North America as a means of increasing the productivity of sodic soils (Cairns, 1962; Mech *et al.*, 1967; Rasmussen *et al.*, 1972; Harker *et al.*, 1977; Lavado and Cairns, 1980; Buckland and Pawluk, 1985a,b; Webster and Nyborg, 1986; McAndrew and Malhi,

1990). Crop responses to deep plowing are extremely variable, and the factors responsible have often not been identified (Harker *et al.*, 1977). Buckland and Pawluk (1985b) reported that cereal yield responses [barley and oats (*Avena sativa*)] to deep plowing at five locations in Alberta ranged from a decrease of 16% at one site to increases of 7–198% at the other sites. A six-year study at two sites in Alberta (Webster and Nyborg, 1986) showed that alfalfa yield on deep plowed plots was 2.5–5.5 times that on conventionally tilled plots. Barley was less responsive to deep plowing than alfalfa. Benefits of deep plowing can be long lasting. At two sites that had been plowed 11–12 years previously, McAndrew and Malhi (1990) obtained grass-legume yield increases of 1.8–2.9 Mg ha^{-1} at one site and of 1.7–2.6 Mg ha^{-1} at the other. Harker *et al.* (1977) concluded that the main factor contributing to improved growth on deep plowed soil was improved water intake and reduced waterlogging, which allowed better root growth and more efficient use of soil and applied N. On the other hand, Buckland and Pawluk (1985b) suggested that yield responses to deep plowing were associated with reductions in moisture stress.

The impact of deep plowing on soil chemical properties can be quite variable. Most workers report a significant, though usually not very large, decrease in sodicity and increases in Ca availability (Webster and Nyborg, 1986; McAndrew and Malhi, 1990). Buckland and Pawluk (1985a) observed no decrease in ESP of the Ap horizon of five deep plowed sites. In contrast, Rasmussen *et al.* (1972) reported almost total reclamation of an irrigated sodic soil in Oregon within 3–4 years of deep plowing to 90 cm. The incorporation of material from the C horizon, which is rich in gypsum and lime, increases the pH and salt concentration of the A and B horizons (Buckland and Pawluk, 1985a; Webster and Nyborg, 1986).

7.6.2 Sulfuric Acid

The use of sulfuric acid to reclaim sodic soils was investigated in laboratory and field studies at the University of Arizona in the 1970s (Stroehlein *et al.*, 1978). Sulfuric acid applied to calcareous, sodic soils reacts with lime to produce gypsum, the soluble Ca form which can then replace exchangeable Na. Acid may also react with other soil constituents such as calcium phosphates and iron compounds, making nutrients more available to plants.

The ability of sulfuric acid to lower soil pH and increase the availability of certain nutrients (Fe, Mn, Zn, P) is a major difference between acid and gypsum amelioration strategies (Ryan *et al.*, 1974, 1975). Growth studies (Overstreet *et al.*, 1951; Stroehlein *et al.*, 1978) show that, in certain nutrient-

TABLE 7.8 Growth of sorghum and EDTA-extractable Fe as influenced by application of gypsum, sulfuric acid, and Fe chelate to a sodic (ESP 21) Pima clay loam from Arizona [Stroehlein *et al.*, Sulfuric Acid for Improving Irrigation Waters and Reclaiming Sodic Soils, Dept. of Soils, Water and Engineering, University of Arizona, Tucson, (1978)]

Treatment	Yield	Extractable Fe
	(g pot^{-1})	(mg kg^{-1})
Control	0.2	2.5
Gypsum	0.6	2.3
Sulfuric acid	4.2	7.0
Fe-chelate	4.1	---

deficient soils, acid may be superior to gypsum as an amendment. Sorghum (*Sorghum bicolor*) failed to respond significantly to gypsum on a Fe-deficient sodic soil in Arizona, but sulfuric acid, which increased ethylene diamine tetraacetic acid (EDTA)-extractable Fe in the soil from 2.5 to 7 mg kg^{-1}, produced a large yield increase (Table 7.8). Sorghum also responded well to Fe chelate. These results strongly suggested that the acid-induced yield increase was due to a correction of Fe deficiency. Wallace and Muller (1978) claimed that the acidification of a small (<1%) part of the mass of calcareous soil effectively prevents Fe chlorosis in soybean (*Glycine max*).

Because soil pH values in excess of 9, which are sometimes found in sodic soil containing Na_2CO_3, can be injurious to plant roots (Thorup, 1969), Stroehlein *et al.* (1978) suggested that the alfalfa yield response to sulfuric acid on an Na-saturated soil from Arizona was due to a reduction in pH from a toxic value of 10 to 8.2. Alfalfa failed to respond to gypsum on the Na-saturated soil. Other potential benefits of acidification are accelerated reclamation of soils high in B (Prather, 1977) and reduction of N losses by ammonia volatilization (Stroehlein *et al.*, 1978).

The Arizona research has clearly shown that acid is an effective amendment for calcareous, sodic soils and that it may be superior to gypsum for soils dominated by Na_2CO_3 or when the availability of Fe or P is low. The success of an acid treatment depends on proper identification of the problem and correct methods of aplication (Stroehlein *et al.*, 1978). Use of sulfuric acid is restricted to areas where there is a cheap supply such as in the southwesternUnited States, where the production of H_2SO_4 outstrips demand.

7.7 CONCLUSIONS

Although the impacts of sodicity on the structural behavior and the surface properties of soils have long been recognized, information on the role of sodicity on nutrient constraints to plant growth is still scarce. While limited information exists on the availability of macronutrients in sodic topsoils, subsurface sodicity limitations to crop growth are imperfectly known. Even less is understood about the availability of micronutrients in alkaline sodic soils, although basic, fundamental chemistry would predict their presence in unavailable forms. The effect of dissolved organic matter on metal mobility in sodic soil solutions, particularly in relation to plant nutrient availability, is poorly understood. It is, therefore, imperative that increased research efforts be directed to evaluating the micronutrient dynamics in sodic soil solutions in relation to dissolved organic matter.

8

Agricultural Management of Sodic Soils

J. D. Oster and N. S. Jayawardane

8.1 INTRODUCTION

Sodic soils usually have poor physical and chemical properties, particularly when the electrolyte or dissolved salt concentration of the soil solution (salinity) is inadequate to compensate for the effects of exchangeable sodium (Na) on clay swelling and dispersion (Shainberg and Letey, 1984; Rengasamy and Olsson, 1991; Jayawardane and Chan, 1994; Oster et al., 1995; Chapter 3). A commonly encountered physical problem is slow water infiltration which results in low soil water storage and, for irrigated agriculture, the need to irrigate more frequently (Oster and Singer, 1984; McKenzie et al., 1993; Chapter 5). When sodic soils are wet, problems of slow water entry into soil and slow internal drainage, poor aeration, trafficability, and compaction commonly occur because of their low hydraulic conductivities (HC) (Meyer et al., 1985; Ford et al., 1993). In dryland agriculture, low infiltration rates (IR) often restrict water intake during rain. The consequent enhanced runoff on sloping lands can increase erosion and, where surface drainage is poor, waterlogging on the lower lands. Under both irrigated and dryland conditions, the preparation and maintenance of seedbeds with good tilth (or aggregate size distribution) that fosters seed germination (Tyurin et al., 1960; Hoyle et al., 1972; Tisdall and Adem, 1988) is difficult. In addition, rainfall during the emergence stage of crop growth followed by a dry spell can cause strong crusts to develop which may reduce seedling emergence.

Chemical problems related to nutrition (Grieve and Maas, 1988; Naidu and Rengasamy, 1993; Chapter 7) which range from deficiencies [e.g., copper (Cu), zinc (Zn), manganese (Mn), iron (Fe)] to high Na levels that can cause calcium (Ca) deficiency in plants, can also occur. In this chapter, only nutritional problems related to Na effects on Ca deficiencies will be discussed while minor element nutrition is discussed further in Chapter 7.

Problems with the behavior of sodic and saline-sodic soils can be expected to increase in the future. The need to provide more food for an expanding population (Borlaug and Dowswell, 1994) and to dispose of drainage waters (Tanji and Karajeh, 1993) will result in the use of poorer quality waters and soils for food production. Furthermore, supplies of good-quality irrigation water will decrease because the development of new water supplies will not keep pace with the increasing needs of industries and municipalities (Bouwer, 1994). Some irrigation water currently used by agriculture will be used first by municipalities, and a portion of the remaining water with its higher salinity, sodicity, and pathogen content will then become available for use by agriculture. In addition, present allocations of irrigation waters to agriculture are being decreased in developed countries such as Australia and the United States (Westcot, 1988) to reduce and control the inevitable and adverse environmental impacts of irrigated agriculture on the quality of rivers and groundwaters (van Schilfgaarde, 1990), as well as to mitigate existing adverse impacts on the habitats for wildlife and fish (Oster, 1994a; Chapter 9). It follows that the need for understanding how to properly manage sodic and saline-sodic soils will increase in order to maintain and improve crop productivity in many areas of the world.

The emphasis in this chapter is on the application of the basic soil and plant sciences to reclamation and management practices used by farmers in developed and developing countries. Initially, the effects of sodicity and salinity on soil properties and crop production are discussed, followed by the strategies involving tillage, amendment, and cropping practices available for managing sodic soils. Finally, practices are discussed that are currently used by farmers in Australia, California, Canada, Israel, India, and Russia under both irrigated and dryland conditions, which have been deemed successful and consistent with research based knowledge and economic constraints. Recent reviews by Gupta and Abrol (1990a), Sumner (1993a), Jayawardane and Chan (1994), and Oster *et al.* (1995) have been used extensively in developing the current material.

8.2 SODICITY AND SALINITY EFFECTS ON SOIL PROPERTIES AND CROP PRODUCTION

Increasing crop production on sodic soils requires an understanding not only of the adverse impacts of sodicity on soil properties, but also of the consequent effects on root zone conditions for crop growth and of the role of cropping in managing and reclaiming sodic soils. Both water entry and its subsequent redistribution within the soil are essential for root and crop growth. Consequently, the focus of this section is on the interaction between soil hydraulic properties, salinity, and sodicity, which are closely linked, particularly for sodic soils. Initially, these linkages are described for infiltration rate, hydraulic conductivity, crust development, and hardsetting followed by a description of how salinity and sodicity are affected by mineral and exchange equilibria. The final subsection deals with reclamation of gypsiferous and calcareous sodic soils, including topics such as water and amendment requirements but with an emphasis on the positive role of cropping in reclamation.

8.2.1 Physical Properties

Infiltration rates and HC decrease with decreasing soil salinity and increasing exchangeable Na. Research conducted since 1954 has documented many instances in which the tendency for swelling, aggregate failure, and dispersion increases as electrolyte concentration decreases even if the exchangeable Na percentage (ESP) is less than 3, that is, a soil with low electrolyte concentration can behave like a sodic soil (Rengasamy *et al.*, 1984b; Shainberg and Letey, 1984; Sumner, 1993a; Chapter 5). These tendencies increase as ESP increases, requiring increasingly higher electrolyte concentrations to stabilize the soil. However, the boundary between stable and unstable conditions varies from one soil to the next (Pratt and Suarez, 1990). In addition, the stability boundary for water entry *into* the soil (infiltration) is different from that for water movement *through* the soil (unsaturated and saturated HC). The surface is more unstable than the underlying soil at low electrolyte and exchangeable Na (Shainberg and Letey, 1984) and magnesium (Mg) levels (Keren, 1991), because the mechanical impact and the stirring action of the applied water on the soil surface destroy soil aggregates and rearrange soil particles into a densely packed, thin soil layer (seal) on the surface (Chapter 5).

Whether soil structure is affected by electrolyte concentration, exchangeable Na, or Mg depends on hydration and on the balance between repulsive and attractive forces between soil particles, particularly clay sized particles (Sumner, 1993a; Rengasamy and Olsson, 1991; Shainberg and Letey, 1984; Chapter 3). Repulsive forces increase with decreasing electrolyte concentration and increasing exchangeable Na. During hydration, soil clays imbibe water or swell, which decreases the intensity of various types of bonds holding particles together, and when repulsive forces exceed attractive forces between soil particles, dispersion results (Chapter 3). Swelling reduces the radii of soil pores while dispersion after breakdown or slaking (Abu-Sharar *et al.*, 1987) and subsequent clay movement lead to the blockage of soil pores. Both swelling and dispersion reduce the saturated and unsaturated HC of the soil.

Hydraulic Conductivity (HC)

The pioneering work of Fireman and Bodman (1939) first demonstrated that increasing the electrolyte concentration of irrigation water applied to nonsodic soils increased HC. Subsequent work, as reviewed in Chapter 5 and by Sumner (1993a), identified many contributing factors, such as clay mineralogy, soil texture, organic matter, oxide mineral content, pH, and soil weathering, all of which influence the effects of electrolyte concentration and sodicity on soil hydraulic properties. The complex interactions between these factors have limited progress in developing mathematical models to predict the effects of electrolyte concentration and sodicity on HC (McNeal, 1968; Cass and Sumner, 1982a,b,c; Russo and Bresler, 1977). The prediction of electrolyte concentration and sodicity impacts on IR will be even more difficult because soil particles of all sizes at the soil surface are subject to rearrangement by irrigation water, rain, and tillage.

Predictions based on the Equivalent Salt Solution Series (ESSS) concept have been successful in predicting changes in saturated and unsaturated HC values for several soils (Jayawardane, 1979, 1992; Jayawardane and Blackwell, 1991). An ESSS for a given soil is defined as the combination of electrical conductivity (EC) and sodium adsorption ratio (SAR) of the soil water that produces the same extent of swelling and, hence, the same changes in pore geometry and HC. Here it is important to note that the numerical values of SAR and ESP are about equal in the range of 0–40 after equilibrium conditions are established between the soil solution and the exchanger phases. Limited data currently available indicate that the ESSS values are similar for soils with similar clay mineralogy. Increases in clay swelling and decreases in saturated HC brought about by decreasing electrolyte concentration of the soil solution, for a given soil, should be reversible; however, they often are not (McNeal and Coleman, 1966; McNeal *et al.*, 1968; Mitchell and Donovan, 1991). This suggests that an irreversible change in the soil matrix often occurs together with clay swelling. This change could be the result of clay particle movement and lodgment in the conducting pores.

Due to the limited ability to predict HC and the inability to predict IR, water quality guidelines for soil and water management (Table 8.1) are largely based on data obtained from laboratory, lysimeter, and field experiments, as well as from farmer experience as documented by researchers, extension specialists, and farm advisors (Ayers and Westcot, 1985; McKenzie *et al.*, 1993; Ford *et al.*, 1993; Oster *et al.*, 1995).

Quirk and Schofield (1955) introduced the concept of threshold electrolyte concentration (TEC), the salinity or electrolyte concentration of the soil solution at which a 10–15% decrease in HC occurred for a silty loam soil from the Sawyers Field at Rothamsted. A plot of TEC, expressed in units of electrical conductivity (EC) against ESP (SAR) resulted in an approximately linear line for SAR values between 0 and 60 (Fig. 8. 1), which adequately described the data. Electrolyte concentrations to the right of the line being greater than the TEC resulted in stable HC; in other words, decreases in HC were less than 10–15%. Electrolyte concentrations to the left of the line resulted in unstable HC due to swelling and/or dispersion. Since 1955, various authors (Pratt and Suarez, 1990; Sumner, 1993a) have summarized data for different soils where more than

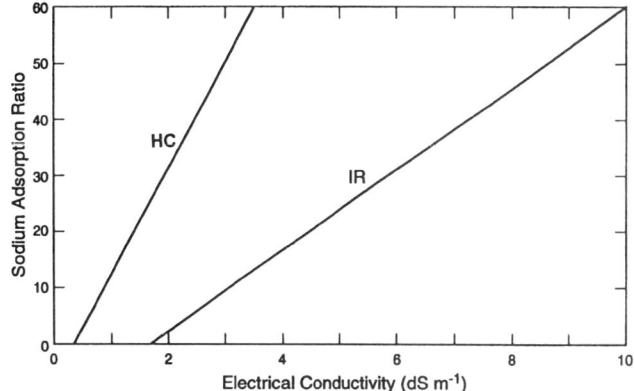

Figure 8.1 Effect of sodium adsorption ratio (SAR) and electrical conductivity (EC) on hydraulic conductivity (HC) and infiltration rate (IR). Lines labeled HC and IR represent combinations of SAR and EC, which resulted in 10–15% reduction in HC for a British Soil [Quirk and Schofield, *J. Soil Sci.* 6:163-178, (1955)] and 25% reduction in steady-rate IR of a cropped North Dakota loam soil (Udic Haploboroll) [Oster and Schroer, *Soil Sci. Soc. Am. J.* 43:444-447, (1979)].

a 20–25% reduction in HC occurred at electrolyte concentrations higher than the TEC for the British soil. Thus each soil responds to EC and SAR in a unique manner. In summary, significant reductions (10–25%) in HC can occur for soils with ESP values of 10–15 if saturated paste extract EC_c values are less than 0.5–5 dS m^{-1}. Based on research conducted since 1966, similar reductions can be expected for soils with SAR (or ESP) values as low as 3 if EC_e is less than 0.2–1 dS m^{-1} (Shainberg and Letey, 1984; Sumner, 1993a).

Infiltration Rate (IR)

When water is applied to the soil surface at a rate that exceeds the IR, whether by rainfall or by irrigation, some penetrates the surface and enters the soil, while the remainder which fails to penetrate either accumulates on the surface or runs off. Generally, the IR is high during the initial stages but decreases exponentially with time to approach a constant value (Fig. 8.2). Two main factors are responsible for this decrease in IR: (a) a decrease in the matric potential gradient which occurs as infiltration proceeds, and (b) the formation of a seal or crust at the soil surface (Chapter 5). In cultivated soils from semiarid regions, where the organic matter content is usually low, soil structure is unstable, and sealing is a major factor determining the steady-state IR (Duley, 1939; McIntyre 1968; Morin and Benyamini, 1977). Seal formation at the soil surface is, in turn, due to two processes: (a) physical disintegration of soil aggregates and soil compaction caused by the impact of water, especially water drops, and (b) chemi-

TABLE 8.1 Water quality guidelines: Combined effects of sodium adsorption ratio (SAR) and electrical conductivity of either a saturation extract (EC_e) or an irrigation water (EC_{iw}) on the likelihood of problems with low infiltration rates (IR) or hydraulic conductivities (HC) [Ayers and Westcot, *Water Quality for Agriculture*, Irrig. Drain. Pap. 29, Food and Agriculture Organization, Rome, (1985)]

SAR soil water	Potential water problem	
	Infiltration, unlikely if EC_e or EC_{iw} (dS m^{-1}) is	Hydraulic conductivity, likely if EC_e or EC_{iw} (dS m^{-1}) is
0–3	>0.7	<0.3
3.1–6	>1.0	<0.4
6.1–12	>2.0	<0.5
12.1–20	>3.0	<1.0
20.1–40	>5.0	<3.0

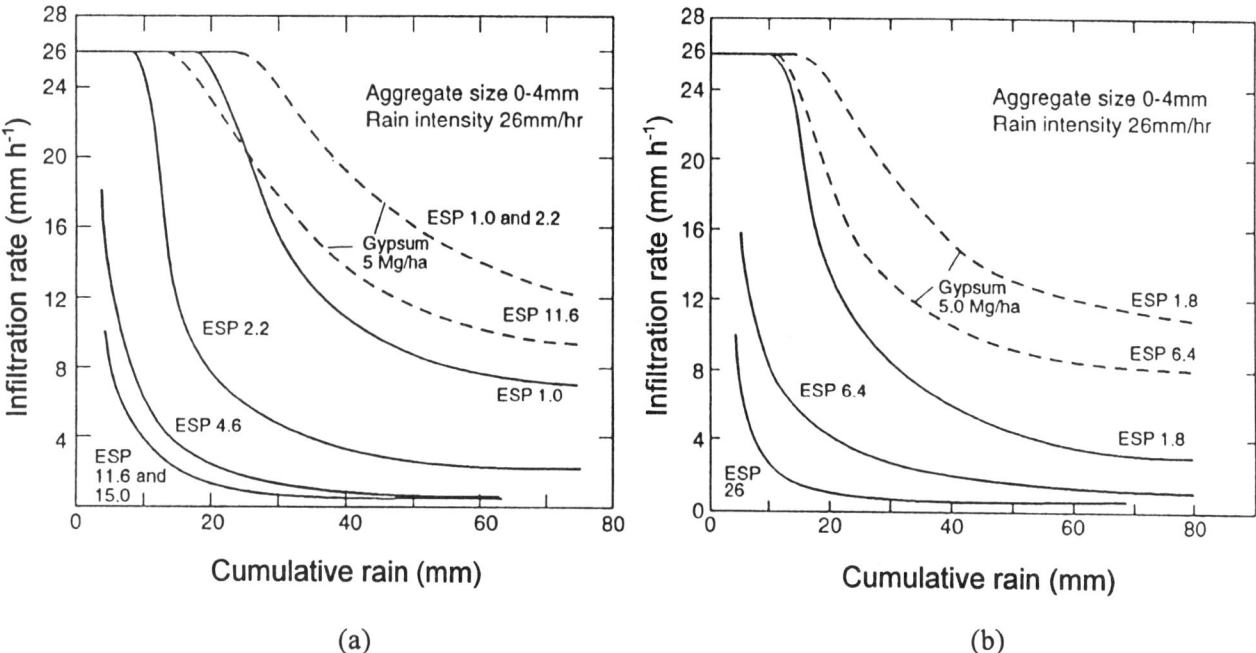

Figure 8.2 Impacts of exchangeable sodium percentage (ESP) and phosphogypsum (PG) on infiltration rate (IR) of (a) Hamra-Netanya sandy loam (Rhodoxeralf) and (b) loessial Nahal-Oz soil (Calcic Haploxeralf) [Kazman *et al.*, *Soil Sci.* 35:184-192, (1983)].

cal dispersion and movement of clay particles and the resultant plugging of conducting pores. Both these processes act simultaneously, with the first enhancing the second (Agassi *et al.*, 1981).

Chemical dispersion becomes severe when electrolyte concentration decreases below the critical flocculation concentration (CFC), at which clay minerals flocculate (Shainberg and Letey, 1984; Goldberg and Forster, 1990). Dispersion is enhanced by the mechanical stirring action of falling water drops, by overland water flow, and by the relative freedom of particle movement at the soil surface (Oster and Schroer; 1979; Rengasamy *et al.*, 1984b; Shainberg *et al.*, 1989). Rainfall can cause electrolyte concentration to fall below the CFC, resulting in enhanced dispersion and severe reductions in IR in irrigated arid regions as well as under humid, rainfed conditions (Sumner, 1993a). Similar conditions occur where nonsaline irrigation waters (~0.10 dS m^{-1}) are used for irrigation, such as along the east side of the Central Valley of California (Doneen, 1948).

It follows that the SAR$_{iw}$ and EC$_{iw}$ of an irrigation water also affect the IR. In studies in which waters of different qualities were applied to cropped columns of a loam soil (Udic Haploboroll), Oster and Schroer (1979) obtained a considerably better correlation between the final IR (FIR) and the SAR and EC of the applied water than those of the soil solution averaged either over the total length of the soil column (530 mm) or for the surface soil (0–76 mm). In

addition, for a given SAR, the boundary between stable and unstable conditions occurred at much higher electrolyte concentrations than that for HC (Fig. 8.1). Under rainfall, similar behavior occurs, as illustrated by the effect of ESP on the IR of a sandy loam in Fig. 8.2a (Kazman *et al.*, 1983). As the ESP increased from 1.0 to 2.2, the FIR fell sharply from 7.5 to 2.3 mm h^{-1} and reached a value of 0.6 mm h^{-1} at an ESP of 4.6. Similar results were obtained with a loessial silt loam (Fig. 8.2b). For both soils, spreading phosphogypsum (PG) on the soil surface which readily dissolves, increasing the EC and the Ca concentration of the soil solution at the soil surface, was effective in reducing seal formation and the associated reduction in IR. Magnesium is not as effective as Ca in improving IR. In fact at a given ESP, replacing exchangeable Ca by Mg enhanced the rate of IR decline and reduced the FIR for the soils used to obtain the data in Fig. 8.2 (Keren, 1991). This specific effect of Mg is due to the difference in size between hydrated Mg and Ca ions, with resulting differences in the strength of attraction to cation exchange sites (Chapter 3). Hydrated Mg, which is larger than hydrated Ca, decreases the linkages between external surfaces within a soil aggregate, decreasing in turn the amount of raindrop energy needed to break down soil aggregates.

Although out of sequence, the subject of HC will be revisited as it is also affected by Mg (McNeal *et al.*, 1968; Emerson and Chi, 1977; Sumner, 1993a). Magnesium enhances the effects of Na on dispersion and HC, particularly at low ESP (Fig. 8.3). The influence of Mg on HC is clear until ESP reaches 12, beyond which the Na effects predominate. Similar results have been reported by Rengasamy *et al.* (1986) and Levy *et al.* (1988). Finally, Alperovitch *et al.* (1981, 1986) compared the effects of Mg and Ca for three noncalcareous soils at SAR (ESP) values ranging from 5 to 20 and found that Mg caused greater clay dispersion and a larger reduction in HC. The same authors also were able to show that Mg enhanced lime dissolution in calcareous soils, thereby partially mitigating the deleterious effects of Mg on HC. Based on these and other data, Sumner (1993a) concluded that Mg is "likely to enhance sodic behavior when very pure waters are applied to highly weathered soils with ESP values < 20."

Crusts, Hardsetting and Soil Strength

Topsoil aggregates that are not water stable, slake or slump and disintegrate during wetting because of stresses set up by rapid water uptake, release of entrapped air, heat of wetting, mechanical disturbance caused by rapidly flowing irrigation water, or impact of raindrops. Slaking results in a reduction in the number and size of large pores at the soil surface as well as in the dispersion of clay and silt particles. Upon drying, a thin, hard surface crust forms, called a structural crust (Fig. 8.4) (Moore and Singer, 1990). A thicker crust may form when sediment-laden water infil-

Figure 8.4 Four stages in structural crust formation of silty clay loam (Typic Chromoxerert) during rainfall: (a) Uncrusted surface, (b) After 10 min, (c) after 25 min, and (d) after 90 min of rainfall [Moore and Singer, *Soil Sci. Soc. Am. J.* 54:1117-1123, (1990)].

trates into the soil (Shainberg and Singer, 1986), causing the sediment to be deposited on the soil surface, and forming what is called a depositional crust. Crusts reduce IR, increase runoff and erosion, and decrease seedling emergence during seed germination and plant establishment phases of crop growth (Chapter 5).

Hardsetting, another type of surface condition with consequences similar to crusting, occurs after wetting soils that do not contain stable aggregates. The major difference between hardsetting and crusting soils is the lack of structural stability in hardsetting soils, which permits complete aggregate breakdown and clay movement within the entire Ap horizon, whereas in crusting soils, clay mobility is manifest only in the top few millimeters of the soil. Hardsetting soils exhibit a massive, compact, and hard surface condition which forms on drying (Mullins *et al.*,1990). The dried surface is not disturbed or indented by the pressure of a forefinger. Soil conditions that facilitate hardsetting include low organic matter content and a texture conducive to high bulk density ρ_b development. This is particularly true for loamy sands, sandy loams, sandy clay loams, and sandy clays that possess a clay mineralogy dominated by kaolinites, micas, or both (Mullins *et al.*, 1990). These soils do not crack on drying. The soil strength that develops during drying of hardsetting soils may arise from the effect of matric potential acting within interparticle and

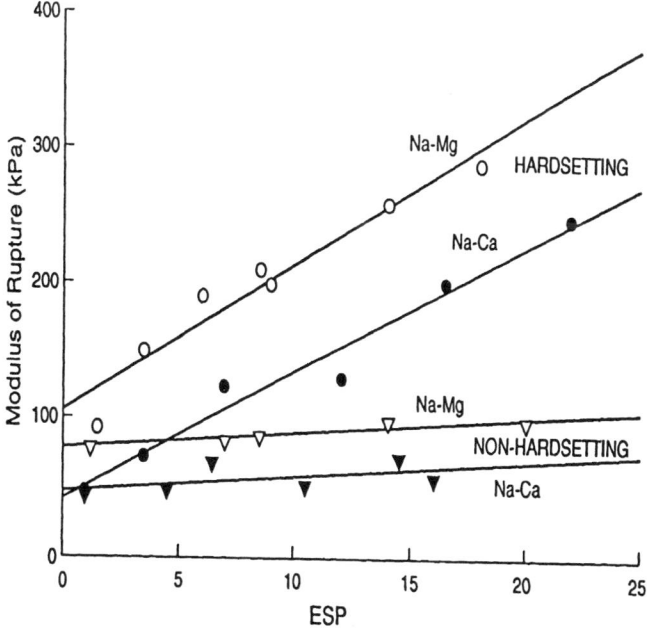

Figure 8.3 Effect of exchangeable sodium percentage (ESP) and Ca/Mg ratios on hydraulic conductivity (HC) of a Vertisol (72% clay) from northern NSW, Australia [Horn, Dip. Agr. Sci. Thesis. University of New England, Armidale, Australia, (1983)].

interaggregate bridges formed from material mobilized during wetting, including silts and clays, and from precipitation of cementing agents such as amorphous silica, imogolitelike aluminosilicates, feldspathoid minerals, and silica-Fe complexes (Chartres *et al.*, 1990).

Exchangeable Na and Mg and electrolyte concentration have been implicated in hardsetting behavior in Australia as demonstrated by Aylmore and Sills (1982), who obtained greater slopes of the modulus of rupture (MOR)-ESP relationship with Mg than with Ca as the dominant cation (Fig. 8.5). Based on a replotting of these data, Sumner (1993a) showed that the MOR decreased with increasing EC. From the perspective of agronomic management, hardsetting of the surface soil invariably results in reduced seedling emergence, as So and Aylmore (1993) clearly demonstrate in Fig. 8.6.

Both irrigation and dryland farmers have used gypsum to alleviate hardsetting and crust formation (Doneen, 1948; Simms and Rooney, 1965; Loveday and Ayers, 1994) which is consistent with research results obtained from studies conducted in Australia, Israel, South Africa, and the United States. Several groups of researchers (Loveday and Bridge, 1983; Shainberg *et al.*, 1989; Sumner, 1993a; Ford *et al.*, 1993) have shown that increased electrolyte concentration and exchange of Ca for Na and Mg resulting from gypsum dissolution are the reasons that gypsum is effective. Infiltration rates are increased, reducing waterlogging on the soil surface, a condition that is not conducive to seed germination and, generally, delays farm operations. Hydraulic conductivities are also enhanced, facilitating water movement to the soil surface as the soil dries and reducing

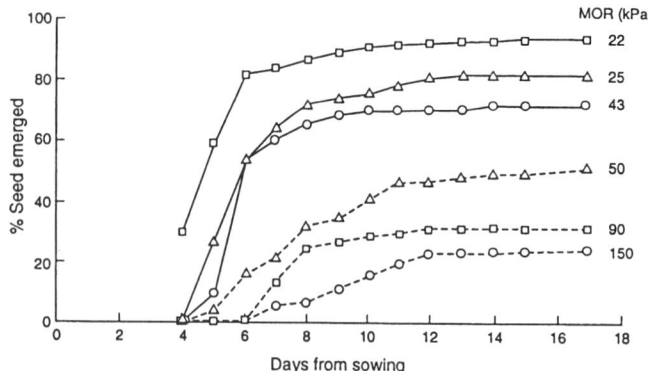

Figure 8.6 Effect of modulus of rupture (MOR) on seedling emergence in hardsetting soils from Western Australia [So and Aylmore, *Aust. J. Soil Res.* 31:761-777, (1993)].

the rate of drying, crust formation, and hardsetting (So and Aylmore, 1993). However, the effects of surface-applied gypsum are temporary because gypsum is leached from the soil surface as it dissolves so that periodic applications (2–5 Mg ha^{-1}) are usually necessary to maintain favorable levels of electrolyte and to reduce sodicity (Oster and Singer, 1984; Jayawardane and Chan, 1994).

The injection of gypsum directly into the irrigation water at rates ranging from 0.17 to 0.34 kg m^{-3} represents an alternative to periodic gypsum application to the soil (Oster *et al.*, 1992). Reliable and reasonably priced equipment is now available to inject gypsum at controlled rates into irrigation water. In addition, appropriate management practices following gypsum application can prolong its benefits, leading to a reduced need for repeated applications. For example, Greene and Wilson (1989) demonstrated that, over a period of 3.5 years, the beneficial effects of gypsum on clay dispersion were lost as a result of leaching, but because the establishment of pasture protected the surface from the impacting raindrops, no loss in HC was recorded. The application of organic matter in conjunction with gypsum is effective in promoting stable aggregation (Muneer and Oades, 1989a,b; Chapter 4). Taylor and Olsson (1987) and Quirk (1978) concluded that increased levels of organic matter arising from pasture root systems stabilize soil structure after gypsum is no longer present at the soil surface. Similar results were obtained in California (Oster, 1994b) with a hardsetting sandy loam soil (thermic Typic Haploxeralf) in a sprinkler-irrigated vineyard. Growing an intercrop of blando brome (*Bromus mollis*), a self-seeding winter annual, proved to be just as effective as surface application of 2 Mg ha^{-1} of phosphogypsum (PG) to the freshly tilled soil surface every other year. The adoption of farming practices such as minimum tillage and direct drilling without tillage lead to increased retention of crop residues in the form of surface mulches. This encourages soil

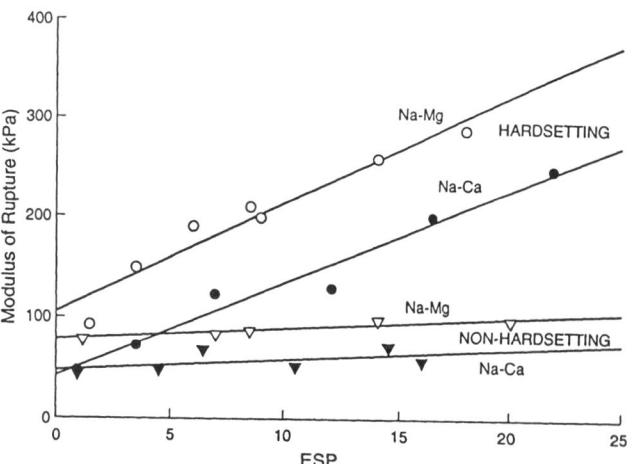

Figure 8.5 Relationship between modulus of rupture (MOR) and exchangeable sodium percentage (ESP) for Na-Ca and Na-Mg systems of hardsetting and nonhardsetting soils from Western Australia [Aylmore and Sills, *Aust. J. Soil Res.* 20:213-224, (1982)].

faunal activity which helps increase and maintain the continuity of biopores, which in turn conduct water and air to subsoils (Tisdall, 1991; Jayawardane and Chan, 1994).

8.2.2 Chemical Properties

Because IR and HC values tend to deteriorate when the electrolyte concentration is insufficient to compensate for the deleterious effects of Na, a pertinent question arises: What are the factors that determine the lowest electrolyte concentration that can occur under field conditions? In irrigated agriculture with no rainfall, the lowest level is determined by the concentration of the irrigation water and usually occurs near the soil surface. If soils receive both irrigation water and rainwater, electrolyte concentrations fluctuate, particularly near the soil surface. Under all of these conditions, reactions between the solid and liquid phases of the soil greatly impact the minimum electrolyte concentration that can occur.

Soil Mineral Equilibria

Leaching with rainfall or with nonsaline irrigation water will not decrease the electrolyte concentration below a level that is strongly affected by soil mineral dissolution. Dissolution, in turn, depends on the soluble minerals present in the soil and on associated chemical equilibria that involve the compositions of the solid, exchanger, liquid, and gaseous phases of the soil (Oster, 1982). Under rainfall conditions, extensively leached noncalcareous soils release from 3 to 5 $mmol_c L^{-1}$ of Ca and Mg to the soil solution as a result of exchangeable cation hydrolysis and the dissolution of plagioclase feldspars, amphiboles, pyroxenes, and other minerals (Rhoades *et al.*, 1968; Oster and Shainberg, 1979; Chapter 5). Dissolution in soils containing only calcite, gypsum, or both can maintain Ca^{2+}, bicarbonate (HCO_3^-), and sulfate (SO_4^{2-}) concentrations at higher levels, depending on the exchangeable ion composition and the partial pressure of carbon dioxide P_{CO_2} (Oster, 1982).

For these soils, the minimum electrolyte concentration, expressed as EC_e, that can be achieved by leaching increases linearly with soil solution SAR for a fixed Mg-Ca ratio and P_{CO_2} (Fig. 8.7). The two sets of solid lines on the left and right sides of Fig. 8.7 are for calcareous and gypsiferous soils, respectively. For these four lines, the only sources of electrolyte are those generated by the dissolution of calcite and gypsum, respectively. The line labeled Mg = 0, P_{CO_2} = 0.032 kPa represents the calculated linkage between the minimum EC_e and SAR of the soil solution at that partial pressure for a calcareous soil that does not contain any chloride (Cl) or SO_4 salts. The next solid line to the right represents the EC_e/SAR linkage where concentrations of

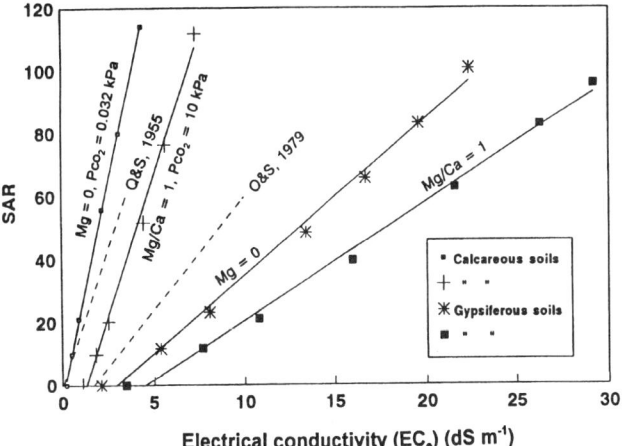

Figure 8.7 Impact of sodium adsorption ratio (SAR), partial pressure of CO_2 (P_{CO_2}), and Mg/Ca ratio on minimum electrial conductivities (EC) for calcareous and gypsiferous soils (solid lines). The dashed line labeled Q&S, 1955 represents the combination of EC and SAR that resulted in a 10–15% reduction in HC of a British soil [Quirk and Schofield, *J. Soil Sci.* 6:163-178, (1955)]; the dashed line labeled O&S, 1979 represents the combinations of EC and SAR that resulted in a 25% reduction in stady-state IR of North Dakota loam soil [Oster and Schroer, *Soil Sci. Soc. Am. J.* 43:444-447, (1979)].

Ca and Mg are equal and P_{CO_2} = 10 kPa. These two lines bracket the likely electrolyte minima for a calcareous soil, which can occur after extensive rainfall. During reclamation with irrigation water, these minima would be increased by an amount approximately equal to the contribution of Cl and SO_4 salts in the irrigation water. Because gypsum is considerably more soluble than calcite, the minimum soil solution concentrations in equilibrium with gypsum (lines labeled Mg = 0 and Mg/Ca = 1 in Fig. 8.7) are greater than for soil solutions in equilibrium with only calcite. Magnesium also increases the minimum electrolyte concentration, but P_{CO_2} has no effect. Because of common-ion effects between Ca, HCO_3, and SO_4, the minimum concentrations for gypsiferous soils are independent of both P_{CO_2} and the presence or absence of calcite. During reclamation with irrigation water, only the Cl salts in the water would increase the minimum electrolyte concentration of a gypsiferous sodic soil.

The pH also depends on equilibrium constraints imposed by the presence of calcite in the same way as for mimimum electrolyte concentration (Gupta *et al.*, 1981). The corresponding pH values for the line labeled Mg = 0, P_{CO_2} = 0.032 kPa in Fig. 8.7 range from 8.4 at SAR = 0 to 9.6 at SAR = 100. For the next solid line to the right, the pH values range from 6.9 at SAR = 0 to 7.6 at SAR = 100. The difference in pH between the two lines results from the higher P_{CO_2} for the lower line (10 kPa vs. 0.032 kPa), as well as from the higher concentration of Mg. Because P_{CO_2} in the root zone is a dynamic parameter which depends on microbial and root respiration and on the soil water content, values may range from 0.32 near the soil surface to about 10 kPa in the lower portions of a rapidly respiring root system (Buyanovsky and Wagner, 1983). Increases in P_{CO_2} benefit the HC of sodic soils for two reasons: (a) the minimum electrolyte concentration is higher, and (b) the pH is lower (Suarez *et al.*, 1984; Chapter 5). Consequently, cropping aids reclamation provided that the amount of infiltrated water exceeds the amount used by the crop.

Physics and Chemistry of HC and IR: A Synthesis

The dashed lines in Fig. 8.7 labeled "Q&S, 1955" and "O&S, 1979" provide quick comparisons of the minimum TEC values for HC and IR for calcareous and gypsiferous soils, respectively. Combinations of EC/SAR to the right of the Q&S, 1955 line are likely to result in less than 10–30% reduction in HC whereas those to the right of the O&S, 1979 line are likely to result in less than 20–30% reduction in FIR. These comparisons can be summarized as follows:

1. Minimum electrolyte concentrations for calcareous soils during irrigation with low-salinity irrigation waters, or after extensive rainfall, may not be adequate to meet TEC requirements for HC unless P_{CO_2} is enhanced by cropping (Robbins, 1986) or the soil contains significant levels of Mg (Alperovitch *et al.*, 1981). Furthermore, minimum electrolyte concentrations for calcareous soils will not meet the TEC requirements for IR for soils where the soil solution SAR near the surface exceeds 5, unless the irrigation water contains sufficient Cl and SO_4 salts to compensate.

2. For soils that contain gypsum, minimum electrolyte concentrations can be expected to meet the TEC requirements for HC, but this may not be true for IR. Electrolyte concentrations at the surface will depend on the soil gypsum content, quantity of gypsum applied to the surface, gypsum dissolution kinetics, and IR (Oster, 1982), as well as on the presence of any Cl salts in the irrigation water.

8.2.3 Reclamation of Sodic Soils

Although salt transport models have provided considerable insight into the chemistry and physics of reclamation (Dutt *et al.*, 1972; Tanji *et al.*, 1972; Robbins *et al.*, 1980; Simunek and Suarez ,1994), suitable mathematical models are not available to describe, on a field scale, the tran-

sitory effects of tillage on IR and HC, and the effects of variable rates of water flow in soil among soil pores of differing sizes on the kinetics of reactions involving cation exchange, mineral dissolution and precipitation, and CO_2 equilibria. Consequently, although many of the basic physicochemical processes are understood and can be described mathematically, reclamation guidelines are also based on experimental relations obtained from field and lysimeter reclamation experiments.

Reclamation of Gypsiferous Sodic Soils

Gypsum, either incorporated into the soil or left on the surface, is the Ca source most commonly used to reclaim sodic soils and to improve low water infiltration due to low electrolyte concentration, high sodicity, or both. Sources include mineral deposits as well as PG and flue gas desulfurization gypsum (FDG), by-products of the phosphate fertilizer and power generation industries. When gypsum is incorporated into the soil, the reduction in ESP upon irrigation and leaching is primarily limited to the soil depth interval in which it is present (Oster and Frenkel, 1980) (Fig. 8.8a), which is a consequence of the greater selectivity of exchange sites for Ca than Na. In addition, the exchanger phase is an effective sink for Ca which replaces exchangeable Na. Consequently, electrolyte concentrations expressed as EC_e in gypsiferous soils depend on exchangeable Na levels and can considerably exceed the EC_e (2.3 dS m^{-1}) of a saturated gypsum solution in distilled water (Fig. 8.8b). The EC_e of the soil solution within and below the gypsum-amended layer decreases as the ESP within the amended layer decreases (Frenkel *et al.*, 1989). In a soil

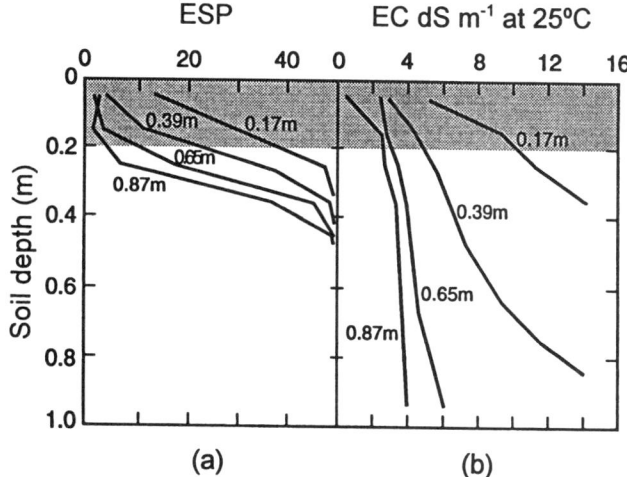

Figure 8.8 Computer-model results (ESP and EC) for reclamation of a soil (initial ESP = 50; CEC = 200 mmol$_c$ hg^{-1}) with gypsum and water (EC = 0). Numbers next to each line are depths of applied water [Oster and Frenkel, *Soil Sci. Soc. Am. J.* 44:41-45, (1980)].

amended with gypsum, only as reclamation approaches completion within the amended soil layer does the gypsum that dissolves begin to replace exchangeable Na at greater depths. During this second phase of reclamation, the required TEC levels in lower soil layers may not be maintained, resulting in decreased HC, which during partial soil reclamation is consistent with the HC data which Robbins (1986) obtained from 1 m long lysimeters containing a calcareous sodic silt loam soil (Xerollic Calciorthid). He found that (a) the greatest reclamation occurred within the upper 0.2 m of the uncropped gypsum-treated lysimeters, the layer in which adequate gypsum was incorporated to reclaim the upper 0.5 m of soil (Fig. 8.9a), and (b) the water flow rate declined to near zero in the gypsum-treated lysimeters after reclamation in the upper 0.2 m was complete and one pore volume of drainage had occurred.

Because of the high EC_e values of sodic soils which contain gypsum, reclamation is usually successful by leaching without additional amendments. For example, IR after 48 h ranged from 7 to 9 mm h^{-1} during reclamation of two clay loam soils (30 < ESP < 40) under ponded conditions using an irrigation water with EC_{iw} of 2.5 dS m^{-1} and an SAR_{iw} of 6 (Reeve *et al.*, 1948). The gypsum content in the upper 600 mm of soil was sufficient to replace most of the exchangeable Na. In the same study, surface application of gypsum increased the IR from 8 to 14 mm h^{-1} for a clay soil which did not contain sufficient gypsum. In a lysimeter experiment, Jury *et al.* (1979) reported similar successes with sandy loam and clay loam sodic soils which contained gypsum, both from continuous applications of water using surface ponding and from daily applications at an unsaturated rate of about 0.4 mm h^{-1}, with EC_{iw} values of 0, 0.5, and 1.3 dS m^{-1}. Infiltration rates were unaffected by EC_{iw}, even in the most extreme case, namely, a clay loam soil reclaimed with distilled water which sustained an IR of 4.5 mm h^{-1} for 3.6 pore volumes of drainage.

Reclamation of sodic soils with underlying gypsiferous layers is possible using deep plowing techniques (Rasmussen *et al.*, 1972; Wetter *et al.*, 1987). In locations where Na-affected surface soils are underlain by soil containing significant quantities of gypsum, deep plowing has been effective in breaking up and mixing the layers that supply soluble Ca to aid in reclamation. The depth of plowing required may vary from 0.5 to more than 1.0 m, depending on the concentration and depths of the Na-and Ca-rich layers. A procedure is available to predict the opti-

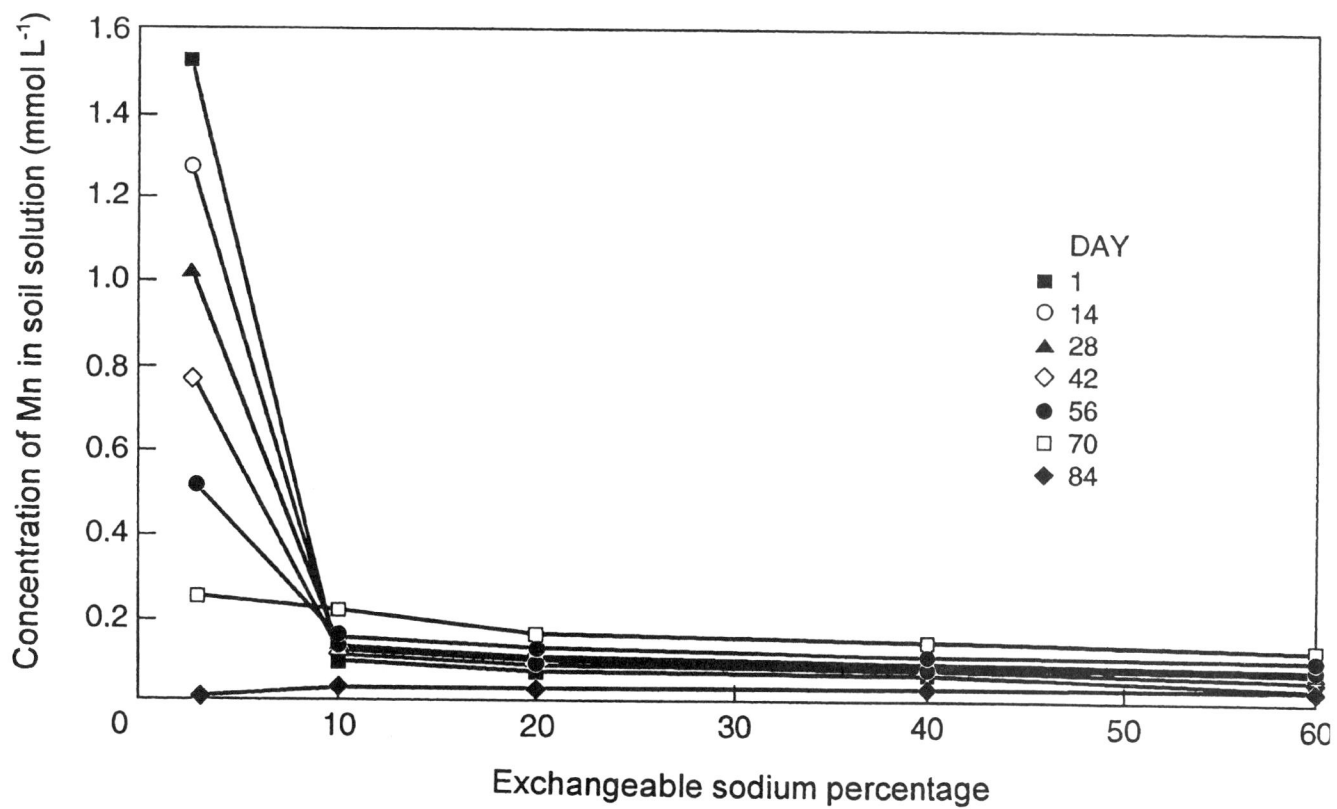

Figure 8.9 Final distribution of exchangeable Na levels with soil depth resulting from reclamation treatments of Xerollic Calciorthid. (a) Uncropped, and (b) cropped treatments [Robbins, *Agron. J.* 78:916-920, (1986)].

mum depth of plowing to maintain adequate permeability during the reclamation process (Rasmussen and McNeal, 1973).

The water requirement for the reclamation of the gypsiferous soils in the experiments of Reeve *et al.* (1948) and Jury *et al.* (1979) was the same as when the method proposed by Hoffman (1986) was used. Hoffman (1986) summarized leaching results of several field reclamation studies and proposed the following relationship between the fraction of initial electrolyte concentration remaining in the profile, C/C_0, and the depth of water infiltrated, D_w, through a given depth of soil, D_s:

$$\frac{C}{C_0} \cdot \frac{D_w}{D_s} = K \qquad [8.1]$$

where K differs with the soil type. Hoffman (1986) obtained a value of $K = 0.3$ for the reclamation of five saline clay loams during ponding. Three of the five reclamation studies were those conducted on gypsiferous sodic soils by Reeve *et al.* (1948), suggesting that gypsum in sodic soils does not affect the water requirement for the reclamation of saline soils under ponded conditions. The K value for the lysimeter experiment conducted by Jury *et al.* (1979) is 0.29 (Oster, 1993), a figure in close agreement with that obtained by Hoffman (1986). Thus a K value of 0.3 applies for gypsiferous soils across a spectrum of soil textures ranging from sandy loam to clay. The data of Jury *et al.* (1979) also suggest that the effects of bypass water flow, hydrodynamic dispersion, and reaction kinetics are the same for this range of soil textures where water application rates ranged from 0.04 to 8 mm h^{-1}.

Reclamation of Calcareous Sodic Soils

In a lysimeter study, Robbins (1986) demonstrated that the solubility of calcite under cropped conditions can be sufficient to reclaim a nonsaline calcareous sodic soil. He measured the P_{CO2} in the root zone during reclamation of a nonsaline ($EC_e = 2.4$), sodic ($ESP = 33$), calcareous silt loam (Xerollic Calciorthid) ($CEC = 210$ mmol$_c$ kg^{-1}) under cropped conditions and demonstrated conclusively the benefits of increased levels of CO_2 in the soil atmosphere to provide sufficient electrolyte and Ca to maintain an adequate HC, with a concurrent reduction of exchangeable Na. In the case of lysimeters cropped with a hybrid sorghum (*Sorghum bicolor* var. Sordan), but not amended with gypsum, the amount of Na removed exceeded that of uncropped lysimeters amended with gypsum. In addition, for a given amount of leachate, the total amount of Na leached in the cropped lysimeter exceeded that of the uncropped lysimeter amended with gypsum. For the lysimeter cropped with alfalfa (*Medicago sativa*) but not amended with gypsum and for a given amount of leachate, the amounts of Na removed were similar to those for the uncropped lysimeter amended with gypsum. Reclamation of the uncropped gypsum-amended lysimeters (Fig. 8.9a) occurred primarily in the zone where gypsum was incorporated, whereas for cropped lysimeters, it occurred throughout the root zone (Fig. 8.9b). Leaching the soil before cropping (Fig. 8.9b, leach + sorghum) resulted in low HC because of inadequate electrolyte concentration and subsequent poor plant growth. The successful use of calcite as a chemical amendment to reduce the impacts of sodicity on the physical properties of soil reported by McKenzie *et al.* (1993) in soils with pH values less than 7.5 is consistent with the findings of Robbins (1986).

Crop Selection

Crops grown during sodic soilreclamation must tolerate both poor soil physical properties (Gupta and Abrol, 1990b) and Na-induced Ca deficiency characterized by necrotic curled leaf tips and heavily serrated leaves (Carter *et al.*, 1979a; Maas and Grieve, 1987; Grieve and Maas, 1988). High Na concentrations reduce the amount of Ca that is available for plant uptake (Cramer and Läuchli, 1986). Calcium concentrations that are adequate under nonsodic conditions (1–2 mmol$_c$ L^{-1}) can thus become inadequate when the Na/Ca ratio of the soil solution is high (Bernstein, 1975). Seedling growth stages of cereals are particularly susceptible to Ca deficiency, though considerable variability occurs among crops (Grieve and Maas, 1988). Variability also occurs among genotypes of sorghum (Grieve and Maas, 1988), rice (*Oryza sativa*) (Grieve and Fujiyama, 1987), triticale (*X. triticosecale*) (Norlyn and Epstein, 1984), and wheat (*Triticum aestivum*) (Kingsbury and Epstein, 1986). The data for rice are of interest because it is often recommended for growing during reclamation of sodic soils in India (Gupta and Abrol, 1990b). For the rice cultivars M9 and M-201, Grieve and Fujiyama (1987) reported severe Ca deficiencies at an osmotic potential of -0.4 MPa ($EC_e = 10$ dS m^{-1}) and at a Na/Ca molar ratio of 78 (SAR = 87), whereas Yeo and Flowers (1985) reported no Ca deficiencies at Na/Ca molar ratios as high as 500 (SAR = 160) for cultivar IR2153. Based on limited data for seedlings of different crops, the general order of increasing susceptibility to Ca deficiency is rice cultivar IR2153, cotton (*Gossypium hirsutum*), barley (*Hordeum vulgare*), wheat, rye (*Secale cereale*), sorghum, and cowpea (*Vigna unguiculata*) (Grieve, 1993. Private communication). The wide diversity among crops and crop genotypes increases the likelihood that field trials conducted under nonsaline sodic conditions will identify local crops that are adaptable to these conditions.

Extensive areas of the Indo-Gangetic plains of northern India have calcareous sodic soils. Field plot studies at the Central Soil Salinity Research Institute near Karnal conducted since 1970 provide the crop tolerance results shown in Table 8.2 (Gupta and Abrol, 1990a). Differential, nonsaline sodic conditions were obtained by applying different amounts of gypsum to the soil and leaching by ponding for about 20 days (Abrol and Bhumbla, 1979). The ESP values ranged from about 10 to 70, with a corresponding range in SAR of about 7–160. Crops were grown under both rainfed and irrigated conditions, with recommended agronomic, fertilization, and plant protection practices used for all treatments. Rankings in Table 8.2 are based on 50% yield reductions, with rankings of rice, barley, and wheat similar to those reported by Grieve and Maas (1988). However, because of crop genotype variations and varying experimental conditions, this similarity could be fortuitous. In addition to the sodicity-tolerant crops listed in Table 8.2, grasses can also be grown on sodic soils. Karnal or Kallar grass (*Diplachne fusca*), Rhodes grass (*Chloris gayana*), and Para grass (*Brachiaria mutica*) have been reported to be highly tolerant of sodicity (Kumar and Abrol, 1986). In addition, Karnal and Para grasses grow well under ponded conditions.

Because cropping facilitates sodic soil reclamation, it is a commonly used practice throughout the world. In developed countries, the farming practices generally include tillage in conjunction with cropping, application of excess water during the cropping season, and application of chemical amendments if necessary and economical. In developing countries, tillage is often minimal and rice is commonly grown under ponded conditions during reclamation with or without the use of amendments,

depending primarily on affordability (Mehta, 1985; Singh, 1985, Gupta and Abrol, 1990a). For calcareous soils, additional amendments are often not used. Benefits from cropping include the following: (a) increased P_{CO2} due to plant root respiration and decomposition of organic matter (Kelley, 1951; Chhabra and Abrol, 1977; Robbins, 1986; Gupta *et al.*, 1989), which in turn increases EC and decreases pH in calcareous soils, (b) *in situ* production of polysaccharides and fungal hyphae, which, in conjunction with differential dewatering at the root-soil interface, promote aggregate stability (Boyle *et al.*, 1989; Kay, 1990; Tisdall, 1991), (c) the physical effects of root action (Chhabra and Abrol, 1977), including removal of entrapped air from the larger conducting pores (McNeal *et al.*, 1966b), generation of alternate wetting and drying cycles, and creation of macropores and release of CO_2 upon decomposition, and (d) financial or other benefits from crops grown during reclamation that can help support farming operations. The choice of which crop to use depends on the salinity levels associated with the soil to be reclaimed as well as on the level of sodicity (Oster *et al.*, 1995).

Gypsum Requirement and Amendment Equivalents
The gypsum requirement (GR) for reclamation, in $Mg\,ha^{-1}$, can be calculated using the equation

$$GR = 0.0086\ FD_s\rho_b(CEC)\ (ESP_i\text{-}ESP_f) \qquad [8.2]$$

where

F (unitless)	=	Ca-Na exchange efficiency factor
D_s	=	the soil depth in m
ρ_b	=	soil bulk density in $Mg\ m^{-3}$
CEC	=	cation exchange capacity ($mmol_c\ kg^{-1}$)
ESP_i, ESP_f	=	initial and final ESP values

The efficiency factor ranges from 1.1 for an ESP_f of 15 to 1.3 for an ESP_f of 5 (Oster and Frenkel, 1980). The SAR_e (saturation extract) can be substituted for ESP in the range of $0\ SAR_e^-50$. The Schoonover laboratory procedure, which determines the amount of Ca required to replace all the exchangeable Na plus any precipitated by soluble HCO_3^- and CO_3^{2-} ions in the soil (USSL Staff, 1954, procedure 22d), has also been used to determine GR, but the values tend to be high. Since these soluble ions are leached during sodic soil reclamation, Abrol *et al.* (1975) proposed a modified procedure, which involves leaching soil with ethanol to remove soluble HCO_3^- and CO_3^{2-}, followed by equilibration with a $CaSO_4$ solution. The difference between the Ca concentration in the $CaSO_4$ solution and the Ca plus Mg concentration in the clear filtrate gives (CEC) (ESP_i-

Table 8.2 Relative tolerance of crops to sodicity [Gupta and Abrol, *Adv. Soil Sci.* 11:223-288, (1990a)]

SP range	Crops[a]
10–15	safflower, mash, pea, lentil, pigeon-pea, curd bean
16–20	bengal gram, soybean
20–25	groundnut, cowpea, onion, pearl millet
25–30	linseed, garlic, guar
30–50	indian mustard, wheat, sunflower, berseem, hybrid napier, guinea grass
50–60	barley, sesbania, saftal, panicums
60–70	rice, para grass
70+	karnal, rhodes, and bermuda grasses

[a] Yields are about 50% of the potential yields in the respective sodicity ranges

ESP$_f$) in Equation [8.2], or (CEC)(ESP$_i$) since ESP$_f$ is zero after equilibration with the CaSO$_4$ solution.

Other amendments used for calcareous sodic soils include sulfuric acid (H$_2$SO$_4$) and acid-forming materials such as sulfur (S), lime sulfur, pyrite (FeS$_2$), and iron and aluminum sulfates [FeSO$_4$, Fe$_2$(SO$_4$)$_3$, Al$_2$(SO$_4$)$_3$]. These amendments react with calcite, providing a soluble source of Ca within the soil. Equivalents of chemically pure amendments relative to a unit of gypsum or S are presented in Table 8.3.

The following chemical reactions illustrate how various amendments react with calcareous sodic soils. In these reactions X represents the soil exchange phase.

1. Inorganic reactions

Gypsum (CaSO$_4$·2H$_2$O),
$$2NaX_{(solid)} + CaSO_4 \cdot 2H_2O_{(} \rightleftharpoons$$
$$CaX_{2(solid)} + Na_2SO_{4(aq)} + 2H_2O_{(aq)} \qquad [8.3]$$

Sulfuric acid (H$_2$SO$_4$)
$$2NaX_{(solid)} + H_2SO_{4(aq)} + CaCO_{3(solid)} \rightleftharpoons$$
$$CaX_{2(solid)} + Na_2SO_{4(aq)} + CO_{2(gas)} + H_2O_{(aq)} \qquad [8.4]$$

Iron sulfate (FeSO$_4$·7H$_2$O)
$$4NaX_{(solid)} + 2FeSO_{4(aq)} + 2CaCO_{3(solid)}$$
$$+ \tfrac{1}{2}O_{2(gas)} \rightleftharpoons 2CaX_{2(solid)} + 2Na_2SO_{4(aq)}$$
$$+ Fe_2O_{3(solid)} + 2CO_{2(gas)} \qquad [8.5]$$

1. Microbiologically mediated reactions

Sulfur (S$_8$),
$$4NaX_{(solid)} + 2S_{(solid)} + 2CaCO_{3(solid)} + 3O_{2(gas)} \rightleftharpoons$$
$$2CaX_{2(solid)} + 2Na_2SO_{4(aq)} + 2CO_{2(gas)} \qquad [8.6]$$

Table 8.3 Amendments equivalent to a unit of either gypsum or sulfur

		Gypsum	Sulfur
Gypsum	CaSO$_4$·2H$_2$O	1.00	5.38
Calcium chloride	CaCl$_2$·2H$_2$O	0.85	4.59
Sulfer	S$_8$	0.19	1.00
Iron sulfate	FeSO$_4$·7H$_2$O	1.61	8.69
	Fe$_2$(SO$_4$)$_3$·9H$_2$O	1.09	5.85
Aluminum sulfate	Al$_2$(SO$_4$)$_3$·18H$_2$O	1.29	6.94
Iron pyrite	FeS$_2$ (30% S)	0.35	1.87

Iron pyrite (FeS$_2$),
$$8NaX_{(solid)} + 2FeS_{2(solid)} + 7.5O_{2(gas)} +$$
$$4CaCO_{3(solid)} \rightleftharpoons 4CaX_{2(solid)} + 4Na_2SO_{4(aq)} +$$
$$Fe_2O_{3(solid)} + 4CO_{2(gas)} \qquad [8.7]$$

The cost, availability, and time required for the reaction with soil determine which chemical amendment will be best for specific circumstances. Although calcium chloride (CaCl$_2$), probably the most costly amendment, is the quickest to dissolve and react with the soil, iron and aluminum sulfates, H$_2$SO$_4$, and gypsum also react quickly when mixed with the soil. The finer the particle size, the poorer the degree of crystallinity of the solid amendment, and the greater the uniformity of soil incorporation, the faster and more efficient the reclamation reactions. For example, because PG and FDG are more porous crystals than mined gypsum, they dissolve faster, and in the case of H$_2$SO$_4$, which reacts quickly with calcite, it has been found to be more effective than equivalent amounts of gypsum (Overstreet *et al.*, 1951; Miyamoto *et al.*, 1975a,b). After the initial calcite dissolution and exchange reactions are complete, any additional calcite dissolution by H$_2$SO$_4$ would result in gypsum precipitation. This gypsum would be relatively noncrystalline, and its uniformity of incorporation could exceed any that could be achieved by mechanical incorporation of finely ground, mined gypsum.

As S and FeS$_2$ must be oxidized by soil microorganisms, both are classified as slow-acting amendments. Since the microbial activity involved increases with decreasing pH, dilution of these amendments by deep incorporation should be avoided. In other words, shallow rather than deep tillage may be preferable. Because these reactions increase with increasing soil temperature (Overstreet *et al.*, 1955), application should be made during the spring and summer.

Some basic chemical and physical principles underlying sodic soil management and reclamation have been presented, including some illustrations of how these principles have been applied to the management of cropped, sodic soils. The next section provides a more detailed discussion of tillage, amendment, and cropping options, which are under development or are recommended for the management of sodic soils.

8.3 TREATMENT AND MANAGEMENT OF SODIC SOILS FOR CROP PRODUCTION

8.3.1 Management Problems

Sodic soils show wide variations in the degree of sodicity, as well as in the nature of the associated problems for crop growth and production. Understanding these problems is important in using these soils for crop production. This is illustrated by soil water content changes obtained by Jayawardane and Blackwell (1986) on a ponded sodic soil (Natric Palexeralf) with an ESP that increased from 2 in the surface to 12 at a depth of 0.5 m (Fig. 8.10). Even with prolonged ponding of irrigation water (EC < 0.1 dS m^{-1}), only a small increase in water content occurred in the subsoil. This reduction in soil water storage (McIntyre *et al.*, 1982a,b) causes crop water stress during prolonged dry periods. In addition, when irrigation or rainfall exceed the sum of evapotranspiration (ET), infiltration, and drainage, the resultant saturated surface layer (Fig. 8.10) restricts aeration (Jayawardane and Meyer, 1985), causing reductions in crop growth and development. The low macroporosity of sodic soils leads to slow internal drainage and water redistribution within the profile. Hence the rate of drying of the saturated surface soil layer is closely related to the evapotranspiration rate. Thus on flat lands, during the winter when ET is low, excess water will remain on the soil surface for extended periods of time, whereas on sloping lands, the excess water results in runoff and erosion.

Soil physical, chemical, and biological characteristics, for example, ρ_b, aggregate stability, IR, HC, organic matter content and composition, SAR, and EC and their changes due to amelioration, often provide useful insights into whether the ameliorative technique used is likely to be successful over the short or long term. However, in any cropping season, the crop responses to a given ameliorative technique, such as deep ripping, deep plowing, and gypsum application, are very variable, even on similar soils. They depend on the climatic conditions and agronomic management during the cropping season as well as on the variability of the soils within the field. Often these differences have not been adequately accounted for. When soil chemical and biological conditions are optimal, root growth and function are directly affected by water availability, soil aeration, soil strength, and temperature (Letey, 1985). These physical factors, in turn, depend on a variety of other variables. For example, water availability is affected by the hydraulic characteristics of the soil, clay swelling, surface crusting, climate, topography, and irrigation. Furthermore, because each of these soil factors changes through the growing season, regular monitoring of root zone conditions is required to explain crop growth and yield responses resulting from different ameliorative treatments. Fortunately, all of these physical properties are closely related to the soil water content, which is relatively easy to measure.

Non-Limiting Soil Water Range (NLWR)

Letey (1985) proposed the concept of nonlimiting soil water range (NLWR) to relate soil physical factors to crop production. Within the NLWR, crop growth and development are not affected by water or aeration stress or by soil strength. In some sodic clay soils, the NLWR is very narrow (Jayawardane *et al.*, 1987a), making it difficult to maintain the water content within this range during the crop season, even on irrigated lands. Crops may suffer aeration stress immediately after irrigation or during prolonged rainfall. In addition, excessive soil strength (Jayawardane and Blackwell, 1990) and water stress become limiting factors early in the drying cycle. Gypsum slotting, a tillage technique described in the next section, increased the NLWR (Fig. 8.11) of a sodic soil (Natric Palexeralf) cropped with wheat (Jayawardane *et al.*, 1987b), making it easier to optimize the irrigation management required to maintain the soil within the NLWR. The NLWR concept satisfactorily explained crop responses to the gypsum-slotting technique. Jayawardane *et al.* (1987a) showed that during the first half of a winter wheat cropping season within the 0.4 m soil depth, soil aeration was limiting for longer periods in the nonameliorated sodic soil due to prolonged rainfall and low ET combined with poor internal soil drainage when compared to the ameliorated soil. This led to marked decreases in the number of tillers per plant in the nonameliorated soil as compared to the gypsum slotted soil. However, during the second half of the season under higher ET and controlled irrigation, soil aeration conditions were favorable under both treatments, and hence, the number of grains per tiller and the weight per grain were similar. It is important to emphasize that the achievement of optimum soil conditions for crop growth,

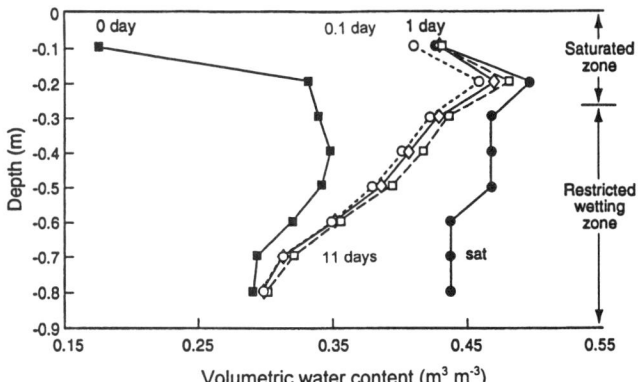

Figure 8.10 Changes in volumetric soil water content profiles of a sodic clay soil (Natric Palexeralf) under prolonged ponding for varying periods of time [Jayawardane and Blackwell, *Soil Use Manage.* 2:114-118, (1986)]

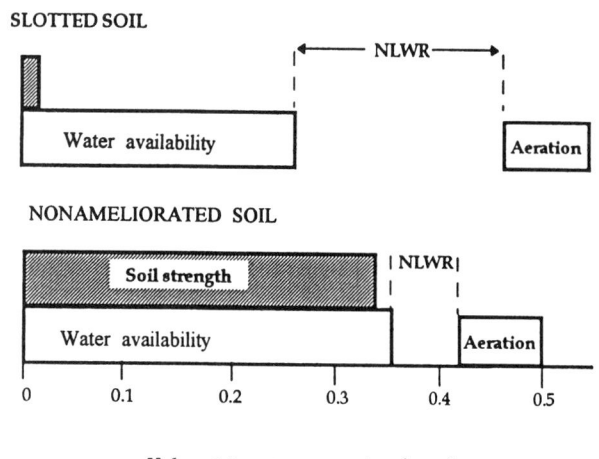

Figure 8.11 Comparison of the nonlimiting water range (NLWR) within a slot in a slotted soil with that of a nonameliorted sodic soil (Natric Palexeralf) under wheat [Jayawardane *et al., Aust. J. Agric. Res.* 38:239-251, (1987b)]

in both ameliorated and nonameliorated soils, is dependent on the interaction between soil properties, climate, and agronomic and irrigation management practices adopted during each cropping season. Thus the differences in crop yields between ameliorated and nonameliorated sodic soils under similar management could be small in seasons with a favorable rainfall pattern, whereas differences could be significant in seasons with excessive or insufficient rainfall, due to the water content of the nonameliorated soil moving outside the NLWR.

8.3.2 Tillage

In some sodic soils, poor crop productivity is often associated with low macroporosity. Tillage increases macroporosity (Klute, 1982; Jayawardane and Prathapar, 1992) and changes the physical and hydraulic properties of soils. Where tillage loosens the soil, ρ_b is reduced and the total porosity is increased, which results in the formation of macropores (Blackwell *et al.*, 1991) that drain at high water potentials (Table 8.4). This usually leads to increases in air-filled porosity and in plant-available water at water contents that occur several days after irrigation. Both saturated and unsaturated HC at high water potentials are increased, although effects on unsaturated HC are smaller at lower potentials. These changes lead to faster infiltration and redistribution of water within the tilled zone and, hence, to improved aeration of sodic soils. However, tillage at unfavorably high soil water contents can cause smearing and compaction, leading to adverse changes in the physical and hydraulic properties and to restricted water movement

and aeration. The following field tests can be used to determine whether the soil water content is satisfactory for tillage: (a) Roll the soil between the palms of the hands into a thread about 3 mm in diameter; if the water content is satisfactory, the soil will crumble. This water content is called the plastic limit (Marshall and Holmes, 1988). (b) If the soil sticks to shoes when walking through the field, it is too wet for tillage (Bauder, 1995).

The physical improvements in sodic soils resulting from tillage are often not permanent. Because wet soils have little strength, they tend to reconsolidate under their own weight which is enhanced by sodicity. Consequently, the hydraulic properties deteriorate and the NLWR becomes narrower. For example, Mead and Chan (1988) found that improvements from tillage were short-lived for hardsetting soils which collapse on wetting.

Shallow Tillage
Shallow tillage, by either disking or chiseling, is effective in reducing problems of soil water penetration caused by crusts, hardsetting, or compaction. Although tillage breaks up the soil, thereby increasing infiltration (Moore *et al.*, 1989), repetition is necessary from time to time (Rawitz *et al.*, 1986). Because crusts reform during rainfall and irrigation, hardsetting soils become hard upon drying, and traffic results in compaction. When tilling to reduce compaction, soils should be chiseled 50% deeper than the lower boundary of the compacted layer, with shanks spaced at 80% of the depth chiseled. For example, if compaction extends down to 0.4 m, rip to 0.6 m with shanks spaced at 0.48 m (Wildman, 1981). A small angle from the horizontal at the lower portion of the shank, which can be achieved with curved or parabolic shanks, enhances the effects of tillage and promotes mixing (Spoor and Godwin, 1978).

Table 8.4 Effects of gypsum application and different tillage practices on macroporosity and intrinsic permeability at −10 kPa water potential in a sodic clay soil (Natric Palexeralf) after one season of irrigated cropping [Blackwell, *et al., Aust. J. Soil Res.* 29:123-140, (1991)]

Soil ameliorative treatment	Macroporosity ($m^3\ m^{-3}$)	Intrinsic permeability (μm^2)
Nonameliorated	0.071	1.6
Gypsum (4 Mg ha⁻¹)	0.078	24.2[a]
Ripping + gypsum (4 Mg ha⁻¹)	0.139[a]	25.4[a]
Slotting	0.093	0.8
Slotting + gypsum (4Mg ha⁻¹)	0.143[a]	15.0[a]
Slotting + gypsum (8 Mg ha⁻¹)	0.190[a]	116.7[a]

[a] Values are significantly different from the nonameliorated soil at $P < 0.05$.

Deep Tillage

Deep tillage (1–2 m) or ripping with straight or parabolic ripper shanks, or plows, can break up hardpans and cemented layers and permanently mix soil layers of different textures. Permanent improvement occurs provided the layer is thin and shallow enough for ripping to shatter it completely and to break into the noncemented layers below. If the layer is not penetrated and destroyed, excessive rainfall or irrigation may pond in the ripper channel and suffocate any roots that have grown into this zone. Evaluation of the depth and thickness of the cemented layer or hardpan is essential before a ripping project is begun. Effective mixing of soil layers with different textures can be achieved by deep rippers, slip and moldboard plows, trenchers, and backhoes (Oster *et al.*, 1992). Mixing of soil profiles with rippers and slip plows is most effective in dry soil which facilitates shattering and mixing. Wildman (1981) recommended a gravimetric soil water content of 10% for clay, 5% for loam, and 2.5% for sand to ensure effectiveness of the slip plow. The slip plow which is frequently used in California, consists of a steel plate box (slip) that is thin (0.1 m), long (3 m), and somewhat wide (0.3 m), made from heavy steel plate. It is attached to a heavy duty ripping shank by a large pin at the bottom and a heavy chain at the top to give a 30–45° angle between shank and slip. The shank and slip are mounted on the hydraulic hitch of a large crawler tractor. Pulled at an angle through the soil, the slip plow lifts and mixes the soil. In California, slip plowing every 4 to 6 years is becoming more common in fields cropped with vegetable, grain, forage, and fiber crops (Colen, 1995).

The slip plow is also often used for the first tillage step in field preparation for planting trees, particularly on soils with different textural layers within the root zone. It is especially effective when trees or vines are planted within the area that has been plowed (Table 8.5) (Wildman, 1985). In 1968, three methods of tillage were compared in the preparation of a claypan soil for planting almonds (*Prunus dulcis*). The ripper and slip plow treatments penetrated to 1.5 m on 2.4 m centers, and the moldboard plow treatment was plowed to 1 m on 1.2 m centers. The slip plow treatment included slip plowing every third ripper track, which coincided with the location of the tree row. The greater the degree of disruption and mixing of the clay layer with the rest of the profile, the better was the root growth and almond yield in 1976, seven years after planting (Table 8.5).

The use of shallow and deep tillage in existing orchards poses a concern, based on reason rather than evidence, that detrimental effects will result from root pruning. In a pecan (*Carya illinoinensis*) grove planted on 9 m centers, Helmers

and Miyamota (1990) chiseled the center 3 m both down the middle of the rows and diagonally to the rows. The purpose was to mix the surface silty clay layer with the underlying fine, structureless sand and to eliminate soil cracking in order to improve leaching. About a year later, after nine irrigations, the trees showed no signs of damage while, in the chiseled area, the average EC_e levels to a depth of 0.9 m decreased from 4.8 to 3.8 dS m^{-1}. Pecan yields can be expected to increase because of these lower salinity levels.

Long-term amelioration of sodic soils requires not only increased macroporosity but also preservation and stabilization of these macropores (Blackwell *et al.*, 1991). The increased macroporosity created by tillage is often subsequently reduced by in-field traffic compaction (van Ouwerkerk and Raats, 1986). Carter (1985), for example, found that cotton fields without traffic had between 21 and 32% better water penetration than those with normal traffic. The use of traffic lanes, sometimes combined with deep tillage and/or bed farming, provides some protection from soil recompaction by traffic. A good example is provided by the tillage and controlled traffic techniques for alfalfa production recently adopted by farmers in the Palo Verde Valley of California, a narrow, alluvial valley next to the Colorado River that has been under irrigation since the early 1900s. The native soils are stratified, with distinct soil layers within the root zone that can vary from clay to sand. For alfalfa production, the fields are slip plowed and then laser leveled (level in one direction, with a slope of 1:1200 in the other) before shallow furrows are formed on a 1 m spacing. These furrows also serve as traffic lanes during planting and harvesting operations. Care is taken to assure that the same furrows are used for traffic lanes for each harvesting operation. The longevity of alfalfa fields using this soil and water management practice is 1–2 years longer than that of similarly tilled and leveled fields that are border irrigated. This increased life is apparently the

TABLE 8.5 Almond yield, trunk circumference, and root count response to mixing a clay pan soil by three tillage methods [Wildman, *Walnut Orchard Management*, University of California Coop. Ext. Spec. Publ. 21410, (1985)]

Tillage method	Yield[a] (Mg ha^{-1})	Trunk circumference[a] (m)	Root count[b] per 0.0847 m^3
None	1.13	0.376	920
Ripper	1.25	0.422	1110
Slip plow	1.32	0.424	1390
Moldboard plow	1.60	0.432	2066

[a] Fourth leaf
[b] Eighth leaf

result of traffic control, which confines soil compaction to a small percentage of the field area, thereby limiting injury to the crop.

8.3.3 Tillage plus Amendments

Deep moldboard plowing (0.5–0.6 m) of a loamy textured soil in Australia broke up the hardsetting surface layer and improved wheat yields, while mixing the clay-rich subsoil with the surface promoted cracking (Hall, 1990). The improvement, however, did not last beyond the second year, unless deep plowing was combined with gypsum applications and crop residue management.

A new technique under study in Australia is gypsum slotting which consists of tillage along narrow slots in the upper soil layers together with the incorporation of chemical amendments at rates sufficient to reclaim the soil in the tilled slot with the rest of the soil remaining undisturbed (Jayawardane and Blackwell, 1985). Gypsum is incorporated at the rate required to correct sodicity in the tilled slots (about 100–150 mm wide, 0.4 m deep, and spaced about 1–2 m apart). The concentration and thorough mixing of the gypsum to the full depth of the slot prevents rapid loss of macropore stability. The resulting preferential water flow pathways facilitate the removal of excess water from the surface layer, long-term reclamation of the soil in the narrow slot, and storage of water in subsoil layers below for use during long dry periods. Soil within the slots is protected from recompaction during subsequent trafficking by bridging implement wheels across the narrow slot with the compressive stresses beneath the wheel being borne by the stronger undisturbed soil on either side of the slot (Blackwell *et al.*, 1989a,b). This results in longer-term benefits than traditional tillage practices on gypsum-amended sodic soils, which usually undergo recompaction under traffic. As described in Section 8.3.1, field studies have also shown that gypsum-slotting leads to improved soil aeration and increased water infiltration and use by crops, leading to higher crop yields (Jayawardane *et al.*, 1987a).

Costs of slotting will vary depending on the slot configuration and the amount of gypsum required. However, while prototype slotters have been constructed, commercial implements are not yet available. The gypsum slotting technique is an excellent example of revising both the tillage and the amendment application to effect localized reclamation and preferential water flow pathways in a manner that should minimize problems inherent in both the chemistry and the physics of long-term reclamation. In soils with shallow water tables, or with low HC values and slow rates of water redistribution, the combination of the gypsum slotting technique with a mole drain located 0.1–0.2 m

below the slots enhanced drainage (Muirhead *et al.*, 1992, 1994). Without drainage, reclamation cannot occur. However on-farm tests with different crops and soils are required to assess the likelihood of improvements on different soils and the economics of the gypsum-slotting method.

Spreading gypsum on the soil surface at rates required to correct sodicity to a specified depth calculated using Equation [8.2] followed by tillage to mix the gypsum into the soil, is a commonly used amendment technique. Following tillage, a final surface application of gypsum (2–5 Mg ha^{-1}) is often made to maintain IR by increasing the electrolyte concentration of the infiltrating water and reducing exchangeable Na and Mg in the surface soil.

After tillage and amendment incorporation, there are two alternatives for water and crop management: (a) ponding water until a predetermined amount has infiltrated (Equation [8.1]) or until it becomes necessary to allow the field to dry to permit land preparation for cropping, or (b) planting the crop and then irrigating, in which case the duration of ponding must be limited to avoid problems with waterlogging and aeration of the seed bed. The latter method has the advantages of using the crop root system to assist in the reclamation process, thereby increasing the probability of maintaining adequate HC. Infiltration rates for subsequent irrigations must be sufficient at least to meet the crop water requirements in order to ensure the maintenance of the reclamation achieved by the initial irrigation. This may require one or two more irrigations than normal for a crop in a given region. One principal advantage of the combined amendment and cropping approach, particularly with grain crops which do not require tillage during the season, is that compaction by equipment is minimized during the first and subsequent crop seasons. The soil in the upper portion of the root zone is dry at the time of harvest and during the preparation of the subsequent seedbed.

Whatever methods are used to apply and incorporate gypsum, the objective is to maintain adequate electrolyte concentration and reduce exchangeable Na and Mg levels to sustain satisfactory IR and HC. However, as explained in Section 8.2 for soils that are not naturally gypsiferous, gypsum incorporation into the soil results in reclamation that is restricted primarily to the depth to which gypsum has been incorporated. Consequently, low HC at some greater depth, particularly on uncropped soils, can restrict water flow into and through the profile, thus limiting the success of reclamation.

8.3.4 Drainage

Deep tile or shallow mole drainage can be used to overcome the problems of surface waterlogging and perched or shallow water tables in sodic soils. Spoor (1995) and MacEwan *et al.* (1992) provide detailed reviews of the design and practical application of tile and mole drainage techniques for the poorly drained duplex soils in Australia. Tile drainage is widely used in fields cropped with high-value horticultural crops in the Murray Basin and for cereal, forage, fiber, and tree crops in California. The main disadvantages of tile drainage are costs of installation and problems associated with the disposal of saline water, often with a high SAR (Oster, 1994a). Muirhead *et al.* (1992) suggested using shallow mole drainage to remove the excess good-quality irrigation water which can then be reused for irrigation. While mole drainage has led to crop yield increases, the instability of moles in sodic soils is a major problem and further studies are needed to develop techniques to cope with this problem. For example, the stability of a mole drain was increased by placing a gypsum-slot above it (Muirhead *et al.*, 1992).

8.3.5 Cropping

Many Vertisols used for cotton production in Australia have sodic subsoils (Stannard and Kelly, 1977). Consequently, deep tillage to below 0.5 m is detrimental because it tends to bring sodic subsoil to the surface which can aggravate waterlogging and infiltration problems (McKenzie *et al.*, 1992). Instead, strong-rooted rotation crops, such as safflower (*Carthamus tinctorius*), have been used to loosen subsoils biologically. Because of the high shrink-swell potential of these soils, extensive cracking to below 1 m results, which increases macroporosity and improves cotton root growth (Hodgson and Chan, 1984; Hulme *et al.*, 1991).

Where previous soil amelioration by tillage or chemicals has been chosen, cropping can also influence root zone conditions favorably. For instance, Jayawardane *et al.* (1994) found that the subsequent irrigation of a sunflower (*Helianthus annus*) crop on a gypsum slotted soil led to more rapid water extraction and increased soil drying and cracking during the drying cycle compared to a non-ameliorated soil (Table 8.6). This enhanced depletion, in turn, led to increased water infiltration during the subsequent irrigation and decreased runoff during the following irrigation and rainfall events compared to non-ameliorated soils (Jayawardane *et al.*, 1994).

Poor aeration and infiltration sometimes occur in the bed farming systems used in Australia. The beds are 1–1.5 m wide and about 0.1–0.15 m higher than the neighboring furrows. Muirhead *et al.* (1994) found that while the air-filled porosity was above a critical limit of about 0.08 mm³ within the bed, aeration conditions were not adequate in

the undisturbed soil below. As the root growth was often restricted to the upper soil layers, reduced water use by the crop during extended drying cycles between irrigations could result. Using more frequent irrigations, water stress and high soil strength can be avoided with the potential for higher crop yields. However, this can lead to increased runoff and reduced irrigation efficiency. Subsoil amelioration below the beds using deep ripping or slotting, combined with gypsum application, could increase the soil water availability and help to extend the time interval between irrigations. Another alternative would be to use ridge instead of bed farming techniques. Plants are located on ridges which are 0.2–0.25 m higher than the neighboring furrow to improve aeration. Well-aerated beds or ridges can at least partially compensate for waterlogged subsoil (Hunter *et al.*, 1980; Hearn and Constable, 1984; Troedson *et al.*, 1989).

In some soils, such as self-mulching grey clays, wetting across the beds from the irrigation furrows usually takes place fairly quickly due to the characteristic extensive and deep cracks (Hodgson and Chan, 1984), which allow rapid wetting of the soil profile. In contrast, red duplex soils crack less and tend to slake easily into fine size particles on wetting which could lead to partial sealing of the furrow wall and restriction of IR. Traffic on wet soil could also lead to furrow wall damage and adversely affect soil structure (McGarry and Chan, 1984). Under irrigation, ponding of water in furrows for long periods can lead to waterlogging and reduced yields of sensitive crops on the beds in red duplex soils (Mason *et al.*, 1987) and self-mulching grey clays (Hodgson, 1982; Hodgson and Chan, 1982; Wright, 1985; Hodgson *et al.*, 1989). Further research to overcome the problems with red duplex soils is under way (Tisdall and Adem, 1988; Tisdall and Hodgson, 1990) and shows considerable promise (Section 8.4) in overcoming problems with aeration and waterlogging.

TABLE 8.6 Water depletion (mm) due to evapotranspiration during a dry cycle and water infiltration (mm) during the subsequent irrigation of a sunflower crop, in nonameliorated and gypsum-slotted sodic clay soil (Natric Palexeralf) [Jayawardane *et al, Subsoil Management Techniques*, Lewis Publishers, Boca Raton, FL (1994)]

Treatment	Water depletion/intake	Irrigation no.					
		1	2	3	4	5	Mean
Nonameliorated	Depletion	18	42	61	87	65	55
	Intake/infiltration	19	49	52	79	63	52
Gypsum-slotted	Depletion	47	72	84	120	73	79
	Intake/infiltration	45	72	81	114	72	75

Describing procedures to reclaim or ameliorate sodic soils which apply to all situations is not possible, since soils are highly variable in their physical properties. These can vary extensively from one region of the world to the next because of differences in the soil forming processes that affect clay mineralogy, soil layering, soil texture, organic matter content, and the distribution of salinity and sodicity within the soil profile. Such differences in soils have been a major cause for the variety of practices developed by farmers which are described in detail for cropping systems in Australia, Canada, India, Israel, Russia, and the United States in the following sections.

8.4 EXAMPLES OF CROP PRODUCTION SYSTEMS USED TO RECLAIM AND MANAGE SODIC SOILS

In this section, the foci are six examples of cropping systems used by farmers to manage problem soils or waters in various regions, ranging from the use of nonsaline, nonsodic waters to irrigate the unstable soils of the Riverine region of central Australia to the use of saline, sodic groundwaters to irrigate salt-tolerant crops in Israel. Also included are examples of farming practices used to reclaim saline sodic soils, both in India, where farming equipment for tillage is a limiting factor, and in Russia and Canada, where it is not. Russia and Canada also represent examples of the reclamation of saline-sodic soils under rainfed conditions. The objective is to describe successful cropping systems and to relate the reasons for their success to the discussions provided in the previous section.

8.4.1 Australia: Irrigation of Unstable Duplex Soils

In the Riverine plain of central Australia, the land is extraordinarily flat, with slopes ranging from 1:500 to 1:2000. The climate is continental, with a winter season during which precipitation exceeds ET and a summer season when rainfall is inadequate. Irrigation waters in the region are nonsaline (EC ~ 0.1 dS m^{-1}) and nonsodic (SAR ~ 1.0). The soils are very unstable due to their sodicity, low organic matter contents and clay mineralogy dominated by illite and kaolinite with only a small amount of smectite. Grain and row crops are traditionally grown after 6 to 8 years of pasture, exploiting the accumulated nutritional and physical benefits conferred by this (ley) system. However, cropping quickly destroys the soil structure. Under cultivation, the surface soil is susceptible to slaking under flood irrigation and heavy rainfall. As the soil dries, it sets hard, and because water does not penetrate deeply, frequent irrigations become a necessity. Cultivation to break up the hard surface only worsens the subsequent slaking. After 1 or 2 years of cropping, yields deteriorate to very low levels and the land must be replanted to pasture for a further 6 to 8 years during which structure improves.

The surface soils (0.15–0.18 m) are dark-brown clay loams to sandy clay loams (usually Typic Halpustalfs), containing less than 2% organic carbon with an abrupt transition to a yellowish red, dense (ρ_b = 1.6 Mg m^{-3}) clay subsoil with a low saturated HC (2–3 mm d^{-1}) and low macroporosity (Stace *et al.*,1968). Even when the surface structure is good, the rooting depth is shallow, the irrigation water rarely penetrates deeper than 0.8 m, and available water is only about 70 mm. Consequently, in the summer, these soils must be irrigated every 5–10 days. Another serious problem is the poor surface and subsurface drainage of the profile; the soil remains waterlogged for several days after rain or irrigation. However, crop production systems have been developed which make it possible to maintain good physical properties without including prolonged periods of pasture in the crop rotation. The Tatura trellis tree and permanent bed systems are examples with proven benefits.

Tatura Trellis Tree System
In this system (Cockroft and Martin, 1981; Olsson *et al.*, 1994), the soil is first analyzed to determine fertility and chemical status. The field is then graded to ensure good surface drainage. Before the trees are planted, the soil is ripped to a depth of 0.6 m along the tree row to improve root penetration into the subsoil. Well before the ripping operation, gypsum is applied as needed to improve the physical properties of the soil so that the subsequent deep tillage of the subsoil results in a low ρ_b. The interrow space is then cultivated to the full depth of the topsoil and banked up to the trees forming a bed. Thereafter, zero tillage is practiced with the use of herbicides and mulches to control weeds. To minimize the collapse of the loose soil, a stabilizing crop, usually ryegrass (*Lolium perenne*), is grown continuously until the bed is fully permeated by tree roots. Thereafter, the stabilizing crop is necessary only during the dormant season. Herbicides and mowing are used to control competition from the stabilizing crop. Use of the stabilizing crop and careful irrigation management through slow wetting and good surface drainage are both essential to control soil structure and maintain adequate soil physical condition. Mini-sprinkler and drip irrigation systems are used daily to apply water to individual trees in amounts required to replace the ET of the previous day. Yields obtained are threefold those under cropping systems that use surface irrigation systems and repetitive tillage.

Tatura Permanent Bed System
This system, which combines laser grading and minimum tillage with mulching practices to maintain adequate IR and surface drainage (Adem *et al.*, 1984; Tisdall and Adem, 1988; Tisdall and Hodgson, 1990), shows considerable

promise for continuous double cropping of cereal and row crops. The aim is to produce and maintain the desired soil properties in three specific zones in the soil (Fig. 8.12): (a) the seeding zone A, which is tilled while wet, but at the water content corresponding to the plastic limit (about $0.2\,g\,g^{-1}$) to develop the desired size distribution of soil aggregates, (b) the traffic- and tillage-free soil between adjacent crop rows, referred to as the water management zone B, and (c) a shallow furrow and traffic lane C. Mulch is obtained by growing a winter cereal crop. Immediately after its harvest, the stubble is shredded and moved onto the water management zone. The bed is then irrigated and allowed to dry to the lower plastic limit. Tillage of the seeding zone, planting, and herbicide application are conducted in a single operation. A tractor-mounted rake is then used to cover the seeding zone. The seedlings emerge without further irrigation. Organic residues from roots and mulched stubble provide sufficient microbial and faunal activity to maintain aggregate stability. Mulching also reduces the maximum temperature at the 30 mm depth by about 8 °C during the summer, which fosters earthworm activity throughout the year. Commercial yields from double cropping with wheat, barley, sunflower, and corn (*Zea mays*) over a period of 6 years (Tisdall and Adem, 1988) ranged from 50 to 90% of the best experimental yields recorded in the region (Cockroft and Mason, 1987).

8.4.2 California: Reclamation and Management of Saline and Sodic Soils

In the Central Valley of California, rainfall (150–460 mm) occurs primarily during the winter months and, the temperatures range from –7 to 40°C with an annual evapotranspiration of about 1400 mm. The native soils are calcareous with a wide range of salinity, sodicity, gypsum, and boron levels, particularly along the west side of the valley with some being nonsaline sodic (Overstreet *et al.*, 1955; Kelley, 1951). Cropping during reclamation was a common practice in this area. Chemical amendments were used selectively, with reclamation being possible without them on many soils (Overstreet *et al.*, 1955). Amendments such as gypsum and H_2SO_4 (on calcareous soils) increased the rate of reclamation, with Overstreet *et al.* (1951, 1955) reporting field data that indicate H_2SO_4 is superior to gypsum. However, recent laboratory studies show that gypsum, $FeSO_4$, $Fe_2(SO_4)_3$, $Al_2(SO_4)_3$, and H_2SO_4 are equally effective when applied to moderately sodic calcareous, silty clay soils (Miyamota and Enriquez, 1990).

Barley, a winter crop, was usually the first crop grown on new ground during reclamation in the 1950s and 1960s, with border irrigation used to supplement annual rainfall. After one or more barley crops, cotton was often added to the rotation. Cotton fields were ripped before planting, amended with gypsum if necessary, listed (a ridge thrown up between two furrows by a lister in plowing) to create furrows and ridges with good tilth, and preirrigated. Large amounts of water (250–350 mm) infiltrated during preirrigation, resulting in considerable leaching and reclamation. The following quote summarizes recommendations provided by the University of California at the time (Dean's Committee, University of California, 1968):

> Field crops, particularly barley, wheat, sorghum, cotton, and sugar beets (*Beta vulgaris*), are real tools for use in reclamation, or as transition crops to "get acquainted" with soils and to get from "where we are" to "where we want to be" from the standpoint of soil salinity, sodium, and boron. By utilizing *more water* on these crops than is actually needed, salts, sodium, and boron can be leached beyond reach of roots, and the soils can be prepared for later plantings of more sensitive high income crops.

Reclamation using such methods required considerable time, since excess water necessary to reclaim the upper 1.5 m of soil ranged from 1.5 m for salinity to 4.5 m for boron (B) (Bingham *et al.*, 1972). Additional water was usually required to compensate for spatial variability of soils and associated IR values (Jaynes and Hundsaker, 1989; Wichelns and Oster, 1990), but with continuing effort success was eventually achieved. In the San Joaquin Valley of California, vineyards planted in the 1970s now occupy fields where reclamation of saline-sodic soils began in the 1950s.

Generally speaking, California farmers tend to practice reclamation over the long term. For example, annual or semiannual gypsum applications of 2–4 Mg ha^{-1} to an entire field are continued for as long as crop growth is uneven and yields are low. Amendment applications to small areas within a field with continually poor plant growth are made by some farmers, although this is not a typical practice. Reclamation by one or two large, annual irrigations, usually before planting

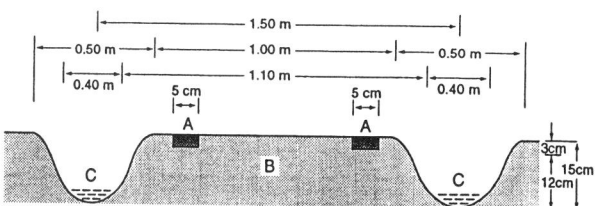

Figure 8.12 Cross section of Tatura permanent bed system on a Lemnos loam showing A seeding zone, B water-management zone, and C furrow [Tisdall and Adem, *J. Agric. Eng. Res.* 40:23-32, (1988)].

or during the first crop irrigation, is a form of intermittent reclamation that has also been successful when practiced over the long term. Farmers along the west side of the San Joaquin Valley of California use ripping tools to till soils to depths of 0.5–0.75 m. Following disking, land planing, and listing to prepare ridges and deep (0.3–0.4 m) furrows, 0.25–0.35 m of water infiltrates during the subse-quent irrigation. Gypsum may be applied at rates of 2–4 Mg ha^{-1}, either before or after tillage, depending on farmer ex-perience with gypsum on a particular field or on soil analysis that indicates it would be beneficial. For cotton, the heaviest irrigation occurs 1–3 months before planting. Then, several days before planting, the tops of the beds are removed ("decapped") and the cotton is planted into the exposed wet soil. In saline fields, the depth of decapitation is increased to ensure that seeds are planted at a depth where the soil is less saline and wetter than if the decapitation were shallower.

8.4.3 California: Use of Gypsum to Maintain Infiltration Rates

The addition of gypsum to the soil surface (1–2 Mg ha^{-1}) or to irrigation water (3–5 mmol$_c$ L^{-1}) for the purpose of maintaining IR is also a common practice in California. Doneen (1948) reported that 270 000 Mg of gypsum were applied in 1945 in the San Joaquin Valley to improve infiltration. The addition of gypsum to Friant-Kern irrigation water (EC < 0.2 dS m^{-1}) or applied on the surface of non-sodic soils irrigated with it was a common practice in the 1950s on the east side of the valley between Fresno and Bakersfield (Ayers, 1980), and such practices continue today. In recent years, machines to inject gypsum slurries into the irrigation water have also become available, and their use is increasing. The gypsum particle size is an important consideration for both surface applications to soil and injection through the irrigation water. For land application, a typical particle size of commercially available gypsum (92% pure) in California is 87% finer than 2.4 mm, 52% finer than 0.3 mm, and 25% finer than 0.07 mm. Finer material is needed for injection into irrigation water, with a typical particle-size distribution of 99–100% finer than 0.15 mm, 93–97% finer than 0.07 mm, and 3–78% finer than 0.04 mm.

8.4.4 Israel: Sustained Irrigation with Saline-Sodic Groundwater

In the western Negev region of Israel, there is a saline aquifer with EC values ranging from 2.5 to 8.5 dS m^{-1} and SAR values of 15–26 (Keren *et al.*, 1990; Frenkel and Hadas, 1981). The soils are silt loams (usually Calcic Haploxeralfs) under a Mediterranean climate, with winter rainfall ranging between 250 and 400 mm. Cotton is the dominant crop. The effect of 16 years of irrigation with water from a well at kibbutz Nahal-Oz (EC = 4.6 dS m^{-1} and SAR = 26) provides a typical example of ongoing reclamation practices. Irrigation during the summer (450 mm) results in ESP values of 20–26 in the upper 600 mm of soil with no hydraulic deterioration during the summer due to the high EC$_{iw}$. However, deterioration can occur during the rainy season due to the low electrolyte concentration in the rainwater. To offset this, PG at a rate of 5 Mg ha^{-1} is spread annually on the soil surface following tillage in the fall, which prevents seal formation and maintains high IR values. This, in turn, provides sufficient leaching of salts from the root zone. Fall application of PG and leaching during the rainy season, coupled with adequate irrigation with the saline sodic water to meet crop needs during the summer months, has resulted in seed cotton yields averaging 5 Mg ha^{-1} between 1979 and 1988. These yields were similar to those obtained when only nonsaline water was used for irrigation.

8.4.5 India: Reclamation of Nonsaline Sodic or Alkali Soils

In the Indo-Gangetic plains of India, most farmers begin reclamation in the monsoon season (July–September; 600–900 mm rainfall) by growing rice. Exchangeable Na levels can be as high as 100%. Excess exchangeable Na, high pH, lack of adequate zinc (Zn) and Ca, and the resulting poor soil physical conditions and nutritional properties of the soil are the chief causes for poor productivity.

Recommended practices for this area (Mehta, 1985; Singh, 1985; Gupta and Abrol, 1990a) include the following: (a) dividing fields into subplots of 0.4 ha each, bordered by 0.4 m high dikes (bunds), (b) leveling, locating high and low spots in the subplots, using a shallow irrigation, and releveling with a slope of 0.1% toward the drainage channel, (c) avoiding tillage while the soils are wet because of their poor physical properties, (d) applying finely powdered gypsum (100% finer than 2 mm, 75% finer than 1 mm, and 35% finer than 0.125 mm) at rates of about 10–15 Mg ha^{-1}, depending on soil sodicity and texture, to reduce the ESP of the surface 150 mm of soil to 15–20, (e) incor-porating gypsum in the top 60–80 mm of soil by shallow tillage, (f) leaching for 15–20 days prior to transplanting recommended varieties rice (P2-21, IR 8, PR 106, and Basmati 370), (g) installing tube wells and utilizing the pumped water for irrigation, particularly where water tables are high, and (h) planting and transplanting three-leaf rice seedlings (35–45 days old) grown on reclaimed soil (rice seedlings are sensitive to sodicity at the 1–2 leaf stage).

If properly fertilized with nitrogen (N) and Zn (Chhabra and Abrol, 1977; Singh, 1985; Gupta and Abrol, 1990a), yields of dwarf rice varieties during reclamation can approach levels achieved in fully reclaimed soils when ESP values for the surface 150 mm are 60 or less (soil solution SAR < 100). The intention of the initial amendment application is to reduce ESP to less than 25–30 so that excellent yields of rice and moderate yields of wheat can be obtained even during the first year. Farmers often do not apply gypsum, but instead resort to prolonged leaching and accompanying application of farmyard manures. When this is done, rice yields are reduced, and rice may have to be grown for 3–5 years before wheat can be grown with even moderate yields. Reclamation is not considered to be complete until the upper 600 mm of soil are fully reclaimed, i.e., soil ESP or soil solution SAR levels are less than 5. At this juncture, wheat yields approach the maximum potential, though the yields of more sensitive crops such as peas (*Pisum sativum*) may still be reduced because of moderate sodicity levels at soil depths below 600 mm.

Karnal or Kallar grass *(Diphlachne fusca* or *Leptochloa fusca)* which grows well under ponded conditions in saline sodic soils, is transplanted into flooded fields using stem cuttings or root stolons (Kumar, 1985; Malik *et al.*, 1986). Because it is perennial, it can be planted at any time of the year, but the best planting time is March since growth is greatest during the summer. If irrigation continues during winter, a further cutting can be obtained in addition to three cuttings between spring and fall. Experiments conducted in Pakistan (Malik *et al.*, 1986) indicate that N and phosphorus (P) fertilizers have little effect on growth. Growing this grass for 3–4 years results in sufficient reclamation to grow rice and wheat, without any need for further reclamation practices. However, market demand for the grass is small, limiting the usefulness of this reclamation method.

8.4.6 Russia and Canada: Sodic Soil Reclamation under Dryland Conditions

Extensive areas of calcareous solonetzic soils with a structural B horizon occur in Hungary, Romania, Yugoslavia, Russia, the North Central Plains of the United States, and the adjoining Canadian prairie provinces (Szabolcs, 1989). These soils have a shallow, dense, sodic claypan, or natric horizon, which varies in depth from 0 to 0.4 m below the surface with ESP values that range from

15 to 50. This layer severely restricts root growth and water penetration because of its high ρ_b (1.5–1.8 Mg m^{-3}) and ESP. A calcareous layer that can be saline, gyspiferous, or both occurs immediately below the sodic layer. These soils often occur in irregularly shaped patches over distances of 1–10 m within a landscape of soils without claypans. Variations in topsoil depth, claypan characteristics, and distribution of solonetzic soils within a field make the preparation of good seedbeds difficult. The patches of solonetzic soils are often too wet for proper tillage, while the rest of the field is too dry. Without reclamation, a wavy growth pattern of crops usually occurs because of differences in crop establishment, rooting depth, and available water.

Deep tillage of solonetzic soils in conjunction with cropping usually results in permanent improvement and increased crop yields. Tyurin *et al.* (1960) summarized the results of reclamation studies conducted between 1930 and 1950 in Russia. These studies included applying various combinations of fertilizer, gypsum, and lime together with cropping and tillage to depths of up to 0.6 m. They concluded that:

> Cultivation of Solonetz soils includes, among its first tasks, the breaking up of the compact columnar horizon B, by its plowing, comminution [pulverization] and thorough admixture with horizon A, so that the two horizons are transformed into a uniform tillage layer.

Similar results were obtained with deep tillage and the use of soil amendments in studies conducted in Canada and the United States beginning about 1950 (Cairns, 1962; Rasmussen *et al.*, 1972; Sandoval, 1978). Deep plowing to depths below the claypan destroys this impermeable layer and mixes it with the calcareous, sometimes also saline and gypsiferous, layer located immediately beneath the sodic claypan, thereby providing a source of Ca for amelioration of the claypan if sufficient water infiltrates for leaching to occur (Rasmussen and McNeal, 1973; McAndrew and Malhi, 1990). In addition to the increased infiltration that results from the physical and chemical amelioration, deep tillage also facilitates deeper root development. Consequently, the available water supply, an important limiting factor for crop production particularly under dryland conditions, increases substantially.

Ripping to depths of 0.4–0.6 m is also effective, although the degree of mixing is reduced. The resulting improved crop growth is largely attributed to the physical shattering of the claypan, but substantial mixing of the soil may occur if ripping is carried out with a curved or parabolic shank when the soil is dry (Wetter *et al.*, 1987). Based on a 14-year field trial begun in 1979, Lickacz and Coy (1991) reported average annual wheat yields of 1809, 2356 and 1350 kg ha^{-1} for ripping, deep plowing, and no deep tillage treatments, respectively on a Typic Natraboroll, respectively (Fig. 8.13). Because deep plowing is both slower and more costly than ripping or subsoiling, Canadian farmers in Alberta rapidly adopted ripping as the preferred deep tillage method for reclamation (Lickacz, 1993). The shank spacing can range from 0.5 to 1.5 m, with shattering decreasing with increasing shank spacing. The power requirement ranges from 50 to 70 horsepower per shank. To prevent structural damage from rocks, shanks are equipped with either a shear pin or a spring trip assembly. During ripping, the extent of shattering should be checked by digging a trench perpendicular to the direction of ripping. Where more shattering is desired, the soil should be ripped a second time at an angle to the first pass. The soil water content should be low when the soil is ripped or deep plowed because if the subsoil is moist, little shattering will occur during ripping and mixing will be reduced if the soil is deep plowed. The ideal conditions for both tillage methods in the northern plains of the United States and Canada usually occur in the fall on stubble land from which subsoil water has been depleted. Similar conditions can also occur during the summer under perennial forages, particularly if rainfall has been limited during the spring and early summer. After ripping or deep plowing, further tillage using heavy disks and/or chisel plows is necessary to level the soil and to prepare a seedbed. Some producers prefer to pull several sections of heavy harrows in tandem behind a ripper or deep plow to facilitate subsequent tillage. If tillage occurs in late fall and there is not enough time available to prepare a seedbed, the soil remains open throughout the winter to allow for maximum infiltration of snowmelt water, often necessitating delayed seeding in the following spring. The first crops after tillage are usually wheat or barley, which have large seeds that can be planted deeply in a seedbed that has poor soil tilth. Several years of cropping can be required before it is possible to prepare a seedbed with a tilth suitable for shallow planting of small-seeded crops.

8.5 CONCLUDING COMMENTS

The major thrust of this chapter has dealt with the scientific basis and the relative effectiveness of different practices for the reclamation of sodic soils as well as examples of successful reclamation practices from different areas of the world. However, there are two additional major factors that lie beyond agronomic considerations, but which cannot be ignored by agricultural scientists, farm advisors, consultants, and farmers.

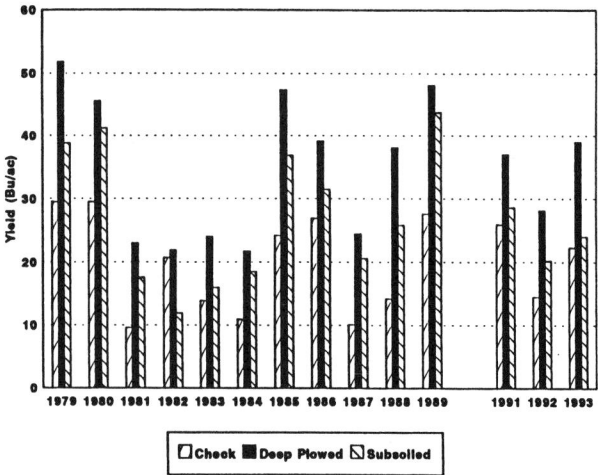

Figure 8.13 Effect of deep plowing and subsoiling on wheat yields on a dark-brown Solodized Solonetz (Typic Natraboroll) in east central Alberta [Lickacz and Coy, Saskatchewan Soils and Crops Workshop, Saskatoon, Alberta Agriculture, Food and Rural Developement, Edmonton, (1991); Lickacz, personal communication, (1994)].

Optimum reclamation practices depend on the particular problems requiring correction, the crops grown in the region, and the equipment available for tillage. In developed countries, mechanical power is available for deep tillage and other land preparation techniques (spreading and incorporation of chemical amendments, land leveling, preparation of high earthen dikes, etc.) required to prepare large areas of land (60–500 ha) so that large amounts of irrigation water can be applied and allowed to infiltrate. Cropping in conjunction with repeated tillage and heavy irrigation of saline soils is a common practice, as is cropping of sodic soils where tillage is combined with the application of amendments. Farmers in developing countries, who have only limited access to mechanical power, must rely more on crops like rice which can be grown in small fields under ponded conditions. Cropping in conjunction with reclamation is a popular method in both types of countries because it provides concurrent income; that is, it is a "pay-as-you-go" option.

Reclamation also has significant environmental consequences (Chapter 9). Improving soil in one place degrades soil and water resources somewhere else. Reclamation requires irrigation or rainfall, drainage, and a place for salt disposal. Whether artificial tile drainage is installed or not, salts will be displaced downward into soil strata and groundwaters beneath the irrigated land. Extensive reclamation often results in shallow saline water tables which may require installation of tile drainage. The resulting saline drainage water generally increases the salinity of receiving surface waters. These negative environmental impacts of reclamation require purposeful and long-term planning and education. Agriculturists working in such areas must know how to reclaim soils with minimum environmental impacts, how to provide the necessary information to local farmers and help them learn how to use it, how to tell others what the tradeoffs are between food production and environmental consequences, and how to inform future generations about what the economic and environmental tradeoffs are and will continue to be.

Acknowledgements
The authors gratefully acknowledge the editorial assistance of Jonathan Langford in the preparation of this chapter.

9

Sodicity, Dispersion, and Environmental Quality

M. E. Sumner, W. P. Miller, R. S. Kookana,
and P. Hazelton

9.1 SODICITY AND DISPERSION: AN ENVIRONMENTAL VIEW

Soil sodicity may have a number of impacts on environmental quality, distinct from its effects on agricultural productivity, which up to now has been the major focus of this book. In the discussion to follow, *environmental quality* will refer to anthropogenic impacts on soil, water, and air that have a negative effect on human and ecological health more or less directly. Some of these effects are obvious, but others are much less predictable.

The existence of sodic and Na-affected soils is a natural phenomenon, and thus in a certain sense, the environmental impacts resulting from the properties of these soils are likewise naturally occurring. Highly dispersive soils observed by Hilgard (1889) in California readily puddled under rainfall and produced high concentrations of soluble humic materials and suspended solids in runoff. That this occurred on virgin soils untouched by humans indicates that such phenomena must be considered within the context of normal pedogenic and hydrologic processes that have occurred for millennia at the earth's surface.

It is clear, however, that human activities in agriculture, forestry, urbanization, and waste disposal have, in many cases, enhanced the potential for Na-affected and dispersive soils to cause or exacerbate environmental degradation. It is largely this anthropogenic component on which this chapter will focus, namely, how human activities result in environmental damage as a result of their interaction with the properties of sodic and dispersive soils.

9.1.1 Sodic Behavior in an Environmental Context

In the context of this review, sodicity is not considered an environmental hazard *per se*. Rather, the interaction of sodicity and the dispersive behavior of soil colloids with human activities such as agriculture and soil contamination result in increased human and ecological risk. It is primarily the effect of colloid dispersion that results in this increased risk, due to a number of interacting factors to be discussed. In sodic soils, as defined in Chapter 1, colloidal particles become highly dispersible at low electrolyte concentrations (Chapter 3). Even at vanishing exchangeable sodium percentages (ESP), sodiclike behavior in the form of appreciable clay dispersion is evident is some soils. This "sodic behavior," in generating stable colloidal suspensions that are mobile in the environment, is a major threat to water quality under certain conditions (Stall, 1972).

Removal of vegetation dramatically increases the potential for raindrop impact and rill processes to enhance the formation of stable colloid suspensions arising from erosion and sedimentation. In addition, any type of soil contamination with metals, pesticides, etc., is more liable to be transported to surface waters as a result of their adsorption on dispersed colloids in runoff. Furthermore, the eroded land is degraded as far as crop production is concerned, with increasingly sodic materials being brought closer to the surface. This can cause waterlogging and may result in crop yield reductions or, in the requirement for increased inputs and management to attain satisfactory productivity.

Engineering properties of soils such as bearing capacity and soil strength are also negatively impacted by sodicity and dispersion. Again, although dispersive soils in the "natural" landscape may result in landslides, tunnel erosion, and other processes related to low soil strength, in the presence of humans, such phenomena are either exacerbated or increased in importance. The use of dispersive soils for urban and agricultural purposes must often be accompanied by special practices to ameliorate or compensate for dispersion-related engineering limitations of such soils.

9.1.2 Sodification and Soil Degradation

Agricultural practices have resulted in secondary sodification of many thousands of hectares of land (Naidu *et al.*, 1995d; Chapters 5 and 8). This increase in the area and extent of sodic soils has a significant potential to lead to enhanced environmental degradation, as well as the more immediate problem of reduced agricultural productivity. One by-product of cultivation and irrigation practices that leads to increased sodification is a deterioration in soil structure and reduced clay flocculation, which inevitably leads to increased dispersion. In many developing countries, the focus on productivity losses due to soil degradation or sodification results in less attention being paid to such environmental effects. In industrial countries, where tillage and chemical inputs may mask any potential yield declines as a result of degradation and there is greater interest in environmental issues, the impacts of dispersion on water quality tend to be obvious, even though these effects may not commonly be linked to the dispersion process.

Salinization of soils is a process related to sodification in semiarid to arid climates where soluble salts mobilized from within the soil profile result in significant land and water quality degradation. Fitzpatrick *et al.* (1994) have discussed salinization in Australia, noting that sodification and salinization by humans have resulted in increased salt levels in surface and groundwaters, expansion of poorly drained and swamp lands due to inhibited soil drainage, and increases in tunnel-related gully erosion. These changes occur even under nonirrigated dryland conditions due to deforestation and replacement of native vegetation, which alters the hydrology of the landscape due to changes in evapotranspiration over the year. While salinization will not be discussed in this review, related increases in ESP will be addressed as they impact the dispersibility of soils and consequently the environment.

9.1.3 Environmental and Engineering Consequences of Sodicity

Sodic behavior in soils, reflected in a high degree of dispersion of the colloid fraction of soil, promotes soil surface sealing, generating more runoff flow and production and delivery of fine sediment to surface water. This is followed by the enhanced capacity of colloids to aid in the transport of any contaminants that may be present in the soil or water column (Chapter 5). In this chapter, this chain of events is considered to be the major environmental impact of sodicity and dispersibility of soils. Chapters 5 and 8 have detailed the effects of sodicity and dispersion on

hydraulic conductivity (HC), infiltration rate (IR), erosion, and productivity. The importance of sediment as a nonpoint-source pollutant of surface waters which is traceable to the dispersible nature of certain soils will be considered here.

Transport of contaminants such as nutrients, heavy metals, anthropogenic organic compounds, and radionuclides by adsorption on colloids has been acknowledged as a major mode of mobility of some contaminants, both in surface water (McDowell *et al.*, 1981; Harter and Naidu, 1995; Helmke and Naidu, 1996) and, in some cases, in the subsurface (groundwater) flow (Buddemeier and Hunt, 1988; McCarthy and Zachara, 1989). Thus where dispersive soils occur in areas of soil contamination, enhanced mobility of these contaminants is likely.

There are other environmental consequences of sodic-related soil properties. In many cases, sodicity is associated with increases in salinity, particularly the presence of chlorides (Cl) in soil solution. Recent work has associated uptake of cadmium (Cd) by plants with increased concentrations of Cl (McLaughlin *et al.*, 1994). While this may to a degree be considered a "natural" result of soil and biological chemistry, its net effect is enhanced plant uptake of contaminants introduced by humans as fertilizer into the soil.

With respect to engineering properties of soils, naturally high levels of sodicity typically result in poor structural stability of soils and an increase in dispersion (Chapter 3). These soils often have low strength and bearing capacity, leading to the settling and cracking of foundations, failures or piping within dams and water impoundments, and related problems. Identifying and compensating for these effects is crucial in the effective use of these soils, and both the causes and potential remedies for sodic related soil engineering problems will be discussed in this chapter.

9.2 SODICITY, EROSION AND SEDIMENTATION

Although soil erosion is a serious concern from the point of view of crop productivity, the increased potential for the pollution of water bodies with sediments is possibly even more important and has certainly received less attention in the literature (Stall, 1972). Erosion is associated with the formation of surface crusts or seals, which markedly reduce the final infiltration rate (FIR) (Chapter 5). Sodicity increases the level of dispersible clay in the soil, causing the blockage of soil pores both at the surface and within the soil profile. This leads to crust or seal formation at the surface (Chapter 5) and the development of highly impervious textural B horizons within the profile (Naidu *et al.*, 1993b) (Chapter 6). These constraints to water movement in the soil lead to waterlogging, surface runoff and erosion, tunnel erosion, and lateral movement of subsurface water above the impervious B horizon. In addition, it should be remembered that sodic behavior can be exhibited in soils even at very low ESP levels provided that the electrolyte concentration is below the critical flocculation concentration (CFC), allowing clay dispersion to take place. Such highly erodible soils produce elevated levels of both organic and inorganic colloids suspended in runoff water during rainfall or irrigation events. The role which Na plays in promoting clay dispersion, crust formation, runoff, and erosion has been extensively reviewed by Shainberg and Letey (1984), Shainberg *et al.* (1989), and in Chapters 1, 3, and 5.

9.2.1 Waterlogging

Soils exhibiting sodic behavior usually have poor structure and associated mobile clay which clogs pores and cracks in the soil fabric, resulting in substantially reduced IR and HC values (Chapters 5 and 6). This process occurs extensively at the top of the Btn1 horizon in sodic duplex soils, which are very poorly drained as a result (Fitzpatrick *et al.*, 1994). An example of this condition is illustrated in Fig. 9.1. Consequently, such soils are prone to severe waterlogging during periods of extended or high-intensity rainfall, when the internal drainage of the soil is unable to cope with the large influx of water. As a result, such soils are extremely difficult to manage as they often alternate between being too wet and too dry. Waterlogging of a Natrixeralf is illustrated in Fig. 9.1. The "slushy" conditions at the soil surface can persist for many weeks, making it very difficult to till and plant fields on time. In addition, waterlogging can take a heavy toll on crop yield as a result of anoxic

Figure 9.1 Effect of sodicity on waterlogging of a Natrixeralf in Australia (Courtesy of Cooperative Research Centre for Soil and Land Management, Adelaide, Australia).

conditions in the root zone which can also promote toxicities of elements such as manganese (Mn) (Chapter 7).

9.2.2 Erosion Processes

While sodicity is not a prerequisite for erosion, clay dispersion certainly exacerbates the problem. Soil erosion involves rill (channel) and interrill (thin flow) processes. The former arises from the action of channelized water cutting into the soil and is responsible for the transport of most of the material off the soil surface while the latter occurs between rills being driven by raindrop impact and is responsible for the detachment and transport of silt- and clay-sized particles (Young and Onstad, 1978; Loch and Donnollan, 1983). Gullies can also be formed by subsurface flow in cracks and micropipes leading to tunnels, which ultimately collapse in highly sodic materials (Bryan *et al.*, 1978; Govers, 1987).

Unfortunately, most literature dealing with erosion has focused on soil loss from productive land rather than on the contamination of water bodies by suspended sediment. There is little doubt that soil loss degrades the land from which it is lost, and in many cases results in yield reductions as high as 50% (White *et al.*, 1984). Although such yield losses could be considered as environmental impacts of sodic behavior, they will not be discussed here as they have been discussed adequately elsewhere (Gilliam and Bubenzer, 1987). Unfortunately, most erosion studies present little chemical information about the soils investigated, and, therefore, it is impossible to evaluate the role that sodicity may have played in many of the erosion studies reported in the literature. Nevertheless, the suspended organic and inorganic colloids generated in the erosion process due to dispersion (sodic behavior) are responsible for the contamination of surface water bodies with particulates and associated pollutants.

9.2.3. Contamination of Surface Waters by Colloids from Erosion

As the contribution of water dispersible clay to crust formation and, subsequently, to runoff and erosion has been fully discussed in Chapter 5, the main focus of this section will be on the suspended colloids produced during the erosion process. As was pointed out in Chapters 1 and 3, the balance between exchangeable Na and electrolyte concentration is important in determining whether colloids will remain in stable suspension. Even many soils that fall in the Nonsodic Nonsaline (Sod$_L$ Sal$_L$) category (Chapter 1) can be highly dispersive, given low electrolyte concentration and inputs of energy. Thus the discussion to follow will present examples that cover the entire range of Na saturations in soils that give rise to suspended colloids. Increases in the level of exchangeable Na merely exacerbate the release of dispersible colloids. Despite this being well known, few field studies have measured the magnitude of stable colloidal suspension formation.

Yousaf *et al.* (1987) demonstrated under laboratory conditions that at low electrolyte concentrations (<10 $mmol_c$ L^{-1}), the concentration of clay that dispersed from soil aggregates with gentle inversion of the suspension increased markedly with the sodium adsorption ratio (SAR) (Table 9.1). Even at SAR values below 5, substantial concentrations of clay appeared in suspension. They also observed that substituting Mg for Ca increased clay dispersion under similar conditions (Chapters 5 and 8). Ali *et al.* (1987) demonstrated that the mineralogy of the clay dispersed from these soils was essentially independent of SAR and electrolyte concentration, with the composition of the clay in suspension reflecting that of the original soil. On Ultisols

TABLE 9.1 Effect of sodium adsorption ratio (SAR) and electrolyte concentration of clay dispersed from aggregates of three soils in water by gentle inversion of suspension [Yousaf, *et al.*, *Soil Sci. Am. J.* 51:920-924, (1987)]

Electrolyte concentration ($mmol_c L^{-1}$)	Clay dispersed at SAR (g kg^{-1})				
	0	5	10	20	40
Bonsall soil (Natric Palexeralf)					
10	-----	3.8	6.6	18.7	31.6
4	-----	9.1	14.1	28.7	49.3
1	6.2	12.7	15.4	30.7	54.6
0	7.5	13.6	16.3	31.9	59.3
Fallbrook oil (Typic Haploxeralf)					
10	-----	0.6	0.8	3.0	24.1
4	-----	3.8	19.8	37.7	114.6
1	0.3	28.2	33.7	58.4	165.0
0	5.0	32.7	40.3	97.6	196.7
Arlington soil (Haplic Durixeralf)					
10	0.5	0.7	3.5	4.8	18.2
4	0.8	11.1	19.4	26.4	57.4
1	5.1	18.2	23.7	35.5	69.8
0	22.7	23.9	30.4	65.9	101.0

TABLE 9.2 Soil loss and concentration of clay in runoff water from three highly weathered soils in small pans under simulated rainfall conditions (50 mm h^{-1} at 23 J m^{-2} mm^{-1}) for the initial dry and three subsequent events when the soil was wet [Miller, *Soil Sci. Soc. Am. J.* 51:1314-1320, (1987)]

Variable	Dry event	Wet event 1	Wet event 2	Wet event 3
Cecil soil (Typic Hapludult) (ESP 1.2)				
Mean IR (mm h^{-1})	29.8	11.3	9.5	8.5
Soil loss (kg ha^{-1})	266	149	155	157
Clay in runoff (g L^{-1})	0.19	0.09	0.09	0.09
EC of runoff (dS m^{-1})	0.011	0.007	0.006	0.005
Worsham soil (Typic Hapludult) (ESP 1.6)				
Mean IR (mm h^{-1})	11.9	3.6	2.3	1.6
Soil loss (kg ha^{-1})	1315	691	871	885
Clay in runoff (g L^{-1})	0.85	0.66	0.86	1.02
EC of runoff (dS m^{-1})	0.010	0.013	0.009	0.010
Wedwee soil (Typic Ochraquult) (ESP 1.2)				
Mean IR (mm h^{-1})	12.5	4.7	3.2	3.6
Soil loss (kg ha^{-1})	1135	521	589	601
Clay in runoff (g L^{-1})	1.03	0.83	0.93	0.88
EC of runoff (dS m^{-1})	0.010	0.009	0.008	0.007

and acid Alfisols with ESP values below 2 (Nonsodic Nonsaline [Sod_L Sal_L]), Miller (1987) demonstrated that erosion losses could be substantial and stable clay suspensions could be formed (Table 9.2) provided that the electrical conductivity (EC) of the runoff water remained exceedingly low (<0.01 dS m^{-1}) under both initially dry and wet surface conditions. The concentration of clay in the runoff remained relatively constant irrespective of the initial soil surface condition. Maximum clay concentrations observed were of the order of 1 g L^{-1}. In a complementary study, Miller and Baharuddin (1987b) obtained values of 3.3–16.4 g L^{-1} for suspended solids in runoff from 15 Ultisols (Sod_L Sal_L) under similar conditions. Under such conditions, surface waters could become polluted with suspended colloids and river and stream beds clogged with the larger size particles in the soil loss. Consequently, the resulting turbid waters become degraded habitats for fish and other aquatic life (Section 9.3.2). Such soils are highly sensitive to any inputs of Na such as that added in sodium nitrate ($NaNO_3$) fertil-

TABLE 9.3 Effect of NaNO₃ equivalent to 100 kg N ha⁻¹ and gypsum (5Mgha⁻¹ on infiltration rate (IR), soil loss, and proportion of clay in suspension of a Greenville soil (Rhodic Paleudult) over a dry and a wet event [Miller and Scifres, *Soil Sci.* 145:304-309, (1988)]

	Dry event			Wet event		
Treatment	IR (mm h⁻¹)	Soil loss (kg ha⁻¹)	Clay (%)	IR (mm h⁻¹)	Soil loss (kg ha⁻¹)	Clay (%)
Control	35.6b[a]	200b	18b	17.0b	211b	18b
NaNO₃	9.0c	3560a	55a	8.0c	480a	42a
Gypsum	46.7a	15b	0c	35.5a	63c	2c
NaNO₃ + gypsum	47.1a	31b	0c	37.4a	37c	2c

[a] Values followed by the same letter in a column are not significantly different at $p = 0.05$.

izer (Table 9.3) (Miller and Scifres, 1988). Addition of NaNO₃ at a rate of 100 kg N ha⁻¹ resulted in increased soil losses and suspended clay in the runoff water. On the other hand, gypsum (5 Mgha⁻¹) effectively reduced clay concentration to zero in both the control and the NaNO₃ treatments, which was also observed by Miller (1987) in the earlier study and by Kim and Miller (1996), who found that runoff water contained no dispersed clay at an EC above 0.5 dS m⁻¹. In a field study on a highly erodible Typic Kanhapludult, Scifres (1989) obtained a clay concentration of 4.2 g L⁻¹ in a runoff event having an EC of 0.05 dS m⁻¹. Shainberg *et al.* (1992b) investigated the effect of varying ESP (2–19) and simulated rainfall energy (0–800 J h⁻¹ m⁻²) and EC (0–0.95 dS m⁻¹) on the sediment concentration in runoff from a Typic Rhodoxeralf under laboratory conditions (Fig. 9.2). Increasing ESP resulted in sharply increased sediment concentrations, particularly for deionized water (simulating rainfall). Even when deionized water mist was used, substantial concentrations of sediment (40 g L⁻¹) after 60 mm of rainfall at ESP = 19 were observed in the runoff. All these laboratory results indicate that many of the world's soils exhibit sodic behavior and, consequently, are prone to generating suspended colloids under rainfall even though a large proportion of soils would not fall into categories considered to be highly sodic. Because most of these soils have suffered a loss of organic matter under cultivation, their structure is unstable and particularly sensitive to any increases in exchangeable Na.

In Australia, Loch and Donnollan (1983) found that the concentration of suspended sediment generated during runoff from simulated rainfall under field conditions was a function of the aggregate stability to raindrop impact and stresses

involved in rill flow. Maximum suspended clay concentrations were of the order of 3 g L⁻¹, despite the fact that the ESP values of the soils studied were below 2. They found that silt- and clay-sized sediment showed less temporal fluctuation than larger sized fractions, indicating that aggregate breakdown by drop impact and overland flow represents a continuous source of dispersible material with its transport being limited only by the rate at which it is supplied. On less stable soils, suspended load decreases with time because once the aggregates have been disrupted, clay is rapidly removed, leaving a less readily transportable sandy surface, as observed by Cummings (1981). Therefore, in general, suspended loads of dispersed clay are unlikely to remain constant over long periods of precipitation on most soils. In a study on southeastern U.S. coastal plain watersheds in which the soils contain dispersible clay but which have low ESP values (1–3), annual mean suspended sediment concentrations were relatively low (11 and 16 mg L⁻¹) because the low-gradient, heavily vegetated drainage systems promoted deposition of sediments (Sheridan and Hubbard, 1987). On the other hand in the Mississippi Delta, mean annual sediment concentrations were much higher (1320 mg L⁻¹) (Murphree *et al.*, 1985).

It should be pointed out that the clay concentrations in suspension generated in these studies are typical for sodic soils, and even the lowest values given are high in terms of the levels that are acceptable for various water uses (Section 9.3.2) (Table 9.4). Thus sodic soils that have been tilled

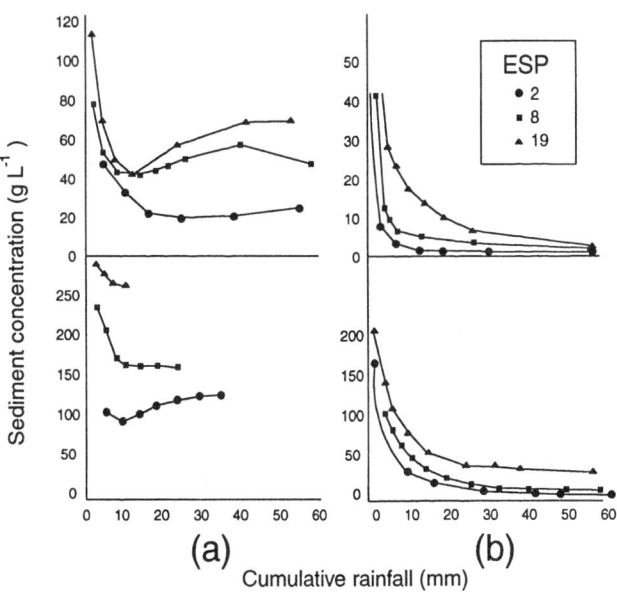

Figure 9.2 Effect of exchangeable sodium percentage (ESP) on sediment concentration in runoff as a function of (a) cumulative rain and, (b) mist for tap water (EC = 0.95 dS m⁻¹) (top) and dionized water (bottom) applied to a Typic Rhodoxeralf on a 35% slope [Shainberg *et al., Soil Sci. Soc. Am. J.* 56:278-283, (1992b)].

TABLE 9.4 Tolerable suspended solids in water for various uses [Stall, *J. Environ. Qual.* 1:353-360, (1976); Kundall and Rasmussen, *Erosion and Sedimentation: Scientific and Regulatory Issues,* Georgia Board of Natural Resources, Atlanta, GA (1995)]

Water use	Tolerable suspended solids (mg L^{-1})
Drinking	> 5
Industry	5–50
Irrigation	?
Swimming	10
Boating	20
Livestock	?
Aquatic life	0–25
Navigation	?
Hydro power	?

and subjected to rainfall are likely to contribute high concentrations of suspended clay in runoff water. Even sodic soils under natural vegetative cover contribute sediment to runoff water (Jenkin, 1986).

All this suspended sediment is readily transported to adjacent water bodies, leading to their degradation. However, great difficulty exists in estimating sediment loads delivered to surface water bodies (Novotny and Chesters, 1989). The concept of the sediment (and pollution) delivery ratio DR (Roehl, 1962) has been used,

$$Y = DR(A) \qquad [9.1]$$

where

Y = basin sediment yield
A = upland erosion or pollution potential estimated from models such as CREAMS (Knisel, 1980)
DR = delivery ratio

For example, Wolman (1977) concluded:

> The relationship between quantities eroded from the land surface and the quantities delivered to some distant downstream location is exceedingly tenuous. The sediment delivery ratio provides a cover for real physical storage processes as well as for errors in estimates of the amount eroded and for temporal discontinuities in the process.

Novotny and Chesters (1989) believe that delivery ratio estimates should only be used with extreme caution in nonpoint pollution studies:

> The sediment delivery process and its components are known only qualitatively. Breaking down the process into its com-

ponents–overland flow, vegetative filtration, and channel processes–and developing quantitative models and descriptions for each component are the most feasible approaches.

Accurate predictions of sediment and pollution yields will have to await such developments.

9.2.4. Subsurface Erosion

Piping or tunneling, which is the development of subsurface channels that may enlarge and collapse to form gullies, is a naturally occurring phenomenon in a variety of soils, including sodic soils. Tunnel erosion occurs widely in Africa (Stocking, 1977), Australia (Ford *et al.,* 1993; Fitzpatrick *et al.,* 1994), New Zealand (Hosking, 1967), and North America (Bryan *et al.,* 1978). In Victoria, over 82% of all recorded tunneling occurs on sodic duplex soils (Ford *et al.,* 1993). There have been a number of investigations of natural tunnel formation in sodic soils (Downes, 1946; Heede, 1971; Crouch *et al.,* 1986; Yaalon, 1987; Ford *et al.,* 1993; Fitzpatrick *et al.,* 1994), all of which agree that the presence of dispersible clay, a hydraulic gradient for water flow, and preferential flow paths are prerequisites for tunnel formation, although a layer of limited permeability remains a common characteristic (Boucher, 1990). The diameters of tunnels formed in Sodic Nonsaline (Sod Sal$_L$) (SAR$_{1:5}$ > 3, EC$_e$ < 4 dS m^{-1}) soils are usually much larger (0.7–1.8 m) than those formed in Nonsodic Nonsaline (Sod$_L$ Sal$_L$)(SAR$_{1:5}$ < 3, EC$_e$ < 4 dS m^{-1}) or in Sodic Saline (Sod Sal) (SAR$_{1:5}$ > 3, EC$_e$ > 4 dS m^{-1}) soils (0.01–0.11 m) (Fitzpatrick *et al.,* 1994). Generally tunneling is confined to soils of illitic and kaolinitic rather than mont-

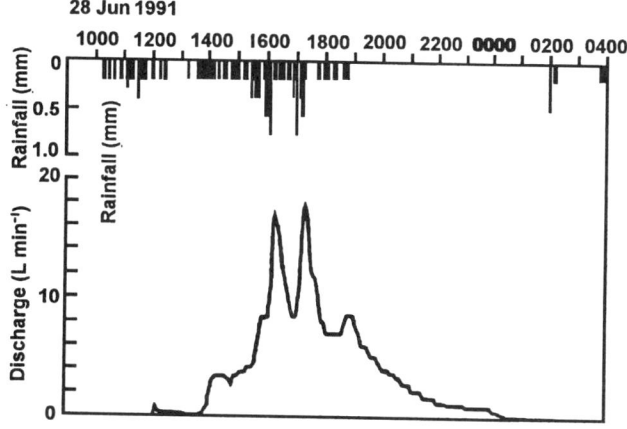

Figure 9.3 Rainfall and tunnel discharge from upper hill slope on Natrixeralf [Fitzpatrick *et al., Australian Sodic Soils: Distribution, Properties, and Management,* Commonwealth Scientific and Industrial Research Organization Publications, Melbourne, Australia, (1995)].

morillonitic mineralogy in Australia. Because of the dispersible nature of the soils in which large tunnels form and the rapid response of flow in these tunnels to rainfall (Fig. 9.3), large quantities of sediment are likely to be mobilized in the process, contributing to the contamination of water bodies. In addition, because these tunnels eventually collapse, open gullies further exacerbate erosion.

9.3 SUSPENDED SEDIMENTS, COLORANTS (TURBIDITY), AND SALTS AS POLLUTANTS

9.3.1. Sources and Characteristics of Turbidity

Basically, eroded material from land surfaces is the major contributor to water turbidity (a measure of water clarity) measured in nephelometric turbidity units (NTU) (Lettenmaier, 1991). In addition, under certain circumstances common in sodic soils (high pH and presence of elevated levels of Na), organic material can dissolve in runoff and percolating water, ultimately being discharged into surface waters (Naidu *et al.*, 1995a). This dissolved organic carbon (DOC) enhances clay dispersion and mobilization (Naidu *et al.*, 1993a) as illustrated by the strong relationship between DOC and suspended colloidal matter (Fig. 9.4). Sediment production, transport, and deposition can vary both

temporarily and spatially, making characterization of mobilized material difficult. Because past erosion has been the major contributor to sediment in streams and rivers, large quantities of sediment are currently stored in these water bodies. In terms of sediment transport, three time scales are involved: (a) the long-term migration of deposited material toward the ocean, which may take centuries, (b) short-term movement over short distances in a river, which occurs annually during high flow, and (c) rapid transport from the soil surface to the water body during storm events. In practice, it is difficult to specifically relate current levels of turbidity in a water body to any one of these potential sources. As a result of improved soil conservation practices, contributions from erosion have been reduced substantially in recent years, and consequently, turbidity in many rivers has fallen steadily over the past 60 years. Nevertheless, turbidity in many rivers is still above acceptable levels. Acceptable levels of suspended solids in water for various uses were presented in Table 9.4.

9.3.2 Effects of Turbidity on Environmental Health

Once eroded material reaches a water body, it makes contributions to both turbidity and bedload solids (material transported along or near the streambed). The turbidity of water is increased by colloidal organic and inorganic particles in suspension and by phyto- and zooplankton. In addition, nutrient inputs which promote the development of these organisms further impact water turbidity (Section 9.4.2). Turbidity adversely affects aquatic systems by light absorption and scattering, gill clogging of fish and crustaceans, and decreasing the water depth to the compensation point at which oxygen (O_2) use exceeds production. The smaller the particle size of the suspended material, the larger the relative effect on light absorption (Cuker *et al.*, 1990). Bedload solids reduce potential spawning sites and cause abrasion of eggs. Thus habitats for aquatic life become more restricted which, in turn, can cause changes in the size and distribution of populations.

For example, suspended silt and clay can seriously affect both filtering and assimilation rates of zooplankton which, in turn, impact planktivorous fish populations (McCabe and O'Brien, 1983). In addition, clays adhere to zooplankton, which can adversely affect feeding, molting, and swimming (Cuker and Hudson, 1992). On the other hand, while organic colloidal inputs can serve as food or nutrient sources for plankton (Arruda *et al.*, 1983; Lind *et al.*, 1992), increased total concentrations can adversely affect ingestion and incorporation by zooplankton. When sediments reach the ocean, offshore marine benthic communities suffer adversely and sea grass and live bottom beds can be smothered.

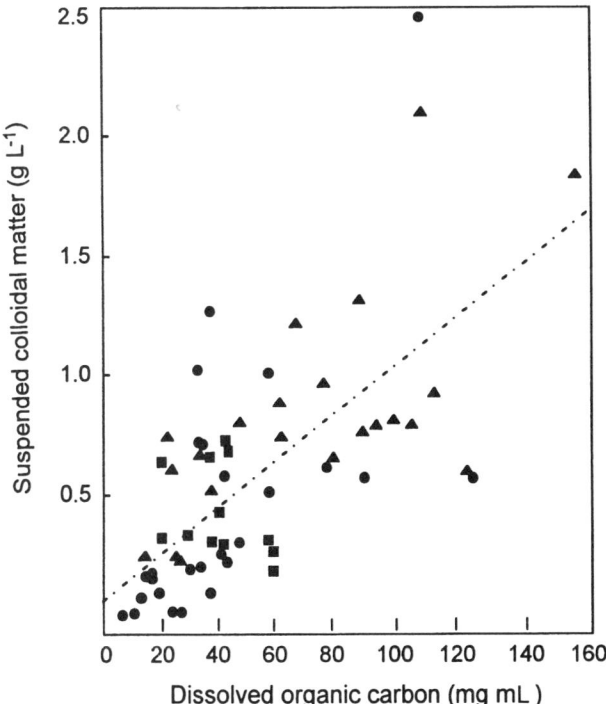

Figure 9.4 Relationship between dissolved organic carbon (DOC) and sediment loading in subsurface throughflow water under pine (▲), native woodland (•), and pasture (●) [Naidu *et al.*, *Aust. J. Exp. Agric.* 33:239-244, (1993a)].

Turbidity reduces fish feeding efficiency as well as habitat (Stuber *et al.*, 1982; Barrett *et al.*, 1992). Although some species are tolerant of sediment, many are very sensitive. For example, over a short section of a river in Georgia, where the total suspended solids increased from 10 to between 25 and 96 mg L^{-1}, as many as six endangered species of fish disappeared as a result of the effects of turbidity. Little is known about the effects on eggs and juveniles of tolerant species, which are likely to be more sensitive. Organic sediments deposited in water bodies can increase the sediment O$_2$ demand, resulting in considerable degradation of the aquatic environment which, consequently, might fail to provide suitable habitat for desired species or meet water quality standards.

Both organic and inorganic sediments can serve as carriers of eutrophying nutrients, toxic metals, and other pollutants (Novotny and Chesters, 1989) (Section 9.4), all of which can lead to a degradation of the aquatic environment, details of which are discussed by Sylvester *et al.* (1994). In addition to environmental health aspects, such as hazardous microorganisms or toxins transported with sediments, decreased aesthetic attraction of the water resource and associated beaches, and increased potential danger to swimmers and divers often result. The costs of purifying water for domestic use increase with increasing turbidity of the raw water (McCutcheon *et al.*, 1993). Finally, transported sediments can adversely affect power generation as a result of increased wear on turbine blades and reduced usable reservoir storage capacity. Sediment removal can be a significant operational cost for utilities that require water for thermoelectric purposes or industries that require water of high quality for manufacturing processes. Channel infilling resulting from sedimentation increases the costs of dredging and reduces water transport capacity during floods (US Bureau of Reclamation, 1977). Ribaudo and Young (1989) have estimated that erosion reduction in the United States could result in savings of over $500 million in terms of off-site benefits.

9.3.3 Water Coloration from Sodic Soils

Water coloration from sodic soils has not been extensively monitored, but clearly organic matter which readily dissolves under sodic conditions is sufficiently mobile to leave the soil and enter adjacent water bodies. In sodic soils, the B horizon is usually relatively impermeable, which enhances surface and subsurface horizontal flow (usually at the top of the B horizon) (Chittleborough *et al.*, 1992; Naidu *et al.*, 1993a). For example, DOC generated by sodicity has been found to enhance dispersion, and hence result in the mobilization of colloidal materials. In soils from the

Mount Lofty ranges of South Australia, Naidu *et al.* (1993a) found high levels of DOC (up to 20–25 mg L^{-1}) and obtained a good relationship between DOC and total colloidal loadings in the throughflow collected immediately above B horizons of duplex sodic soils (Natraqualfs) (Fig. 9.4). Considerable amounts of DOC are released throughout the year. The DOC loads peaked after the dry summer season as a result of the removal of dry, hydrophobic organic matter from the soil surface in overland flow. During the wet winter, DOC loads remained high as a result of subsurface flow through macropores to sodic B2 horizons and then laterally to streams (Smettam *et al.*, 1991). The DOC conveyed into stream water subsequently impaired the water quality in urban water supply reservoirs. Large concentrations of soluble Fe were associated with the DOC concentrations further exacerbating the coloration of the water (Fig. 9.5).

Recent studies (Nelson *et al.*, 1990, 1993) have established that the water quality with respect to DOC is controlled by the capacity of the soil to retain organic matter, preventing it from entering the soil solution for further transport. This, in turn, is governed by the clay content of the A horizon on which the organic matter is adsorbed. The effect of clay content on the mean DOC concentration of waters

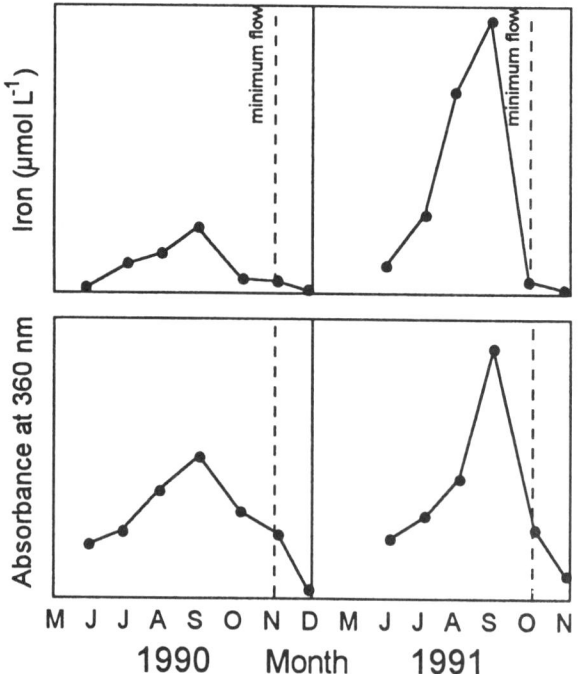

Figure 9.5 Seasonal variations in dissolved carbon index (absorbance at 360 nm) and soluble Fe in Tungali subcatchment, South Australia [Naidu et al., *Australian Sodic Soils: Distribution, Properties, and Management*, Commonwealth Scientific and Industrial Research Organization Publications, Melbourne, Australia (1995a)].

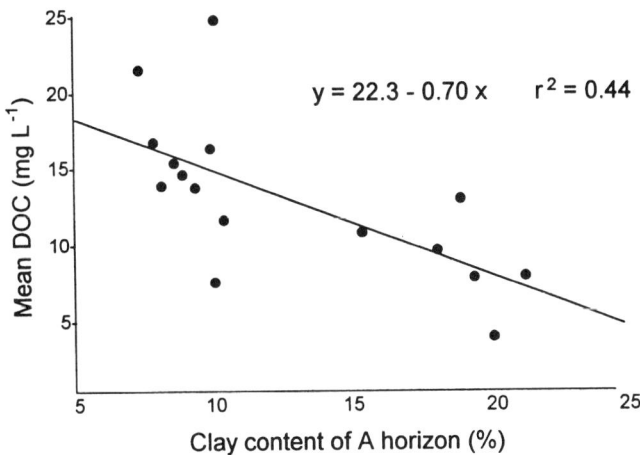

$$y = 22.3 - 0.70 \, x \qquad r^2 = 0.44$$

Figure 9.6 Relationship between A horizon clay contents of soils in various catchments in South Australia and Victoria and dissolved organic carbon (DOC) concentrations of associated stream waters [Oades, *Environmental Impact of Soil Component Interactions: Natural and Anthropogenic Organics*, Lewis Publishers, Boca Raton, FL (1995)].

from a number of catchments in South Australia and Victoria is illustrated in Fig. 9.6 showing that the DOC from sandy soils (5–10% clay) was much higher than that from loams (20% clay). The adsorption of organic matter on the clay fraction is also affected by sodicity and electrolyte concentration of the water, as would be expected from the considerations discussed in Chapter 3. Thus there is a good relationship between DOC concentrations and the SAR of stream waters (Fig. 9.7) showing that even small increases in Na levels (SAR = 1–3) resulted in elevated DOC concentrations, which contribute to increased turbidity in the water. Both N and P contents in the stream waters were correlated with turbidity, illustrating that organomineral complexes in catchments play an important role in determining water quality.

Similar levels of DOC (13–25 mg L^{-1}) to those presented have been reported by Lobartini *et al.* (1991) in surface waters in the southeastern United States where fulvic acids comprised 70% and humic acids 30% of the total. Such movement of clay and organic colloids can result in significant off-site migration of other sorbed contaminants into waterways (Section 9.4.2).

9.3.4 Transfer of Salts to Water Bodies

Many sodic soils contain appreciable levels of salts, which under natural conditions are slowly transferred to adjacent water bodies. When such soils are developed for agricultural purposes, the rate of transfer of salts is usually increased (Fitzpatrick *et al.*, 1995). In Australia and North America, large areas of soils with saline water tables have been deforested and planted to crops or pastures. Because of the lower evapo-transpiration of these systems relative to forest, watertables

continue to rise causing saline seeps which not only result in severe soil degradation but also contribute significantly to the salt load of rivers (Sharma and Williamson, 1984; Halvorson, 1990).

Drainage, which is usually a prerequisite to the irrigation of many sodic soils, produces large volumes of water several times more saline than the original irrigation water. This can have a great impact on the surface water quality, in some cases rendering it unsuitable for further irrigation or domestic and stock uses. Often the impacts of the saline water are greater on industrial and municipal users (damage to plumbing and water heaters, increased costs of water treatment, increases in objectionable taste characteristics) (Holburt, 1984) than on agriculture (Jones, 1984). In certain situations, elements such as selenium (Se) are leached from the irrigated environment, and when these waters are allowed to evaporate in closed basins, severe reproductive failures have been observed in waterfowl and aquatic organisms (National Research Council, 1989):

> Inevitably, irrigation . . . over time cannot avoid causing an adverse off-site effect. This effect must be acknowledged; it can be minimized, internalized, or rejected, but it cannot be ignored. If irrigation is a desired use of water, then its waste waters must be treated and/or disposal provided for.

9.4 COLLOID-ASSISTED TRANSPORT OF CONTAMINANTS

The factors governing the production of stable suspensions of inorganic and organic colloids under sodic conditions have

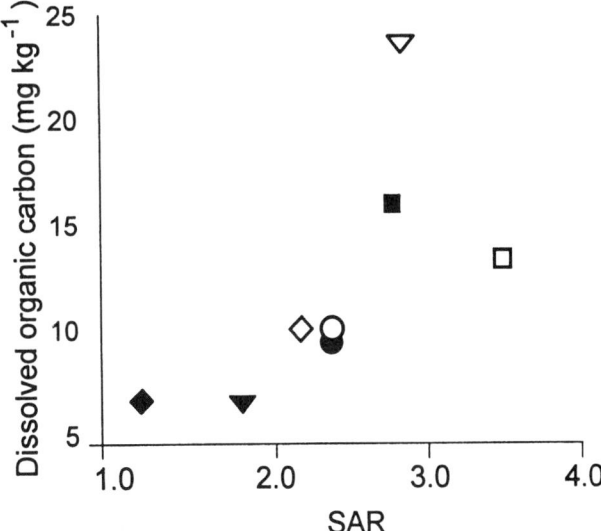

Figure 9.7 Relationship between dissolved organic carbon (DOC) concentrations and sodium adsorption ratios (SAR) of stream waters in Mt. Lofty ranges of South Australia [Skene and Oades, *Soil Sci.* 159:65-73, (1995)].

been discussed in detail in Chapters 3 and 4. With the movement of water of low electrolyte concentration from rainfall through such soil materials, colloids can readily become mobile. These colloidally stable particles move preferentially through macropores and cracks associated with soil structure, ultimately entering groundwater, as illustrated in Fig. 9.8 (McCarthy and Zachara, 1989). However, under strongly sodic conditions, such movement downward may be somewhat curtailed due to the highly impervious nature of many sodic subsoils (Chapter 6). Nevertheless over long time periods, there is substantial evidence (Gschwend *et al.*, 1988; Chittleborough *et al.*, 1992; Kaplan *et al.*, 1993; Naidu *et al.*, 1993a; Fitzpatrick *et al.*, 1994) to suggest that appre-

ciable quantities of organic and inorganic colloids can become mobile within the soil.

9.4.1 Adsorption of Nutrients and Contaminants on Soil Colloids

Metals and Radionuclides

Although most of the work on the sorption of metals and radionuclides has been conducted on soils which are not particularly sodic, it is pertinent to the present discussion because increasing the level of Na in the soil simply promotes colloid dispersion, and consequently, sorbed contaminants will exhibit increased mobility in the environment. In

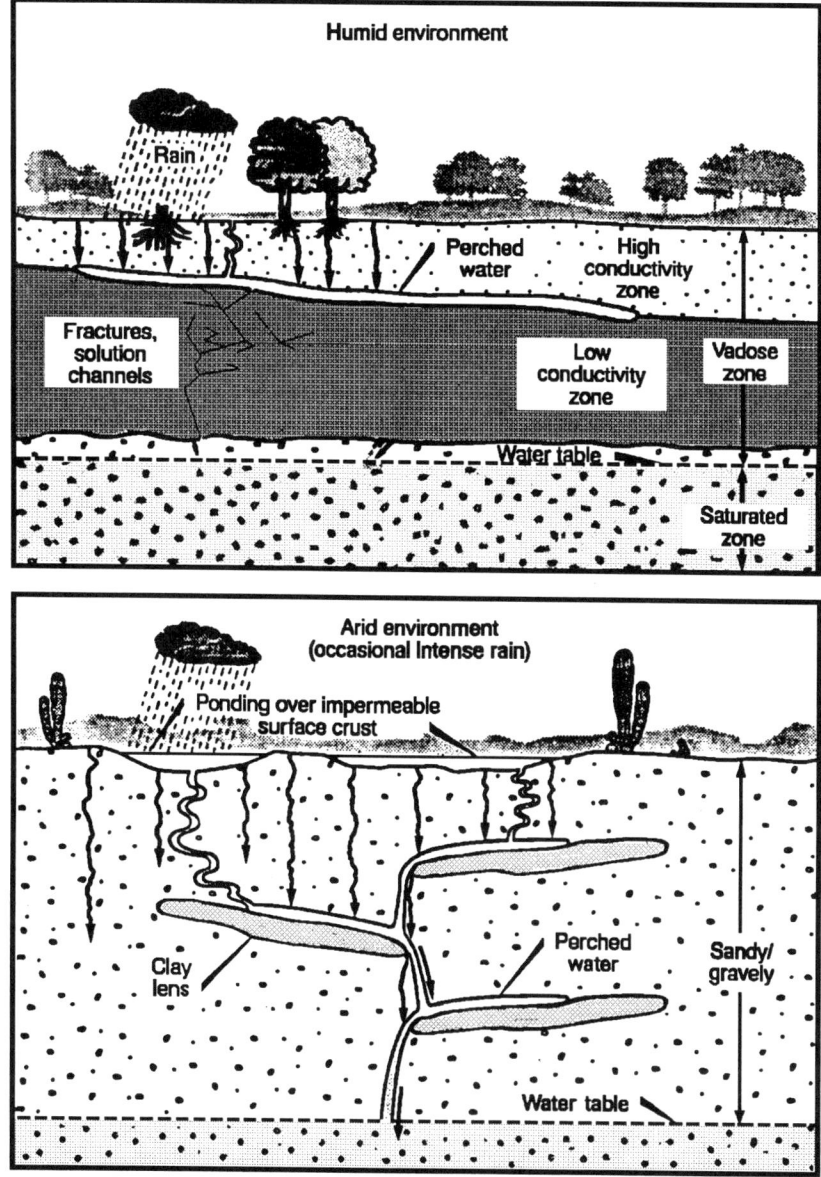

Figure 9.8 Illustration of colloid transport pathways through soil under humid and arid environments [modified from McCarthy and Zachara, *Environ. Sci. Tech.* 23:496-502, (1989)].

aquatic environments, many toxic trace metals (Cd, Pb, Cu) are strongly associated with particulate phases emanating from soils (clays, metal oxides, and organic materials) (Hoffman *et al.*, 1981; Lion *et al.*, 1982). In addition, sodic soils produce elevated levels of DOC, which can complex metals and assist in their transport. Naturally occurring organic matter in water has been shown to bind Cu in direct proportion to the DOC levels (Newell and Sanders, 1986), while toxic radionuclides such as americium (Am), plutonium (Pu), and europium (Eu) are strongly sorbed by natural organic ligands (Nelson *et al.*, 1985; Moulin *et al.*, 1989).

In soil, metals undergo several reactions with aqueous and colloidal phases, depending on the nature of the metal. Metal ions in solutions are closely associated with an enveloping sheath of water molecules (hydration) which, upon dissociation, releases protons and modifies the charge characteristics of hydrated metal ions. Because the pK_a values for most potentially toxic metal pollutants (Pb, Cu, Zn, Co, Ni, and Cd) lie in a narrow range from 7.7 to 10.08, they are mostly present as divalent cations at pH values common in many sodic soils, making them highly reactive with negatively charged surfaces. Metal cations that remain highly hydrated (e.g., Na, Mg) are readily exchangeable, whereas those which hydrolyze, form inner sphere complexes with the sorbing surfaces (e.g., Cu, Cd) and resist desorption back into the soil solution. Changing soil pH from 3 to 5 increased metal sorption from 20–100% due to hydrolysis and increased negative surface charge on soil colloids (Naidu *et al.*, 1994). The relative affinities of metals for soil surfaces tend to follow the Irving-Williams order, i.e., Hg > Pb > Cu > Zn > Ni ~ Co > Cd. Under reducing conditions common in sodic soils which promote the dissolution of sesquioxides, sorbed metals are readily released into solution, increasing their mobility in the liquid phase. Sorption of metals on colloids is also affected by competing cations and ligands and by the ionic strength of the solution, which can be as high as 0.1 *M* in saline sodic soils. Metals form complexes of varying stability with both inorganic and organic ligands, determined by their respective properties. For example, humic acid (HA) and fulvic acid (FA) strongly complex certain metals such as Cu, significantly influencing metal solubility and consequent environmental impacts.

Nutrients

From an environmental standpoint, P is the nutrient of greatest concern as it is usually limiting in natural water bodies and colloids assist in its transport. The sorption of P on soil particles has been well documented (Barrow, 1987). The reaction is rapid at first and continues for a long time (Sparks, 1986; Barrow, 1987), which makes sorption kinetics of great importance in contaminant transport through soils. Immediately following the application of soluble P to soil, stable (highly insoluble) complexes are formed with metal cations (Fe, Al, Ca, and Mg) and sesquioxide surfaces. Such sorption can lead to charge reversal, further promoting dispersion (Chapter 3). However, sodic soils are usually low in sesquioxides, which means that P immobilization will be less severe than in many other soils, allowing higher soil solution P concentrations to be subtended. If P-enriched colloids become mobile, substantial quantities of P can be transported to surface water bodies where the potential exists for the sorbed P to become bioavailable as a result of reducing conditions. Eutrophication usually results (Sharpley and Withers, 1994).

Pesticides and Other Organic Contaminants

Sorption of organic contaminants on colloidal particles is the first and, perhaps, the most important process that affects several other processes including bioavailability, degradation, volatilization, leaching, and runoff. As many of them are subsequently transported attached to mobile colloids, sodicity will promote the process. Commonly used pesticides include ionic (both cationic and anionic) and nonionic types. While interactions of pesticides with model materials (homoionic clays, ion exchange resins, organic matter fractions) are relatively well understood (Mortland, 1970), in natural soils, the complexity is much greater (Hamaker and Thompson, 1972). Sorption of cationic pesticides, such as diquat and paraquat, which is governed by the cation exchange capacity (CEC), will be high in sodic soils and affected by clay type and amount (Kookana and Aylmore, 1994). For nonionic pesticides, the amounts sorbed are more directly related to organic matter than to clay contents (Hamaker and Thompson, 1972; Green and Karickhoff, 1990) because the retention mechanism involves the partitioning of the chemical between the aqueous phase and the hydrophobic organic matter (Chiou *et al.*, 1979), akin to partitioning of an organic compound in octanol from octanol-water systems. Consequently, the sorption coefficient K of a pesticide in soil is generally expressed on the basis of the organic matter content in the soil,

$$K_{oc} = K f_{oc} \qquad [9.2]$$

where

f_{oc} = fraction of organic carbon in soil

K_{oc} = sorption coefficient per unit of organic matter

K can be measured directly by batch or column methods or estimated through semiempirical equations derived from physical properties such as solubility, melting point, and octanol-water partition coefficients (Briggs, 1973, 1981; Green and Karickhoff, 1990). Because the organic matter contents of sodic soils are somewhat limited, nonionic pesticides are likely to be less strongly adsorbed. Weber (1994), who has recently compiled the K_{oc} values for a large number of pesticides (Table 9.5), has summarized the sorption behavior of pesticides on soils in general as follows. The pesticides strongly retained by soil colloids are quaternary nitrogen (N) pesticides, organic arsenic (As) and P acid pesticides, organometalic fungicides, dinitroaniline herbicides, some organophosphorus pesticides and their metabolites, pyrethroids, and nonionic pesticides of very low water solubility. Weakly sorbed pesticides belong to the classes of carboxylic and aminosulfonyl acid herbicides; amide and anilide herbicides, hydroxy acid, carbamate, carbanilate, and highly water soluble nonionic pesticides.

Because most pesticides have a greater affinity for organic than for inorganic colloids, the nature and amount of

TABLE 9.5 Values of K_{oc} and soil reactivity for a number of classes of pesticides [Weber, *Mechanisms of Pesticide Movement into Ground Water*, Lewis Publishers, Boca Raton, FL (1994)]

Chemical class	K_{oc} Range	Soil reactivity
Quaternary N pesticides	$10^5 - 10^6$	Very high
Pyrethroid pesticides	$10^4 - 10^6$	High - very high
Dinitroaniline herbicides	$10^4 - 10^5$	High
Organic As and P pesticides	$10^4 - 10^5$	High
Organometallic fungicides	$10^3 - 10^5$	Moderate - high
Chlorinated hydrocarbon pesticides	$10^3 - 10^5$	Moderate - high
Phenylurea herbicides	$10^2 - 10^4$	Low - high
Organophosphate insecticides	$20 - 10^5$	Very low - high
Carbamate and carbanilate pesticides	$5 - 3000$	Very low - moderate
Thiocarbamate herbicides	$200 - 3000$	Low - moderate
Basic pesticides	$20 - 2500$	Very low - moderate
Amide and anilide herbicides	$150 - 1500$	Low
Aminosulfonyl acid herbicides	$30 - 2000$	Very low - low
Hydroxy acid pesticides	$10 - 2000$	Very low - low
Carboxylic acid herbicides	$1 - 100$	Very low - low
Fumigant pesticides	$10 - 60$	Very low

organic matter are of importance (Stevenson, 1982) in determining their transport both on the surface and in subsurface environments. For example, humic materials from grassland soils are rich in HA whereas in forest soils, FA dominates. The sorption capacities of HA and FA have been found to be different (Hayes *et al.*, 1978). Both the type of organic material and its stage of decomposition are important in determining the adsorption of some pesticides (Walker and Crawford, 1968). Furthermore, the rapid decline of organic matter content with depth controls not only the sorption, but also the degradation of pesticides. Because the organic matter content decreases rapidly with depth, particularly in sodic soils, pesticide mobility and degradation are either enhanced or reduced, creating a major impact on the groundwater pollution potential (Kookana and Aylmore, 1994).

Soil pH often has a significant bearing on the sorption affinity of pesticides because the CEC and the ionization of basic (*s*-triazines) and acidic (2,4-D, dinoseb, sulfonylurea) pesticides are pH dependent (Weber, 1994; Singh *et al.*, 1990). The sorption of simazine, chlorsulfuron, trisulfuron, and to a lesser extent linuron and fenamiphos, decreases with increasing pH (Kookana, 1989; Blacklow and Pheloung, 1992). Consequently, acidification, which is common in many Australian sodic top soils, and management practices such as liming result in soil pH changes which can affect pesticide sorption (Kookana, 1989). In addition, the pH at the colloid surface or in the rhizosphere soil may differ substantially from that of the bulk soil solution (Nye, 1986), causing a certain degree of uncertainty.

The reversibility of sorption reactions determines whether the soil solid phase actually acts as a permanent sink or merely provides temporary storage for the pesticide. Sorption-desorption hysteresis, which is not well understood, the phenomenon yet having to be incorporated into transport models, has frequently been noted for pesticides (Brusseau and Rao, 1989).

9.4.2 Contaminant Transport in Surface Runoff

During periods of heavy rainfall, surface-applied or soil-incorporated nutrients, pesticides, and other chemicals which are insoluble in water but have a high affinity for soil colloids can be transported with runoff sediment to adjacent water bodies. Both sodicity (Section 9.2) and organic compounds (Shanmuganathan and Oades, 1983a) (Chapter 4) promote colloid mobilization, leading to increases in colloid and contaminant loadings in streams (Fitzpatrick *et al.*, 1994). Any increase in the Na content of the soil will exacerbate this process. In addition to reducing input efficiency, the suspended particles with their contaminant load pose a threat to environmental quality. Detachment of sediments and their associated pollutants from the original soil is selective for dissolved pollutants, clay, and organic matter (Novotny and Chesters, 1989). Consequently, these materials occur in higher concentrations in the runoff than in the original soil, and this phenomenon is termed enrichment (Table 9.6). The enrichment in P and several metals is quite marked. Because of the association of many organic chemicals and pesticides with the colloidal fraction of soils, they are usually greatly enriched in suspended sediments (Novotny and Chesters, 1989). In general, as the delivery ratio decreases, the enrichment ratio increases, which means increased turbidity and pollutant load. When these pollutants reach ponds and lakes, they may be released under appropriate redox conditions and may move even faster than if they were transported in infiltrating water through the soil.

Rao *et al.* (1983) showed that total edge-of-field pesticide losses may be important in determining the long-term impacts of nonpoint source loadings on water quality in adjacent water bodies. Herbicide concentrations are usually lower than the lethal concentration (LC_{50}), but for insecticides, this value is often exceeded.

Metals

Due to the strong affinity of certain metal contaminants for soil colloids, metals which are usually relatively immobile in soil have been found to move in association with mobile colloids (Kaplan *et al.*, 1993). The role of humic substances, mainly HA

TABLE 9.6 Enrichment of suspended sediments in clay and associated pollutants relative to contents in the original soil [Novotny and Chesters, *J. Soil Water Conserv.* 44:568-576, (1989)]

Constituent	Units	Content in Soil	Content in Suspended sediment	Enrichment ratio[a]
Clay	%	27	91	3.4
Total P	$\mu g\ g^{-1}$	810	1,700	2.1
Pb	$\mu g\ g^{-1}$	19	39	2.0
Cd	$\mu g\ g^{-1}$	0.30	0.31	1.0
Zn	$\mu g\ g^{-1}$	69	280	4.0
Cu	$\mu g\ g^{-1}$	25	45	1.8
Al	$\mu g\ g^{-1}$	22 000	49 000	2.2
Fe	$\mu g\ g^{-1}$	21 000	46 000	2.2
Mn	$\mu g\ g^{-1}$	700	730	1.0
Cr	$\mu g\ g^{-1}$	29	56	1.9
Ni	$\mu g\ g^{-1}$	17	45	2.6

[a]Enrichment ratio = C_r/C_s, where C_r and C_s are concentrations in runoff sediment and original soil, respectively.

and FA, in the transport of radionuclides in natural aquifers is also well recognized (Kim *et al.*, 1984).

Madak *et al.* (1992) reported high levels of mobile trace metals in association with HA and FA in the river Ganges in India. In rivers in the southeastern United States, total Al concentrations correlate well with dissolved organic matter (DOM) (Shuman, 1992) whereas Naidu *et al.* (1995a) found a strong correlation between DOC and Al and Fe concentrations in the throughflow water collected at the top of the B horizon in duplex sodic soils in South Australia. Korentajer *et al.* (1993) investigated the transport of cadmium (Cd) in surface runoff from three South African soils of varying degrees of erodibility. Losses of Cd in runoff varied from about 5 to 50% of that applied, with over 93% being associated with the colloid transported in the runoff. Thus for metals that have a high partition coefficient to the solid phase (Cu, Pb, Cd) colloid-assisted transport would be a factor of major importance.

Elevated levels of trace metals (Cr, Cu, Pb, Zn) in various river systems in Australia have been reported (Cooper *et al.*, 1994; Helmke and Naidu, 1996) frequently exceeding the water quality guidelines set by ANZECC/NHMRC (1992). In many cases, the elevated levels of metals from irrigated agricultural areas have resulted from increased soil loss and sediments, highlighting the importance of particulate transport in loading the water bodies with trace elements. Storm event monitoring revealed a good relationship between high metal loadings and high flow rates, implicating sediment and particulate transport pathways from diffuse sources in metal inputs into waterways (Fig. 9.9). While the soils in the monitored catchments were not of a strongly sodic nature, the pathways of contaminant transport are relevant to sodic soils.

Nutrients

Nutrient runoff from agricultural land and other nonpoint sources is widely recognized as the major cause of surface water eutrophication throughout the world. Agricultural runoff has been identified as the cause of water quality problems in 55–58% of the surveyed river lengths or lake areas in the United States (US EPA, 1990). During runoff, colloids (clay, organic matter) which have higher P and other contaminant contents than coarse fractions are preferentially eroded leading to enrichment factors for P ranging from 1.2 to 6.0 (Sharpley *et al.*, 1994). Both the P and N supply in surface water bodies contribute to eutrophication, with the particulate fraction for P constituting the major portion (75–95%) reaching water from conventially tilled lands (Sharpley *et al.*, 1994). Subsequent release of P from sediment surfaces can lead to eutrophication. For example, nutrient loadings in Australian river systems received much

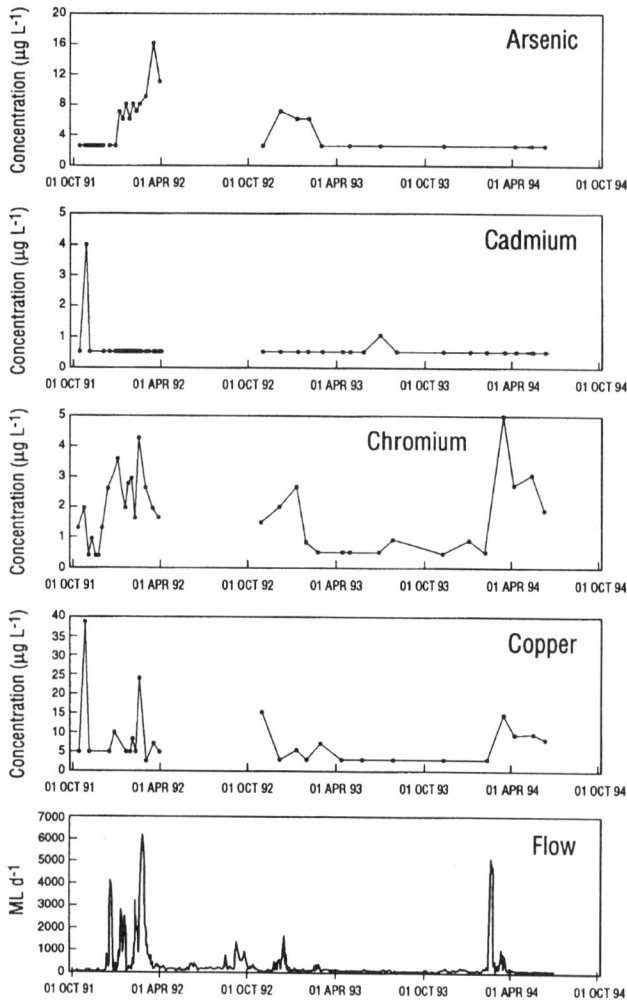

Figure 9.9 Time series plots of arsenic, cadmium, chromium, and copper concentrations and flow rate in the Barwon River, Mungindi, Australia, October 1992 to June 1994 [Cooper *et al.*, *Report on Trace Metal Monitoring*, New South Wales Department of Water Resources TS94.090, Sydney, Australia (1994)].

attention when a poisonous blue-green algal bloom which caused the death of fish and cattle, occurred in the Murray-Darling river system. Such heightened awareness in the community led to a comprehensive monitoring program for nutrients and other contaminants throughout Australia. As a result of this effort (Cooper, 1994), the Namoi River basin has been identified as the major contributor of nutrients to the Darling River, accounting for 42.5 and 154 Mg P per year in 1993 and 1994, respectively. The concentration of P in the water paralleled turbidity (Fig. 9.10), demonstrating that most instream P was associated with particulate matter, with only 20–45% of the total P being present as orthophosphate. Furthermore, storm event analysis showed that the highest turbidities and nutrient levels were associated with major storms. Indeed 48 Mg N and 16 Mg P were lost during a ten-day period of storms in the Namoi basin. The N:P ratio

was generally found to be low (<10:1) in all water samples monitored.

Sharpley (1995) has developed a site vunerability index based on transport (erosion and runoff) and source factors (soil test P, rate and method of P application, organic or inorganic), which is closely related to observed total P losses from watersheds in Oklahoma and Texas (Fig. 9.11), making it a valuable tool in identifying potential sites contributing to the pollution of water bodies.

Phosphogypsum (PG), which often contains appreciable levels of P, has often been used as an ameliorant for soil crusting (Shainberg *et al.*, 1989). The possibility that the P contained in PG which is usually applied on the surface could lead to water pollution was investigated by Korentajer *et al.* (1991). They found that P exports were determined by the effects of the PG rate on the quantity of P available for transport, as well as its effect on runoff volume and soil loss. Concentrations of P in runoff frequently exceeded 0.025 mg L^{-1}, the concentration at which algal proliferation begins in surface waters. Thus the use of PG while reducing sediment load (Chapter 5), may contribute to the eutrophication of surface water bodies.

Pesticides

Strongly sorbed pesticides (K > 100 mg L^{-1}) are relatively immobile in terms of transport in solution. However, due to their high affinity for soil surfaces, colloid-assisted transport is the major mechanism for their transport. Conversely, pesticides which are weakly sorbed are more prone to leaching than to transport associated with colloids. Pesticides are generally lost in the water phase of runoff, except for highly sorbed pesticides or where sediment loading is high (> 10 g L^{-1}) (Leonard, 1990). Pesticides with solubilities below 2 mg L^{-1} (cationic pesticides excepted) are preferentially transported on colloids in runoff, whereas those with solubilities greater than 10 mg L^{-1} are transported mainly in the water phase. However, the relative importance of sediments in pesticide transport is dependent on scale and the sediment-to-water ratio. Generally, the total loss of pesticides in runoff is strongly correlated with the total amount present in the top 0.1 m of soil (Leonard, 1990).

Because nonionic pesticides and organic compounds have high affinities for organic matter, erosion and transport of organic matter largely determine movement. Due to what is known as "raindrop stripping," eroded sediments have a much higher proportion of fine particles (clay and organic matter) than the parent soil, and consequently, are enriched from one- to five-fold in sorbed pesticides with the enrichment ratio being lower in fine-textured and highly erodible soils (Sharpley, 1985; Rose and Ghadiri, 1991). In addition, because aggregate surfaces have higher concentra-

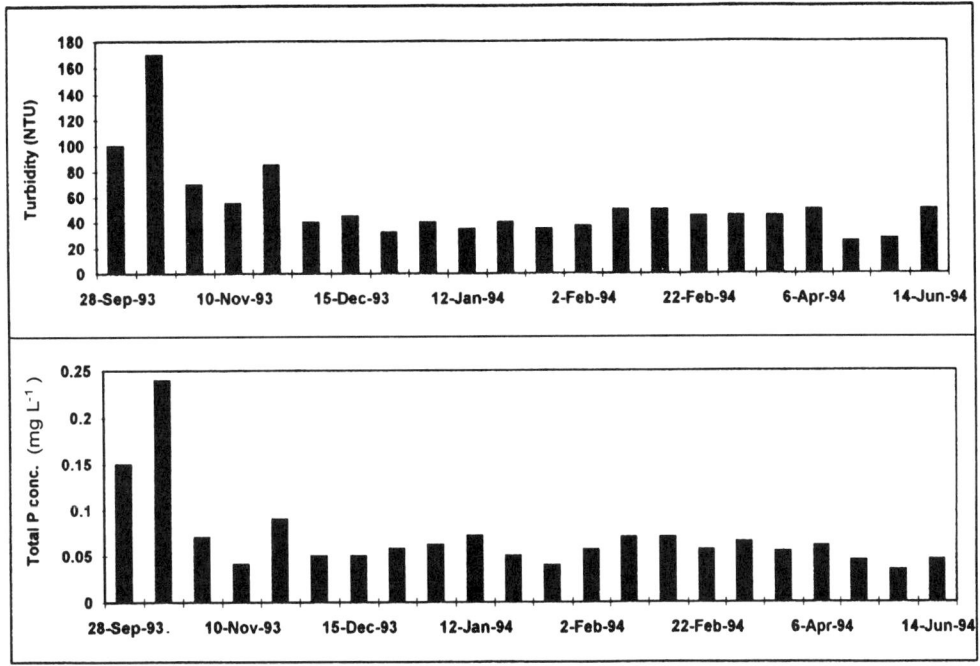

Figure 9.10 Turbidity and total P concentrations in Macquarie River basin, Oxley, Australia [Daly, 1993/1994 *Report on Nutrients and General Water Quality Monitoring*, New South Wales Department of Water Resources TS94.088, Sydney, Australia (1994)].

tions of pesticides than their interiors (Fig. 9.12), preferential transport is exacerbated further. Thus far, most work has focused on evaluating the enrichment of nutrients in sediments with little work on the pesticides being conducted. However, surface transport models such as CREAMS (Knisel, 1980) do include this process, but only in a very simplistic manner.

Reports on colloid-assisted transport of pesticides in surface waters originating from sodic soils are not available in

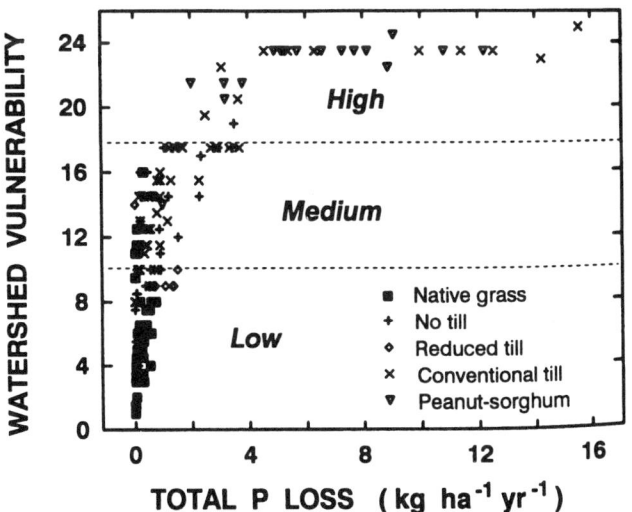

Figure 9.11 Relationship between P index rating of watershed vulnerability to P loss in runoff and measured total P loss in runoff [Sharpley, *J. Environ. Qual.* 24:947-951, (1995)].

the literature. However, there are several examples of pesticide transport from nonsodic soils that highlight the role of colloids in contaminant transport. Some of these examples are considered next.

The monitoring of pesticide residues in eastern Australia in areas where sodic soils are common (Cooper *et al.*, 1994) has shown that several pesticides (endosulfan, atrazine, metalochlor, and prometryn) frequently occur in surface waters, especially those associated with irrigated agriculture (mainly cotton). The levels of some pesticides exceeded the guidelines for the protection of aquatic ecosystems (ANZECC/NHMRC, 1992). In the case of endosulfan, 80% of the samples from irrigated areas exceeded the limit. Although the transport pathways of these pesticides from fields to rivers were not investigated, colloid-assisted transport is highly likely given that they have moderately high affinities for soil colloids. For example, in a storm, atrazine, prometryn, fluometuron, and diuron behaved similarly, showing initial high loadings falling away, and as the hydrograph peak arrived, the load rose again (Fig. 9.13). This bimodal pattern of export is probably due to colloid-assisted transport, resulting in an increase in loading as the hydrograph peaked. At these sites, a strong relationship between turbidity and P export was also observed. Peterson and Batley (1989) found that a large proportion (66%) of endosulfan and its metabolites was associated with suspended sediments in irrigation lagoons and canals in cotton fields in New South Wales. Agassi *et al.* (1995) dem-

Sodic Soils

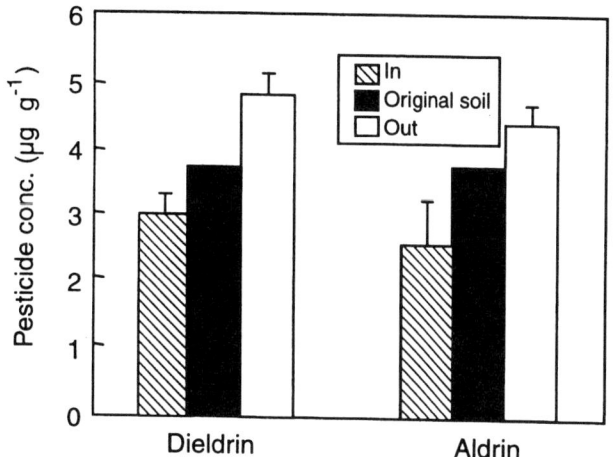

Figure 9.12 Pesticide concentrations in outer layers and inner parts of aggregates compared with concentration in original whole aggregate [Rose and Ghadiri, *Modelling the Fate of Chemicals in the Environment,* CRES, Australian National University, Canberra, Australia (1991)].

onstrated that reductions in the erosion of fine particulates will reduce the transport of sorbed pesticides from furrow irrigated soil.

9.4.3 Subsurface (Groundwater) Contaminant Transport

Subsurface Colloid Mobilization

The occurrence of colloids in groundwater and their role in facilitating the transport of contaminants is widely acknowledged. Commonly occurring colloids in the groundwater may be of detrital or authigenic origins or are produced through geochemical or biological processes (McCarthy

and Zachara, 1989). While layer silicates and sesquioxides may be present in the original geological material (detrital), the alterations of thermodynamically unstable primary minerals can also produce aluminosilicates, silica, and hydrous oxides. Colloidal transport is influenced by several factors, including ionic strength, solution composition, flow velocity, quantity, nature and size of suspended colloids, geological composition and structure, and groundwater chemistry (Puls and Powell, 1992). Once mobilized, the colloids do not necessarily flocculate as a result of increased ionic strength or the presence of excess Ca over Na because of kinetic limitations. Not only the mobilization but also the subsequent deposition and attachment of colloids in porous media are sensitive to solution chemistry (Elimelech and O'Melia, 1990; McDowell-Boyer, 1992).

In light-textured sodic soils subjected to low electrolyte conditions, clay becomes mobile and contributes to high turbidity (Rowell *et al.*, 1969; Felhendler *et al.*, 1974). Even at very low ESP levels, clay can be mobilized provided the electrolyte concentration of the leaching water is low enough (Chapter 3). Chiang *et al.* (1987) demonstrated that clay could be mobilized in Ultisols and Alfisols even at SAR values below 3 when the soils were subjected to water of low EC. Colloids can also be generated by dispersion from aggregates caused by the infiltration of waters of low ionic strength dominated by Na (Gschwend *et al.*, 1988). In sodic soils, Na dissolves organic matter in the form of soluble Na-humates that contribute to DOC, which is highly mobile. Under conditions which favor leaching, colloids move from the soil surface or root zone and reach the groundwater, especially through preferential pathways (Jardine *et al.*, 1989). Layer silicate clays mobilized from surface soils have caused turbidity in wells considerable

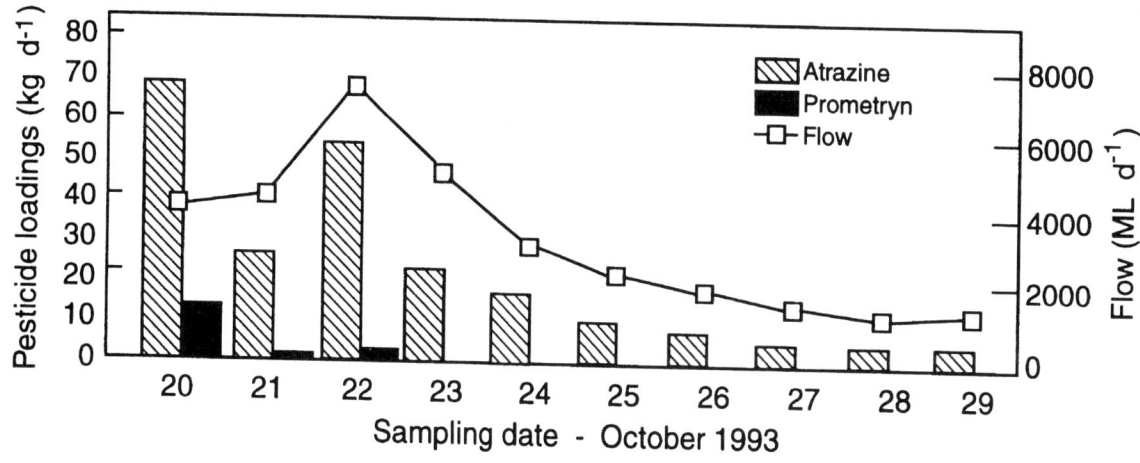

Figure 9.13 Atrazine and prometryn loadings in the Namoi River, Tarriaro, Australia, during 1993 storm event [Cooper, 1993/1994 *Report on Pesticide Monitoring,* New South Wales Department of Water Resources TS94.087, Sydney, Australia (1994)].

distances from recharge sites (Nightingale and Bianchi, 1977). Dissolved organic C as well as substantial numbers of bacteria originating from the surface have also been observed in groundwaters.

These colloids play a significant role in the subsurface transport of contaminants (McCarthy and Zachara, 1989). For example, humic substances because of their affinity for organic contaminants such as pesticides, metals, and radionuclides, can enhance the mobility of the contaminant, and consequently, its transport through soils (Ballard, 1971; Kile and Chiou, 1989). In addition, organic matter coatings on layer silicate clays can alter the reactivity of clays with contaminants (Dalang et al., 1984). In recent years, the association between the transport of contaminants such as radionuclides and inorganic colloids has been established in some field studies (Buddemeier and Hunt, 1988). For example, uranium (U) and its daughter products sorbed on Fe- and Si-rich colloidal particles were transported several hundred meters downgradient from ore deposits in Australia (Short et al., 1988).

Colloids are often highly mobile in aquifers. For example, Gschwend and Reynold (1987) showed that ferrous phosphate colloids became suspended and possibly mobile in a sand and gravel aquifer. Tannin and lignin components of pulp liquor, which is highly sodic, were found by Robertson et al. (1984) to migrate through a sand aquifer at nearly the groundwater flow velocity. McCarthy et al. (1993) studied field scale transport of natural organic matter through a sandy coastal plain aquifer. They found that the smaller sized (<3000 MW) and more hydrophilic component of the organic matter was more mobile than the larger sized hydrophobic components. Colloidal viruses have also been found to travel long distances through soils into groundwater (Wellings et al., 1975). Because natural organic matter potentially has considerable mobility through an aquifer, the transport of associated contaminants in groundwater becomes possible. Puls and Powell (1992), who studied the movement of iron oxide colloids through columns packed with sand and gravelly aquifer material, found that the colloids were not only significantly mobile but, under certain hydrogeochemical conditions, moved faster than tritiated water, the conservative tracer.

Organic Contaminants

Sodic soils can subtend substantial concentrations (200–300 mg L^{-1}) of DOC in the soil solution which can affect processes such as bioaccumulation, degradation, volatilization, solubility, sorption, and transport behavior of toxic organic compounds (TOC) (Senesi and Chen, 1989; McCarthy and Zachara, 1989). Solubilities of TOCs, such as DDT, PCBs, and 1,2,3-trichlorobenzene, can increase

many times in the presence of humic substances (Chiou et al., 1986). The affinity of DOC for hydrophobic organic contaminants was demonstrated by Poirrier et al. (1972), who noted that the coloring colloids in natural waters adsorbed and concentrated trace amounts of DDT present in water to levels far in excess (by a factor of 15 800) of that expected in pure water (Shuman, 1992). Wershaw et al. (1969) noted that the solubility of DDT was 20 times higher in a 0.5% aqueous solution containing Na humate than in water. Carter and Suffet (1982) also showed that a significant fraction of the dissolved DDT in natural waters is associated with DOC, with the extent of such binding depending on the source of the humic material, pH, ionic strength, and DOC concentration. The solubilities of other organic contaminants, such as polychlorinated biphenyls (PCB), 1,2,3-trichlorobenzene, and several pesticides, increase manyfold in the presence of humic substances (Chiou et al., 1986; Madhun et al., 1986).

In column studies, DOC even at low concentrations (5 mg C L^{-1}) increased the mobility of PCBs and Cd (Dunnivant et al., 1992b). At 20.4 mg C L^{-1}, the mobilities of a PCB (a hexachlorobiphenyl) and Cd were three- and twofold greater than when no DOC was present. Aqueous phase natural organic matter (NOM) has been shown to facilitate polyaromatic hydrocarbon (PAH) transport in soil column studies (Liu and Amy, 1993) by forming NOM-PAH complexes. Furthermore, the presence of NOM also affected the kinetics of interaction between PAH and the solid phases. Because the binding of TOCs on DOC increases with hydrophobicity, the presence of DOC is expected to increase the mobility of the most hydrophobic TOCs. Indeed, enhanced transport of hydrophobic TOCs (log K$_{ow}$ > 4) in the presence of DOC has been observed (Hutchins et al., 1983; Enfield and Yates, 1987).

Since hydrophobic constituents of DOC have strong affinities for many TOCs, rapid movement of DOC in the unsaturated zone (Jardine et al., 1989) may facilitate the transport of highly hydrophobic TOCs into groundwater. Kan and Tomson (1990) demonstrated, both theoretically and experimentally, that for a specific soil and DOM level, all organic compounds with K$_{ow}$ above a certain value move at the same rate in the groundwater (Fig. 9.14). They also reported that, in some situations, DOC can increase the movement of highly hydrophic compounds (such as DDT) by a factor of 1000 or more.

In addition to sorption and transport, DOC can affect the degradation of TOCs by dealkylation, photosensitization, and photodegradation, resulting in products with different mobility, toxicity, and persistence than those formed in the absence of DOC (Senesi and Chen, 1989). Pulp effluents, which are sodic and have relatively high DOC loadings

(Collins and Allen, 1991), may significantly affect TOC behavior in soil. However, these processes are not well understood and constitute a very important area of further research.

Radionuclides, Metals, and Metalloids

Mobilization of colloids can have a major influence on the transport behavior of metals in soils. For example, Pu and Am disposed of in liquid form at a seepage site in the United States were predicted to move only millimeters in the absence of colloid-facilitated transport, while, in fact, these radionuclides migrated up to 30 m (Nyhan *et al.*, 1985). In another study (Nelson *et al.*, 1985), Pu and Am were detected in monitoring wells located several kilometers away from a liquid waste outfall at Los Alamos in New Mexico. Ultrafiltration studies demonstrated that the transported radionuclides were indeed present on colloids. Similarly, Penrose *et al.* (1990) found detectable concentrations of Pu and Am some 3.4 km downgradient from the source, irreversibly associated with particles between 0.025 and 0.45 μm in size. In Australia, U and thorium (Th) sorbed on colloids were found in groundwaters hundreds of meters away from U ore deposits (Short *et al.*, 1988). In Germany, the presence of mobile ferric hydroxide colloids in groundwater facilitated the transport of some radionuclides (Kim *et al.*, 1984). Transport of cesium (Cs) was primarily controlled by the migration of clay colloids in soils (Torok *et al.*, 1990). Arsenate transport through columns packed with aquifer material was enhanced 21-fold over that of dissolved arsenate by iron oxide colloids on which it is strongly sorbed (Puls and Powell, 1992). Guggenberger *et al.* (1994), who studied the movement of Cd, Cu, and Cr through a Spodosol, found that Cr and Cu moved in the form of stable complexes with hydrophilic organic acids while Cd moved in inorganic form. These examples demonstrate that although metals and radionuclides are expected to be essentially immobile because of their high affinity for soil particles, these contaminants can be transported by mobile colloids in soil and groundwater over significant distances. However, although there is little information in the literature that deals specifically with transport under sodic conditions, a similar pattern of behavior can be expected.

9.5 EFFECT OF SODICITY OF PESTICIDES AND METALS ON TOXICITY AND FATE

9.5.1 Pesticide Fate

Because sodic soils are often strongly alkaline (pH 8–9.5), the behavior of several pesticides is affected, the best example of which is the sorption, persistence, and leaching of sulfonylurea herbicides. In several countries, including Australia, these herbicides have become very popular due to their high potency (grams per hectare rather than kilograms per hectare for other common herbicides) and low toxicity, among other factors. Under acid conditions, sulfonylurea is rapidly hydrolyzed and short lived, whereas in sodic soils, persistence is extended because the only remaining mechanism for degradation is microbial. This extended persistence of as little as 1% of the initially applied sulfonylurea can cause injuries to crops, especially legumes, in the following season (Ferris *et al.*, 1995). A further problem arises because these herbicides are weak acids (pK_a 4.0) existing predominantly in anionic form in sodic soils, and consequently, are prone to rapid leaching into the subsoil, where the lack of organic matter and associated microbial activity promote longevity. For example, in a Victorian sodic soil (surface pH 8.5), 55 mm of rain was sufficient to leach these herbicides into the more alkaline subsoil (0.3–0.4 m) (Ferris *et al.*, 1995) where suitable acid conditions for hydrolysis and microbial activity are absent. In relatively dry regions, rainfall may be insufficient to leach the material from the root zone resulting in "carryover" injury, while in wetter regions, where leaching takes place, groundwater contamination may result. Long-term computer simulations by Ferris *et al.* (1995) show that substantial leaching of chlorsulfuron is possible, but whether this will pose any toxicological risk in groundwater is unclear.

Sodicity has indirect consequences in terms of modifying pesticide behavior. For example, because the accumulation of organic matter is not favored under sodic conditions, pesticide behavior may be different than in nonsodic soils, where most of the pesticide testing takes place. In

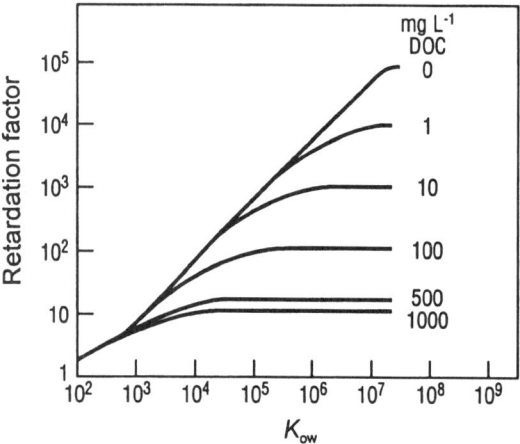

Figure 9.14 Theoretical effect of dissolved organic carbon (DOC) concentrations on relative retardation factor for compounds of different octanol water partition coefficients K_{ow} [Kan and Tomson, *Environ. Toxicol. Chem.* 9:253-263, (1990)].

addition, the landscape hydrology and water transmission properties of a soil are affected by sodicity, which, in turn, can affect the persistence and potential leaching of pesticides through soil and also runoff from a catchment.

Properties such as pH have been found to affect pesticide fate and behavior in soils as well as in surface waters. For example, Gaynor and Volk (1981) compared the surface losses of atrazine and terbutyrin in runoff wter and sediments from limed and unlimed plots under artificial rainfall conditions. They found that losses of atrazine and terbutryn in both runoff water and sediments from unlimed plots were higher than that from limed plots during a simulated rain after 12 h of herbicide applications, presumably due to an increase in infiltration and reduction in runoff due to liming. While pH had less effect on terbutryn persistence, rapid chemical degradation of agrazine was noted at lower pH of unlimed soil. Higher terbutryn adsorption was also a factor that contributed to a greater loss of terbutryn than agrazine under comparable conditions.

9.5.2 Salt Effects on Metal Toxicity

Because many sodic soils contain appreciable levels of salts, the role they may play in releasing and solubilizing toxic heavy metals and other contaminants and increasing their availability to biological systems, both terrestrial and aquatic, needs to be discussed.

Gambrell *et al.* (1991) found that increasing salinity from 2 000 to 10 000 mg L^{-1} in a wetland sediment contaminated with heavy metals increased the levels of Cd, Cu, and Zn in solution as a result largely of exchange reactions. Provided that substantial contamination had taken place, these increased concentrations of metals in solution could adversely impact aquatic life if salinity levels increased. On the other hand, Kunishi and Glotfelty (1985) found that freshwater inflow in an estuary of the Chesapeake Bay reduced salinity sufficiently to lower divalent cation concentrations, allowing the sediment to release bioavailable P. However, although readily detectable, the P contribution from this source was more than matched by that from the mineralization of organic sources.

Recent research (McLaughlin and Tiller, 1994; McLaughlin *et al.*, 1994) has identified soil salinity as a major factor contributing to the increased uptake of Cd by crops such as potatoes (*Solanum tuberosum*) and sunflower (*Helianthus annus*) grown on soils fertilized with phosphatic fertilizers containing Cd. The formation of considerable concentrations of the ion pair CdCl$^+$ (Smolders *et al.*, 1995) increases the availability of Cd to plants, as illustrated in Fig. 9.15, where Cl$^-$ concentrations in soil are directly related to Cd concentrations in the tubers.

9.6 ENGINEERING CONSIDERATIONS IN SODIC SOIL MANAGEMENT

9.6.1 Soil Strength and Bearing Capacity

The property used to assess the ability of the soil to support heavy loads without shear failure is the shear strength of a soil. For example, in road cuts, the shear strength of exposed faces dictates the maximum angle achievable before embankment failure occurs. Shear strength also governs the ultimate bearing capacity, which is the pressure at which shear failure will occur in the soil beneath a foundation (Brink *et al.*, 1984). Because the bearing capacity of the soil determines slope stability, increasing levels of stress cause rupture. Consequently, because soils with elevated levels of Na disperse or swell, or both on wetting, resulting in aggregate breakdown, embankments often fail on sodic soils. Sodic soils pose problems in building earthworks such as dams, which often are susceptible to tunneling failure (Section 9.6.3). When sodic soils swell and shrink with changing moisture content, road pavements, foundations, and underground services including water pipes, optic cables, etc. can be damaged (Palmer and Hazelton, 1994). According to Aitchison and Wood (1965), there is a clear interaction between the level of compaction achieved during construction and the consequent permeability of the material, which determines the potential for postconstruction deflocculation in such material. Deflocculation has been shown to be an important antecedent of piping (Section 9.6.3) failure in earthen dams.

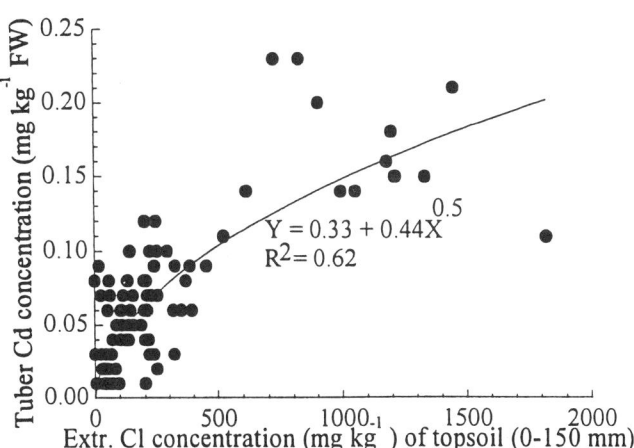

Figure 9.15 Relationship between soil Cl concentrations and Cd content of potato tubers [McLaughlin *et al.*, *J. Eviron. Qual.* 34:1013-1018, (1994)].

9.6.2 Sodicity Effects on Engineering Uses of Soils

To illustrate the effects of sodicity on the use of soils for various construction purposes, a number of examples will be selected.

Urban Development

A large area of the hinterland surrounding Sydney, Australia, is underlain by shales with a dispersed siderite content and a variable (often high) proportion of unstable sodic clay in the subsoil. On these sodic soils, slope failures at gradients between 7 and 10% are prevalent, and this is especially evident on slopes with a south to southeasterly aspect. These slopes which face the sea probably receive increased accessions of Na. The exposure of the subsoil to rainwater combined with the elevated ESP levels results in low soil shear resistance. Consequently, mass movement and gully erosion occur frequently. The proximity of these areas to Sydney has placed them under pressure for residential development. In the predevelopment phase of the land, the gullies, which have formed as a result of erosion of the highly dispersible subsoils, are often mechanically filled in, compacted, and smoothed so that the landforms are remolded and watercourses diverted. During the preconstruction phase, any sediment that becomes mobile as a result of the action of rain on the dispersible materials is collected in ponds (detention basins), which are landscaped into the housing estates. To clarify these ponds, gypsum is used as a flocculant, but problems associated with the dissolution of the large quantities required are often encountered. The rates of gypsum required depend on the solubility of the material, pH, and the clay mineralogy of the suspended sediment.

Densely vegetated "buffer" strips on gently sloping land surrounding construction sites have been found to be highly effective in removing sediment, allowing clear water to exit the area (Wilson, 1967; Jackura, 1980).

To overcome the low bearing strength of these soils, foundations require special engineering specifications and are usually built on piles or on rafted or waffle slabs to resist any effects of soil movement that might subsequently take place.

Stormwater and Sediment Basins

Because sediment derived from dispersible soils remains in suspension for long periods of time, substantial efforts have been made to clarify the water using various flocculants, such as gypsum or alum (Weaver, 1969). In highly dispersible soils in New South Wales, Australia, sediment basins have been treated and water cleared using gypsum. However, after a subsequent large rainfall, the water quality (turbidity) deteriorated to well below the En-

vironmental Protection Authority (EPA) standard of 50 mg L^{-1} of sediment or 30 mg L^{-1} of sediment in sensitive areas such as national parks. As a result, retreatment was required (Hazelton, 1992). Because gypsum appears to be somewhat ineffective over long periods and requiring 72 h for complete flocculation, the use of alternative flocculants such as alum, ferric chloride, or polymers has been suggested. For a more persistent treatment, Ferguson (1994) recommends the use of a "riffle" of gypsum rock above entry points to the basin at elevations only submerged by significant storm flows. Alternatively, an automatic proportional dispenser at the inlet can be used to meter sufficient flocculant in to order achieve adequate flocculation of suspended clay.

Alum has been used in the treatment of stormwater in several areas in Florida and Washington (Livingston *et al.*, 1994). Because the sediment from highly erodible soils in the Florida catchment did not settle within the typical residence time of most management practices for urban development, a stormwater treatment system was designed using sonic flow meters and variable-speed injection pumps to automatically inject liquid alum into the storm sewer lines upstream of the lake at a flow-proportioned rate regardless of the incoming stormwater flow. This system promotes mixing due to turbulence in the storm sewer line, and the floc produced settles on the lake bottom, providing the added benefit of promoting nutrient (P) inactivation in the sediments.

ALFLOC 8103, a highly charged liquid cationic polymer flocculant of moderate molecular weight, has been evaluated on residential sites on the north coast of New South Wales (Towler, 1994) and has been effective in removing suspended clay from sediment basins (20 ML). However, the basin required to be redosed after the next runoff event several weeks later as there was no residual effect from the initial dosing. This product, which meets the USEPA drinking water standard, is effective over a wide pH range.

Filter fabrics used in conjunction with gabions, rock check dams, and perforated standpipes have been shown to be effective in removing all but the finest suspended colloidal particles from throughflowing water (Ferguson and Gonnsen, 1993).

Roads

Road batter failures are common in sodic soil areas in Australia. For example, total slope failures with slumps of up to 200 m in length due to clay dispersion along desiccation cracks and root holes often result when cuts are made through sodic soils with low wet bearing strength. Once the original vegetation is removed, the dispersible soil material restricts seedling emergence and plant growth, resulting in poor stabilization. Several large storms are sufficient to cause severe rilling, resulting in embankment slumping. After several attempts to stabilize slopes by re-shaping and compaction, use was made of a paragrid geotextile. It was necessary to excavate the clay and back-fill with rock ballast. After compacting the section, the geotextile was positioned on the slope as the fill was re-placed and compacted. Seeded topsoil retained by the geotextile promoted vegetation growth, resulting in a stable slope. Sea facing aspects close to the ocean, which pre-sumably received greater Na accessions, result in more frequent batter failure than those facing away, although both can contain sodic soil materials (Road Transport Authority, 1992).

Excavation of Drains

Excavation of a drain through a residential subdivision in Albury, New South Wales, during realignment of a creek to allow the maximum use of the site, induced tunnel erosion (Crouch, 1977). Twelve months after excavation, tunnels extended 7 m from the drain wall and threatened adjacent residential blocks. A survey of a major tunnel by excava-tion and probing indicated that tunnel formation occurred along soil cracks around large structural units (0.5–1 m in diameter). The soil, which had a high ESP (6.5–18.6) with a volume expansion of less than 5% and Emerson Class of 1–3, (Emreson, 1967) was relatively impermeable.

The following remediation measures were suggested. The tunnels should be collapsed by ripping the channel length-wise to prevent further enlargement. By battering the sides of the drain, the hydraulic gradient should be reduced to less than 3:1 and gypsum should be applied. According to Crouch (1977), effective reclamation has resulted if the minimum amount of gypsum required to increase the Dis-persion Index (ratio of suspended particles <0.004 mm with dispersing agent relative to the same particles in water) to 3 is doubled. By spreading suitable topsoil over the area and revegetating, the extent of soil cracking and the rapid infiltration of water along preferred pathways can be re-duced. Regular maintenance includes filling any cracks that develop before they enlarge by erosion.

Flood Control and Irrigation Channels

Sherard et al. (1977) described several examples where dispersive clay piping had badly damaged or completely destroyed concrete canal linings used for flood control and irrigation. Runoff at the top of the canal bank percolated downward into cracks when the canal was empty and then flowed into the canal through the construction joints in the lining. In an extreme case, after a few years, the concrete lined canal was transformed into an earthen canal about twice as wide and much deeper with only disconnected remnants of the concrete slabs lying around. It appears that concrete-lined canals and dispersive clay were not com-patible.

Conclusions

Dispersive soils are subject to surface sealing and crusting, which restricts revegetation of sites under construction. Increased runoff from these soils results in severe rill and tunnel erosion. Batter failures and the undermining and clogging of roadside drains often create a flood hazard in low-lying residential areas. Sodic soils with low wet bear-ing strength are usually pliable and deform easily under pressure when wet. In addition to making site access diffi-cult, they are generally unsuitable for foundations without specific engineering design modifications Soil strength, settlement, and swelling behavior are properties that affect local roads and streets because they influence the ease of excavation, grading, and traffic supporting capacity. Clay dispersion from these soils results in suspended sediments which cannot be removed in gross pollutant traps. This suspended material exacerbates the turbidity of surface waters, negatively impacting aquatic ecosystems.

9.6.3 Piping and Related Phenomena

In addition to being a natural phenomenon (Section 9.2.3), piping or tunneling frequently occurs in earthworks con-structed on sodic soils. For example, in New South Wales, Australia, where sodic soils are common, the following tests are routinely carried out to determine the suitability of soil for earthwork construction: particle size analysis, Atterberg limits, dispersion percentage, linear shrinkage, volume expansion, and the Emerson aggregate test (Emerson, 1967). The relationship of these tests to suscep-tibility to tunneling in earthen construction is illustrated in Table 9.7. Thus soils which slake readily (Emerson Class 1 or 2) and contain more than 33% dispersible clay are susceptible to tunneling.

The colloidal dispersibility of soil can also be directly measured by the pinhole test in which distilled water flows through the soil under a specific head (Sherard, 1976). Because this test simulates a leak in a clay dam, it is used directly to predict soil behavior in such constructions and is widely recognized as giving the most reliable and repro-ducible results (Statton and Mitchell, 1977). However, prior to testing, samples should be maintained at their natural moisture content because air-drying can cause some nor-mally non-dispersive clays to disperse in the pinhole test (Shafer, 1978). Nickel (1977) indicated that only the Emerson crumb test and the pinhole test directly model the condition for clay dispersion where shear stress applied by hydraulic flow must exceed the shear strength of the zone of expansion. Heinzen and Arulanandan (1976) have shown that a critical shear stress of zero defines a dispersive clay system. They also found that dispersion was independent of the nature of the clay when the electrolyte concentration was low and the SAR high. In the engineering context, Sherard *et al.* (1972) concluded that the permeability, shear strength, and compressibility of the subfoundation materials and compacted core, and the presence of dispersive clays are the properties that promote piping in earthen structures. These properties most significantly influence the performance of an earthen dam or embankment. The soil mass permeablility within an embankment or a small earth dam is greater and desiccation cracks may be deeper and larger than in bigger structures, where the weight of the dam increases the compressive stresses, tending to close the cracks. Differential settlement cracks may provide preferred water flow channels where piping could initiate. Consequently, an excessive hydraulic gradient in a narrow core may result in internal erosion or piping. These properties are thought to have contributed to the failure of the 93 m high Teton Dam in Idaho, United States on first filling. Fourteen lives were lost and the total damage was estimated at $400 million. Instrumentation installed in earth dams and other large embankment structures to monitor their long-term performance has led to improved design concepts (Brink *et al.*, 1984).

According to Sherard *et al.* (1976), a concentrated leak such as a crack from any source or a leakage channel left open by a construction deficiency, is required for a dam to fail by dispersive clay piping. However, failure could result if a leak developed later under unusual conditions such as cracking induced by earthquake or dessication and subsequent cracking of the wall during long periods of low water level. Crouch *et al.* (1991) concluded that, even when dispersion limits are set, a range of uncertainty exists in the prediction of earthwork performance because limits are based on average conditions of construction and filling. Sherard *et al.* (1976) speculated that many such failures have not been diagnosed as dispersive clay piping. They contend that it is reasonable practice to build major embankment dams using materials containing dispersive clay provided appropriate design criteria are used. Piping in dispersive clay in dam foundations is unlikely to occur below the depth at which the clay is continuously saturated with water.

Richie (1963) set an arbitrary limit equivalent to 30% dispersion of particles smaller than 5 m, above which the soil would be considered to be susceptible to tunneling failure. According to Crouch *et al.* (1991), subsequent experience with many hundreds of soils and case histories shows no reason to change this limit. Soils suitable for earthwork construction should contain enough dispersible clay to seal, but not enough to allow tunnel failure. To be effective, the clay must disperse sufficiently to move and fill voids between the larger particles.

The formation of settlement cracks can be reduced by ensuring that there are no sharp changes in slope in the earthwork foundations. A high level of compaction should be achieved throughout the wall and not only in the upper layers. Watering to increase the level of compaction is difficult in dispersible soils, and construction may need to be deferred until the soil moisture returns to a more suitable level (Chapman, 1978). For structures built of dispersible soil, sand filters combined with gypsum stabilized zones have been successful in reducing tunneling failure (Sherard *et al.*, 1977). The Soil Conservation Service of New South Wales uses gypsum or hydrated lime at rates of approximately 1 Mg ha^{-1} for amelioration. Under alkaline conditions, lime is unsuitable because of its lower solubility.

TABLE 9.7 Relationship between soil properties and tunneling susceptibility [Murphy, Upper and Lower Macquarie Areas Tech. Pap. 13/80, Soil Conservation Service, Wellington, NSW, Australia, (1981)]

Dispersion percentage (%)	Emerson aggregate class[a]	Particle size analysis	Volume expansion	Tunneling susceptibility
>33	1 or 2(3)	15–30% clay with >50% silt and fine sand	<0 or incomplete wetting	Extreme
>33	1 or 2(3)	>30% clay with >30% coarse sand	<0 or incomplete wetting	Very high
>33	1 or 2(3) or 2(2)	15–30% clay and/or >50% silt and fine sand	<0 or incomplete wetting	High
<33	2(1), 3, 4, 5, or 6	>30% clay with <50% silt and fine sand	>0 and correlates well with clay content	Minimal

[a]Classifies aggregates based on their cohesion in water with stability increasing from Class 1 to 8.

Contour banks built on dispersible soils often develop tunnels in the foundations, the frequency of which can be substantially reduced by ripping the banks in susceptible areas prior to construction (Chapman, 1978).

Examples of Failures Due to Piping

Thailand: Case histories of failure due to piping include four earth dams in northeast Thailand reported by Cole *et al.* (1977). These dams were all earthfill embankments built of silty sand and sandy silt soils of low plasticity with upstream and downstream face slopes of 1 on 2. The Lam Sum Lai Dam failed during the first filling, and sections (about 50 m wide) of the embankment, which had been subjected to rainfall over 5 years prior to filling, causing very severe tunnel erosion, were washed out. The entrances to the erosion shafts were visible along the crest of the dam and the higher parts of both faces. In some locations, these covered almost the full width of the crest, tending to merge into each other as a result of surface erosion. The exit holes in the lower portions of both faces and the associated tunnels ranged up 3 or 4 m in width. Repair of the dam involved excavation to remove all damaged areas, which were replaced with nondispersive or lime-treated material and compacted so that permeability did not exceed 10^5 cm s^{-1}.

Brazil: In Brazil, the construction of the Sobradinho Dam involved the use of dispersive clays occurring mainly in the alluvium along the river. Although no failure occurred, difficulties in compacting the dispersive clay led to the utilization of nondispersive clay in the zones encompassing the concrete structures (Bourdeaux and Imaizumi, 1977). A compacted sand filter built according to the conventional criteria (Terzaghi's filter criterion) prevented the migration of the dispersive clay particles. A thin skin of silt-size particles was formed on the face of the sand. The highly dispersive clays were stabilized by the addition of 6% alum.

9.7 SODICITY MANAGEMENT FOR ENVIRONMENTAL QUALITY

9.7.1. Reclamation of Sodic Soils

The use of gypsum and/or other amendments to reclaim sodic soils as described in Chapters 7 and 8 will reduce the production of dispersed colloids, both organic and inorganic, leading to lower loads being transported to surface waters and groundwaters. However, as has been clearly demonstrated throughout this book, colloids can become mobile in soils with low levels of exchangeable Na provided that the electrolyte concentration is low enough to promote dispersion. Consequently, the mere fact that the

physical and chemical conditions in a sodic soil have been improved does not necessarily mean that the transfer of colloids will cease. What is most important in effecting a reduction in colloid mobilization and transport to water bodies is the maintenance of adequate electrolyte concentrations in the soil solution to promote clay flocculation and reduce mobilization. While this is possible in certain circumstances where the costs of applying ameliorants can be absorbed, in the case of most sodic soils which have relatively low productivities and where farming systems thereon cannot support increased costs, colloid mobilization and transport are likely to be a continuing process.

9.7.2 Sodic Soil Managment in Engineering Applications

Because engineering design criteria to cope with construction on sodic soils are outside the scope of the present treatise, only management strategies to reduce the loss of suspended sediments from construction sites will be discussed here. Many of the same principles used for the management of sodic soils in agricultural systems (Chapter 8) also apply here. These include the use of vegetative cover, surface stabilization, which reduce water velocity.

Vegetative Cover

Vegetative cover, which intercepts and captures sediment in runoff, is less expensive and easier to maintain than engineered barriers or other systems such as settling ponds (Section 9.6.2). Rational scheduling of clearing activities to maintain maximum vegetative cover will effectively reduce runoff and sediment load. Planted buffer strips that surround the site and associated water courses are also effective in reducing sediment load (Sheridan and Hubbard, 1987). The use of surface applied mulches composed of straw or pine needles increases infiltration (Meyer *et al.,* 1970), thereby reducing runoff and associated sediment load. While mulches are only temporary solutions and are generally not as effective as complete vegetative cover, they do create an environment conducive to the germination and growth of seeded vegetation.

Surface Stabilization

Other more expensive alternatives to vegetative cover are available to stabilize disturbed surfaces, including the use of gypsum, geotextiles, and polyelectrolytes. The benefits accruing to the use of gypsum and polyelectrolytes such as polyacrylamide (PAM) to reduce crusting and the consequent runoff produced, have been extensively discussed in Chapter 5. In the engineering context, these materials would be applied to the disturbed surfaces at intervals sufficiently frequent to provide the required protection of the surface. This technique is particularly effective if used as a precursor to seeding of a vegetative cover. Injection of a polyelectrolyte stream into runoff water from a 500 ha surface mining site was highly effective in reducing off-site transport of sediment from 1140 to 23.5 mg L^{-1} for a 10-year, 24-h storm event (Foree and Tapp, 1979).

Woven geotextile sediment mats, hydraulic mulch fibers, plastic erosion control meshes and blankets, silt fences, and fabric-formed revetments have also proved to be effective in erosion control (Theisen, 1994). Most of these are placed on the surface to reduce raindrop impact and hence crust formation or are used to intercept sediment that has become mobile. Sediment mats made of burlap, excelsior, and jute mesh have also been used very effectively in diverted stream beds to trap sediment. When full, they are removed, anchored to the stream bank, and seeded to provide streambank stabilization. Comparative studies have shown that this technique is as effective as diverting the stream by piping and pumping around the construction site.

Bibliography

Aarstad, J.S., and D.E. Miller. 1973. Soil management practices for reducing runoff under center-pivot sprinkler systems. *J. Soil Water Conser.* 28:171-173.

Abdou, F.M., T. El-Kobbia, and L.H. Soerensen. 1975. Decomposition of native organic matter and ^{14}C labelled barley straw in different Egyptian soils. *Beitr. Trop. Landwirtsch. Veterinärmed.* 13:203-209.

Abrol, I.P., and D.R. Bhumbla. 1979. Crop responses to differential gypsum applications in a highly sodic soil and the tolerance of several crops to exchangeable sodium under field conditions. *Soil Sci.* 127:79-85.

Abrol, I.P., I.S. Dahiya, and D.R. Bhumbla. 1975. On the method of determining gypsum requirement of soils. *Soil Sci.* 120:30-36.

Abrol, I.P., J.S.P. Yadav, and F.I. Massoud. 1988. *Salt-Affected Soils and Their Management.* FAO Soils Bull. 39. Food and Agriculture Organization, Rome, Italy.

Abu-Sharar, T.M., F.T. Bingham, and J.D. Rhoades. 1987. Stability of soil aggregates as affected by electrolyte concentration and composition. *Soil Sci. Soc. Am. J.* 51:309-314.

Adam, M., N. Miljkovic, and N. Plamenac. 1988. Solonetz and Solod soils of Yugoslavia. In: Adam, M. (Ed.). *Solonetz Soils.* Proc. Int. Symp. Solonetz Soils. Osijek, Yugoslavia.

Adams, F. 1974. Soil solution. In: Carson, E.W. (Ed.). *The Plant Root and Its Environment.* pp. 441-480. University of Virginia Press, Charlottesville, VA.

Adem, H.H., J.M. Tisdall, and P. Willoughby. 1984. Tillage management changes size distribution of aggregates and macrostructure of soils used for irrigated row corps. *Soil Tillage Res.* 4:561-576.

Adriano, D.C. 1986. *Trace Elements in the Terrestial Environment.* p. 533. Springer-Verlag, New York, NY.

Agassi, M., J. Letey, W.J. Farmer, and P. Clark. 1995. Soil erosion contribution to pesticide transport by furrow irrigation. *J. Environ. Qual.* 24:892-895.

Agassi, M., J. Morin, and I. Shainberg. 1985. Effect of raindrop impact energy and water salinity on infiltration rates of sodic soils. *Soil Sci. Soc. Am. J.* 49:186-190.

Agassi, M., I. Shainberg, and J. Morin. 1981. Effect of electrolyte concentration and soil sodicity on infiltration rate and crust formation. *Soil Sci. Soc. Am. J.* 45:848-851.

Agassi, M., I. Shainberg, and J. Morin. 1986. Effect of powdered phosphogypsum on the infiltration rate of sodic soils. *Irrig. Sci.* 7:53-61.

Agriculture Canada Expert Committee on Soil Survey. 1987. *The Canadian System of Soil Classification,* 2nd ed. Agriculture Canada Publ. 1646.

Ahmad, M., B.H. Niazi, and G.R. Sandhu. 1988. Effectiveness of gypsum, HCl and organic matter for the improvement of saline sodic soils. *Pak. J. Agri. Res.* 9:373-378.

Ahmad, N., R.H. Qureshi, and M. Qadir. 1990. Amelioration of a calcareous saline-sodic soil by gypsum and forage plants. *Land Degrad. Rehabil.* 2:277-284.

Ahmad, R., and S. Ismail. 1993. Provenance trials in Pakistan: A synthesis. In: Davidson, N., and R. Galloway (Eds.). *Productive Use of Saline Land.* pp. 62-66. Australian Council for International Agricultural Research, Canberra, Australia.

Aitchison, G.D., and C.C. Wood. 1965. Some interactions of compaction, permeability and post-construction deflocculation affecting the probability of piping failure in small earth dams. *Proc. 6th Int. Conf. Soil Mech. Found. Eng.* 2:442-446.

Alam, S.M. 1994. Nutrient uptake by plants under stress conditions. In: Pessarakli, M. (Ed.). *Handbook of Plant and Crop Stress.* pp. 227-246. Marcel Dekker, New York, NY.

Alconada, M., O.E. Ansin, R.S. Lavado, V.A. Deregibus, G. Rubio, and F.H. Gutierrez. 1993. Effect of retention of run-off water and grazing on soil and on vegetation of a temperate humid grassland. *Agric. Water Man.* 23:233-246.

Ali, O.M., M. Yousaf, and J.D. Rhoades. 1987. Effect of exchangeable cation and electrolyte concentration on mineralogy of clay dispersed from aggregates. *Soil Sci. Soc. Am. J.* 51:896-900.

Allison, F.E. 1973. *Soil Organic Matter and Its Role in Crop Production.* Elsevier Scientific, New York, NY.

Allison, L.E. 1956. Soil and plant responses to VAMA and HPAN soil conditioners in the presence of high exchangeable sodium. *Soil Sci. Soc. Am. Proc.* 20:147-151.

Allison, L.E., and D.C. Moore. 1956. Effect of VAMA and HPAN soil conditioners on aggregation, surface crusting, and moisture retention in alkali soils. *Soil Sci. Soc. Am. Proc.* 20:143-146.

Alperovitch, N., I. Shainberg, and R. Keren. 1981. Specific effect of magnesium on the hydraulic conductivity of sodic soils. *J. Soil Sci.* 32:543-554.

Alperovitch, N., I. Shainberg, R. Keren, and M.J. Singer. 1985. Effect of clay mineralogy and aluminum and iron oxides on the hydraulic conductivity of clay-sand mixtures. *Clays Clay Min.* 33:443-450.

Alperovitch, N., I. Shainberg, and D. Rhoades. 1986. Effect of mineral weathering on the response of sodic soils to exchangeable magnesium. *Soil Sci. Soc. Am. J.* 50:901-904.

Al-Saleh, I. 1991. Talajviszonyok és a szikesedési folyamatok Szíriában. *Agrokém. Talajtan* 40:243-262.

Altamore, R., R.F. Torres, R.S. Lavado, and J.E. Gimenez. 1983. Efecto del subsolado sobre un suelo salino-alcalino del Oeste Bonaerense. *Cien. Suelo* 1:45-51.

Aly, S., and J. Letey. 1990. Physical properties of sodium-treated soil as affected by two polymers. *Soil Sci. Soc. Am. J.* 54:501-504.

Amrhein, C., and D.L. Suarez. 1987. Calcite supersaturation in soils as a result of organic matter mineralization. *Soil Sci. Soc. Am. J.* 51:932-937.

Amrhein, C., M.F. Zahow, and D.L. Suarez. 1993. Calcite supersaturation in soil suspensions. *Soil Sci.* 156:163-170.

Anderson, D.W. 1987. Pedogenesis in the grassland and adjacent forests of the Great Plains. *Adv. Soil Sci.* 7:53-93.

Anon. 1979. *Glossary of Soil Science Terms.* Soil Science Society of America, Madison, WI.

ANZECC/NHMRC. 1992. *Australian Water Quality Guidelines for Marine and Fresh Water.* National Water Quality Management Strategy, Canberra, Australia.

Arora, H.S., and N.T. Coleman. 1979. The influence of electrolyte concentration on flocculation of clay suspensions. *Soil Sci.* 127:134-139.

Arruda, J.A., G.R. Marzolf, and R.T. Faulk. 1983. The role of suspended sediments in the nutrition of zooplankton in turbid reservoirs. *Ecology* 64:1225-1235.

Arshad, M.A., and A.R. Mermut. 1988. Micromorphological and physicochemical characteristics of soil crust types in Northwestern Alberta, Canada. *Soil Sci. Soc. Am. J.* 52:724-729.

Aslam, M., R.C. Huffaker, and D.W. Rains. 1984. Early effects of salinity on nitrate assimilation in barley seedlings. *Plant Physiol.* 76: 321-325.

Aslam, Z., M. Mujtab, J. Alshtar, R. Waheed, K.A. Malik, and M.Naqvi. 1993. Biological methods for economically utilizing salt-affected soils in Pakistan. In: Davidson, N., and R. Galloway (Eds.). *Productive Use of Saline Land.* pp. 29-31. Australian Council for International Agricultural Research, Canberra, Australia.

Atwell, B.J. 1990. The effect of soil compaction on wheat during early tillering. I. Growth development and root structure. *New Phytol.* 115:29-35.

Avnimelech, Y., M. Kochba, Y. Yotal, and D. Shkedi. 1990. On the use of municipal solid waste compost for the reclamation of saline and alkaline soils. *Trans. 14th Int. Congr. Soil Sci.* IV:186-191.

Ayers, R.S. 1980. Personal communication.

Ayers, R.S., and D.W. Westcot. 1976. Water quality for agriculture. Irrig. Drain. Pap. 29. Food and Agriculture Organization, Rome, Italy.

Aylmore, L.A.G., and J.P. Quirk. 1959. Swelling of clay-water systems. *Nature* 183:1752-1753.

Aylmore, L.A.G., and I.D. Sills. 1982. Characterization of soil structure and stability using modulus of rupture-exchangeable sodium percentage relationships. *Aust. J. Soil Res.* 20:213-224.

Baize, D. 1990. Référentiel Pédologique. AFES, Plaisir, France.

Bakker, A.C., and W.W. Emerson. 1973. The comparative effect of exchangeable calcium, magnesium and sodium on some physical properties of Red-Brown Earth subsoils. III. The permeability of Shepperton soil and comparison methods. *Aust. J. Soil Res.* 11:159-165.

Baldock, J.A., M. Aoyama, J.M. Oades, Susanto, and C.D. Grant. 1994. Structural amelioration of a South Australian Red-Brown Earth using calcium and organic amendments. *Aust. J. Soil Res.* 32:571-594.

Baldwin, M., C.E. Kellogg, and J. Thorp. 1938. Soil classification. In: *Soils and Men.* pp. 979-1001. U.S. Dept. of Agriculture Yearbook, U.S. Govt. Printing Office, Washington, DC.

Baligar, V.C., V.E. Nash, M.L. Hare, and J.A. Price. 1975. Soybean root anatomy as influenced by soil bulk density. *Agron. J.* 67:842-844.

Ballard, T.M. 1971. Role of humic carrier substances in DDT movement through forest soil. *Soil Sci. Soc. Am. J.* 35:145-147.

Banerjee, S. 1959. Some aspects of salt-affected soils of West Bengal. *Soil Sci.* 88:45-50.

Barley, K.P., and J.T. Hutton. 1956. Effects of a lowland rice crop and of gypsum on a saline alkali clay. *Aust. J. Agric. Res.* 7:110-126.

Bar On, P., I. Shainberg, and I. Mochaeli. 1970. The electrophoretic mobility of Na/Ca montmorillonite particles. *J. Colloid Interface Sci.* 33:471-472.

Barraclough, P.B., and A.H. Weir. 1988. Effects of a compacted sub-soil layer on root and shoot growth, water use and nutrient uptake of winter wheat. *J. Agric. Sci. Camb.* 110:207-216.

Barrett, J.C., G.D. Grossman, and J. Rosenfeld. 1992. Turbidity-induced changes in reactive distance of rainbow trout. *Trans. Am. Fish. Soc.* 121:437-443.

Barrow, N.J. 1984. Modelling the effect of pH on phosphate sorption by soil. *J. Soil Sci.* 35:283-297.

Barrow, N.J. 1985. Reactions of anions and cations with variable-charge soils. *Adv. Agron.* 38:183-230.

Barrow, N.J. 1987. *Reactions with Variable-charge Soils.* Martinus Nijhoff Publishers, Dordrecht, Netherlands.

Bar-Yosef, B., S. Fishman, and H. Talpaz. 1980. A model of zinc movement to single roots in soils. *Soil Sci. Soc. Am. J.* 44:1272-1279.

Barzegar, A.R., P.N. Nelson, J.M. Oades, and P. Rengasamy. 1996. Organic matter, sodicity and clay type: Influence on aggregation. *Soil Sci. Soc. Am. J.* In Press.

Barzegar, A.R., J.M. Oades, P. Rengasamy, and L. Giles. 1994. Effect of sodicity and salinity on disaggregation and tensile strength of an Alfisol under different cropping systems. *Soil Tillage Res.* 32:329-345.

Bauder, J. 1995. Personal communication.

Bazilevich, N.I. 1965. *The Geochemistry of Soda Soils.* Israel Program for Scientific Translations, Jerusalem, Israel.

Beater, B.E. 1957. *Soils of the Sugar Belt: Part 1. Natal North Coast, Natal.* Regional Survey Rep. 3, Oxford Univ. Press, Oxford, UK.

Beater, B.E. 1959. *Soils of the Sugar Belt: Part 2. Natal South Coast, Natal.* Regional Survey Rep. 4, Oxford Univ. Press, Oxford, UK.

Beater, B.E. 1962. *Soils of the Sugar Belt: Part 3. Zululand, Natal.* Regional Survey Rep. 5, Oxford Univ. Press, Oxford, UK.

Beckmann, G. 1983. Development of old landscapes and soils. In: *Soils: An Australian Viewpoint.* pp. 107-117. Commonwealth Scientific and Industrial Research Organization Publications, Melbourne/Academic Press, London, UK.

Ben-Hur, M., M. Malik, J. Letey, and U. Mingelgrin. 1992. Adsorption of polymers on clays as affected by clay charge and structure, polymer properties, and water quality. *Soil Sci.* 153:349-356.

Ben-Hur, M., I. Shainberg, D. Bakker, and R. Keren. 1985. Effect of soil texture and $CaCO_3$ content on water infiltration in crusted soils as related to water salinity. *Irrig. Sci.* 6:281-294.

Benjamin, M.M., and J.O. Leckie. 1981. Multiple-site adsorption of Cd, Cu, Zn and Pb on amorphous iron oxyhydroxide. *J. Colloid Interface Sci.* 79:209-221.

Bernstein, L. 1967. Quantitative assessment of irrigation water quality. In: *Water Quality Criteria*. pp. 51-64. American Society for Testing and Materials Spec. Publ. 416.

Bernstein, L. 1974. Crop growth and salinity. In: van Schilfgaarde, J. (Ed.). *Drainage for Agriculture*. pp. 9-54. American Society of Agronomy, Madison, WI.

Bernstein, L. 1975. Effects of salinity and sodicity on plant growth. *Ann. Rev. Phytopath*. 13:295-312.

Bernstein, L., and G.A. Pearson. 1956. Influence of exchangeable sodium on the yield and chemical composition of plants: 1. Green beans, garden beets, clover, and alfalfa. *Soil Sci*. 82:247-258.

BGR. 1980. *Etude de reconnaissance des ressources en Afrique à l'aide d'images satellites. Etudes de recherche multidisciplinaire à l'aide d'images satellites dans la République de Haute-Volta: Géologie, hydrogéologie, pédologie et utilisation de l'espace*. Bundesanstalt für Geowissenschaften und Rohstoffe, Hanover, Germany.

Bhargava, G.P. 1977. Classification of salt-affected soils: Some problems. In: *Management of Salt-Affected Soils*. pp. 31-56. Proc. Indo-Hung. Sem. Manage. Salt Affected Soils. Central Soil Salinity Research Institute. Karnal, India.

Bhargava, G.P. 1979. Soil and water quality survey. In: *A Decade of Research*. pp. 22-34. Central Soil Salinity Research Institute, Karnal, India.

Bhumbla, D.R. 1977. Alkali and saline soils of India. In: *Management of Salt-Affected Soils*. pp. 14-19. Proc. Indo-Hung. Sem. Manage. Salt Affected Soils. Central Soil Salinity Research Institute, Karnal, India.

Biggar, J.W., and M. Fireman. 1960. Boron adsorption and release by soils. *Soil Sci. Soc. Am. Proc*. 24:115-120.

Bingham, F.T., A.W. Marsh, R. Branson, R. Mahler, and G. Ferry. 1972. Reclamation of salt-affected high-boron soils in Western Kern County. *Hilgardia* 41:195-211.

Black, A.S., and B.M.S. Abdul-Hakim. 1985. Effect of pasture or wheat cropping on the permeability of a weakly weathered soil to sodic conditions. *N. Z. J. Agric. Res*. 28:275-278.

Blacklow, W.M., and P.C. Pheloung. 1992. Sulfonylurea herbicides applied to acidic sandy soils: Movement, persistence, and activity within the growing season. *Aust. J. Agric. Res*. 43:1157-1168.

Blackmore, A.V., and R.D. Miller. 1961. Tactoid size and osmotic swelling in calcium montmorillonite. *Soil Sci. Soc. Am. Proc*. 25:169-173.

Blackwell, J., R. Horn, N.S. Jayawardane, R. White, and P.S. Blackwell. 1989a. Vertical stress distribution under tractor wheeling in a partially deep loosened Typic Paleustalf. *Soil Tillage Res*. 13:1-12.

Blackwell, P.S., N.S. Jayawardane, J. Blackwell, R. White, and R. Horn. 1989b. Evaluation of soil recompaction by transverse wheeling of tillage slots. *Soil Sci. Soc. Am. J*. 53:11-15.

Blackwell, P.S., N.S. Jayawardane, T.W. Greene, J.T. Wood, J. Blackwell, and H.J. Beatty. 1991. Subsoil macropore space of a transitional Red-Brown Earth after either deep tillage, gypsum or both. I. Physical effects and short-term changes. *Aust. J. Soil Res*. 29:123-140.

Blair Rains, A., and A.D. McKay. 1968. *The Northern State Lands, Botswana*. Land Resources Study 5. Land Resour. Div., Directorate of Overseas Surveys, Tolworth, Surrey, UK.

Bloom, P.R. 1981. Metal-organic interactions in soil. In: *Chemistry in the Soil Environment*. pp. 129-150. ASA Spec. Publ.40. American Society of Agronomy, Madison, WI.

Bocquier, G. 1964. Présences et charactères des solonetz solodisés tropicaux dans le bassin tchadien. *Trans. 8th Int. Congr. Soil Sci*. V:687-695.

Bocquier, G. 1968. Biogéocénoses et morphogenèses actuelles de certains pédiments du bassin tchadien. *Trans. 9th Int. Congr. Soil Sci*. IV:605-614.

Bocquier, G. 1971. Genèse et évolution de deux toposéquences de sols tropicaux du Tchad. Interprétation biogéodynamique. *Mém. ORSTOM*. 62.

Bodman, G.B. 1937. The variability of the permeability "constant" at low hydraulic gradients during saturated water flow in soils. *Soil Sci. Soc. Am. Proc*. 2:45-53.

Bodman, G.B., and G.K. Constantin. 1965. Influence of particle size distribution in soil compaction. *Hilgardia* 36:567-591.

Bodrogközy, G. 1965. Ecology of the halophilic vegetation of the Pannonicum. II. Correlation between alkali ("szik") plant communities and genetic soil classification in the Northern Hortobágy. *Acta Bot. Hung*. 11:1-51.

Bolt, G.H. 1967. Cation exchange equations used in soil science: A review. *Neth. J. Agric. Sci*. 15:81-103.

Bolt, G.H., and R.D. Miller. 1958. Calculation of total and component potentials of water in soil. *Trans. Am. Geophys. Union* 39:917-928.

Borlaug, N.E. and C.R. Dowswell. 1994. Feeding a human population that increasingly crowds a fragile planet. Keynote Lecture. *Trans. 15th World Congr. Soil Sci*. Acapulco, Mexico.

Boucher, S.C. 1990. Field tunnel erosion: Its characteristics and amelioration. Victoria Department of Conservation and Environment, East Melbourne, Australia.

Boulet, R. 1970. La géomorphologie et les principaux types de sols en Haute-Volta septentrionale. *Cah. ORSTOM Sér. Pédol*. VIII:245-271.

Boulet, R. 1978. Toposéquences de sols tropicaux en Haute-Volta. Equilibre et déséquilibre pédobioclimatique. *Cahiers ORSTOM* 85 *Sér. Pédol*. Strasbourg, France.

Bourdeaux, G., and H. Imaizumi. 1977. Dispersive clay at Sobradinho Dam. In: Sherard, J.L., and R.S. Decker. *Dispersive Clays, Related Piping, and Erosion in Geotechncal Projects*. pp. 13-24. American Society for Testing and Materials, ASTM STP 623.

Bouwer, H. 1994. Irrigation and global water outlook. *Agric. Water Manage*. 25:221-231.

Bower, C.A. 1961. Prediction of the effects of irrigation waters on soils. *Proc. UNESCO Arid Zone Symp*. pp. 215-222. Teheran, Iran.

Bower, C.A., W.G. Harper, C.D. Moodie, R. Overstreet, and L.A. Richards. 1958. Report of the nomenclature committee appointed by the board of collaborators of the U.S. Salinity Laboratory. *Soil Sci. Soc. Am. Proc*. 22:270.

Bower, C.A., and C.H. Wadleigh. 1948. Growth and cationic accumulation by four species of plants as influenced by various levels of exchangeable sodium. *Soil Sci. Soc. Am. Proc.* 13: 218-223.

Bowser, E.E., R.A. Milne, and R.R. Cairns. 1962. Characteristics of the major soil groups in an area dominated by Solonetzic soils. *Can. J. Soil Sci.* 42:165-179.

Boyle, M., W.T. Frankenberger, Jr., and L.H. Stolzy. 1989. The influence of organic matter on aggregation and water infiltration. *J. Prod. Agric.* 2:290-299.

Bradfield, R. 1936. The value and limitations of calcium in soil structure. *Am. Soil Surv. Assoc. Bull.* 17:31-32.

Bradford, J.M., J.E. Ferris, and P.A. Remley. 1987. Interrill soil erosion processes: I. Effect of surface sealing on infiltration, runoff, and soil splash detachment. *Soil Sci. Soc. Am. J.* 51:1566-1571.

Bresler, E., B.L. McNeal, and D.L. Carter. 1982. Diagnosis and Properties. In: Bresler, E., B.L. McNeal, and D.L. Carter (Eds.). *Saline and Sodic Soils: Principles, Dynamics, Modeling.* pp. 1-27. Springer-Verlag, New York, NY.

Brewer, R. 1979. Relationships between particle size, fabric and other factors in some Australian soils. *Aust. J. Soil Res.* 17:29-41.

Briggs, G.G. 1973. A simple relationship between soil adsorption of organic compounds and their octanol/water partition coefficients. *Proc. 7th. Br. Insect. Fung. Conf.* 1:83-86.

Briggs, G.G. 1981. Adsorption of pesticides by some Australian soils. *Aust. J. Soil Res.* 19:61-68.

Brink, A.B.A., T.C. Partridge, and A.B.A. Williams. 1984. *Soil Survey for Engineers.* Oxford University Press, New York, NY.

Broadbent, F.E., and G.R. Bradford. 1952. Cation exchange groupings in the soil organic fraction. *Soil Sci.* 74:447-457.

Brooks, R.H., C.A. Bower, and R.C. Reeve. 1956. The effect of various exchangeable cations upon the physical condition of soils. *Soil Sci. Soc. Am. Proc.* 20:325-327.

Brubaker, S.C., C.S. Holzhey, and B.R. Brasher. 1992. Estimating the water-dispersible clay content of soils. *Soil Sci. Soc. Am. J.* 56:1227-1232.

Bruno, V., and G. Pietramellara. 1992. Anaerobic decomposition of organic matter in submerged soils: Influence of sodicity, salinity and alkalinity. *Agricoltura Mediterranea* 122: 18-26.

Brusseau, M.L. and P.S.C. Rao. 1989. Sorption nonideality during organic contaminant transport in porous media. *CRC Crit. Rev. Environ. Control* 19:33-99.

Bryan, R.B., A. Yair, and W.D. Hodges. 1978. Factors controlling the initiation of runoff and piping in Dinosaur Provincial Park Badlands, Alberta, Canada. *Zeit. Geomorph. Suppl.* 29:151-168.

Buckland, G.D., and S. Pawluk. 1985a. Deep plowed Solonetzic and Chernozemic soils: I. Tilth and physicochemical features of the cultivated layer. *Can. J. Soil Sci.* 65:629-638.

Buckland, G.D., and S. Pawluk. 1985b. Deep plowed Solonetzic and Chernozemic soils: II. Crop response characteristics. *Can. J. Soil Sci.* 65:639-649.

Buddemeier, R.W., and R.H. Hunt. 1988. Transport of colloidal contaminants in groundwater: Radionuclide migration at the Nevada Test Site. *Appl. Geochem.* 3:535-548.

BUNASOL. 1991. Notice explicative de la carte d'aptitude des terres du Burkina Faso. Echelle: 1/1 000 000. Projet BKF/87/ 020. Documentations Téchniques 9.

Buyanovsky, G.A., and G.H. Wagner. 1983. Annual cycles of carbon dioxide level in soil air. *Soil Sci. Soc. Am. J.* 47:1139-1145.

Cairns, R.R. 1961. Some chemical characteristics of a Solonetzic soil sequence at Vegreville, Alberta, with regard to possible amelioration. *Can. J. Soil Sci.* 41:24-34.

Cairns, R.R. 1962. Some effects of deep working on Solonetz soil. *Can. J. Soil Sci.* 42:273-275.

Cairns, R.R. 1963. Nitrogen mineralization in Solonetzic soil samples and some influencing factors. *Can. J. Soil Sci.* 43:387-392.

Cairns, R.R. 1968. Various forms of nitrogen for bromegrass on a Solonetz soil. *Can. J. Soil Sci.* 48:297-300.

Cairns, R.R., and J. D. Beaton. 1976. Improving a Solonetzic soil by nitrogen-sulphur materials. *J. Sulph. Inst.* 12:10-12.

Cairns, R.R., W.E. Bowser, R.A. Milne, and P.C. Chang. 1967. Effect of nitrogen fertilization of bromegrass on Solonetzic soils. *Can. J. Soil Sci.* 47:1-6.

Cairns, R.R., R.A. Milne, and W.E. Bowser. 1962. A nutritional disorder in barley seedlings grown on an alkali Solonetz soil. *Can. J. Soil Sci.* 42:1-6.

Campbell, C.A. 1978. Soil organic carbon, nitrogen and fertility. In: Schnitzer, M., and S.U. Khan (Eds.). *Soil Organic Matter.* pp. 173-271. Elsevier Scientific, Amsterdam, Netherlands.

Carter, C.W., and I.H. Suffet. 1982. Binding of DDT to dissolved humic materials. *Environ. Sci. Tech.* 16:735-740.

Carter, D.L., and H.B. Peterson. 1962. Sodic tolerance of tall wheatgrass. *Agron. J.* 54:382-384.

Carter, L.M. 1985. Wheel traffic is costly. *Trans. ASAE* 28:430-434.

Carter, M.R. 1983. Growth and mineral composition of barley and wheat across sequences of Solonetzic soil. *Plant Soil* 74:229-235.

Carter, M.R. 1986. Microbial biomass and mineralizable nitrogen in Solonetzic soils: Influence of gypsum and lime amendments. *Soil Biol. Biochem.* 18:531-537.

Carter, M.R., and J.R. Pearen. 1988. Influence of calcium on growth and root penetration of barley seedlings in a saline-sodic soil. *Arid Soil Res. Rehab.* 2:59-66.

Carter, M.R., J.R. Pearen, P.G. Karkanis, R.R. Cairns, and D.W. McAndrew. 1986. Improvement of soil properties and plant with on a Brown Solonetzic soil using irrigation, calcium amendments and nitrogen. *Can. J. Soil Sci.* 66:581-589.

Carter, M.R., and G.R. Webster. 1990. Use of calcium-to-total cation ratio as an index of plant-available calcium. *Soil Sci.* 149:212-217.

Carter, M.R., G.R. Webster, and R.R. Cairns. 1979a. Calcium deficiency in some Solonetzic soils of Alberta. *J. Soil Sci.* 30:161-174.

Carter, M.R., G.R. Webster, and R.R. Cairns. 1979b. Effect of moisture change and salinity on the Mg/Ca ratio and ratio of Ca/ total cations in soil solutions. *Can. J. Soil Sci.* 59:439-443.

Cartwright, B., B.A.Z. Zarcinas, and L.R. Spouncer. 1986. Boron toxicity in South Australian barley crops. *Aust. J. Agric. Res.* 37:351-359.

Cass, A. 1972. Reclamation of Sodic Soils: A Laboratory Investigation. M.Sc. Agric. Thesis University of Natal, Pietermaritzburg, South Africa.

Cass, A. 1980. The Influence of Pore Structural Stability and Internal Drainage Rate on Selection of Soil for Irrigation. Ph.D. Diss. University of Natal, Pietermaritzburg, South Africa.

Cass, A., and M.E. Sumner. 1982a. Soil pore structural stability and irrigation water quality: I. Empirical sodium stability model. *Soil Sci. Soc. Am. J.* 46:503-506.

Cass, A., and M.E. Sumner. 1982b. Soil pore structural stability and irrigation water quality: II. Sodium stability data. *Soil Sci. Soc. Am. J.* 46:507-512.

Cass, A., and M.E. Sumner. 1982c. Soil pore structural stability and irrigation water quality: III. Evaluation of soil stability and crop yield in relation to salinity and sodicity. *Soil Sci. Soc. Am. J.* 46:513-517.

Cavallaro, N., and M.B. McBride. 1984. Zinc and copper sorption and fixation by an acid soil clay: Effect of selective dissolutions. *Soil Sci. Soc. Am. J.* 45:747-749.

Ceppi, S.B., M.I. Velasco, and C.P. De Pauli. 1993. Influencia de Na$^+$ y Ca^{2+} en la adsorción de ácidos húmicos sobre partículas de suelo bajo diferentes sistemas de manejo. *Agrochimica* 37:134-146.

Chaberek, S., and E. Martell. 1959. *Organic Sequestering Agents: A Discussion of the Chemical Behavior and Applications of Metal Chelate Compounds in Aqueous Systems.* John Wiley, New York, NY.

Chand, M., I.P. Abrol, and D.R. Bhumbla. 1977. A comparison of the effect of eight amendments on soil properties and crop growth in a highly sodic soil. *Indian J. Agric. Sci.* 47:348-354.

Chander, K., S. Goyal, and K.K. Kapoor. 1994. Effect of sodic water irrigation and farmyard manure application on soil microbial biomass and microbial activity. *Appl. Soil Ecol.* 1:139-144.

Chang, C.W., and H.E. Dregne. 1955. Effect of exchangeable sodium on soil properties and on growth and cation content of alfalfa and cotton. *Soil Sci. Soc. Am. Proc.* 19:29-35.

Charman, P.E.V. 1978. *Soils of New South Wales, Their Characteristics, Classification and Conservation.* Soil Conservation Tech. Handbk. 1. Soil Conservation Service, Sydney, Australia.

Chartres, C.J. 1993. Sodic soils: An introduction to their formation and distribution in Australia. *Aust. J. Soil Res.* 31:751-760.

Chartres, C.J., R.S. Greene, G.W. Ford, and P. Rengasamy. 1985. The effects of gypsum on macroporosity and crusting of two red duplex soils. *Aust. J. Soil Res.* 23:467-479.

Chartres, C.J., J.M. Kirby, and M. Raupach. 1990. Poorly ordered silica and aluminosilicates as temporary cementing agents in hard-setting soils. *Soil Sci. Soc. Am. J.* 54:1060-1067.

Chen, J., J. Tarchitzky, J. Morin, and A. Banin. 1980. Scanning electron microscope observations on soil crusts and their formation. *Soil Sci.* 130:49-55.

Chen, Y., A. Banin, and A. Borochovitch. 1983. Effect of potassium on soil structure in relation to hydraulic conductivity. *Geoderma* 30:135-147.

Chen, Y., and P. Barak. 1982. Iron nutrition of plants in calcareous soils. *Adv. Agron.* 35:217-240.

Cheverry, C. 1974. Contribution à l'étude pédologique des polders du lac Tchad. Dynamique des sels en milieu continental subaride dans des sédiments argileux et organiques. Thèse. *Cah. ORSTOM Sér. Pédol.* Strasbourg, France.

Chhabra, R. 1985. Crop responses to phosphorus and potassium fertilization of a sodic soil. *Soil Sci. Soc. Am. J.* 77:699-702.

Chhabra, R., and I.P. Abrol. 1977. Reclaiming effect of rice grown in a sodic soil. *Soil Sci.* 124:49-55.

Chhabra, R., I.P. Abrol, and M.V. Singh. 1981. Dynamics of phosphorus during reclamation of sodic soils. *Soil Sci.* 132:319-324.

Chhabra, R., S.B. Singh, and I.P. Abrol. 1979. Effect of exchangeable sodium percentage on the growth, yield, and chemical composition of sunflower (*Helianthus annus* L.). *Soil Sci.* 127:242-247.

Chi, C.L., W.W. Emerson, and D.G. Lewis. 1977. Exchangeable calcium, magnesium and sodium and the dispersion of illites in water. I. Characterization of illites and exchange reactions. *Aust. J. Soil Res.* 15:243-253.

Chiang, S.C., D.E. Radcliffe, W.P. Miller, and K.D. Newman. 1987. Hydraulic conductivity of three southeastern soils as affected by sodium, electrolyte concentration, and pH. *Soil Sci. Soc. Am. J.* 51:1293-1299.

Chiou, C.T., R.L. Malcolm, T.I. Brinton, and E.E. Kile. 1986. Water solubility enhancement of some organic pollutants and pesticides by dissolved humic and fulvic acids. *Environ. Sci. Tech.* 20:502-508.

Chiou, C.T., L.J. Peters, and V.H. Freed. 1979. A physical concept of soil-water equilibria for non-ionic organic compounds. *Science* 206:831-832.

Chittleborough, D.J. 1992. Formation and pedology of duplex soils. *Aust. J. Exp. Agric.* 32:815-826.

Chittleborough, D.J., K.R.J. Smettem, E. Cotsaris, and F.W. Leaney. 1992. Seasonal changes in pathways of dissolved organic carbon through a hillslope soil (Xeralf) with contrasting texture. *Aust. J. Soil Res.* 30:465-476.

Cho, C.M., L. Sakdinan, and C. Chang. 1979. Denitrification intensity and capacity of three irrigated Alberta soils. *Soil Sci. Soc. Am. J.* 43:945-950.

Chorom, M., and P. Rengasamy. 1995. Dispersion and zeta potential of pure clays as related to net particle charge under varying pH, electrolyte concentration and cation type. *Eur. J. Soil Sci.* 46:657-665.

Chorom, M., P. Rengasamy and R.S. Murray. 1994. Clay dispersion as influenced by pH and net particle charge of sodic soils. *Aust. J. Soil Res.* 32:1243-1252.

Chretien, J., and E.B.A. Bisdom. 1983. The development of soil porosity in experimental sandy soils with clay admixtures as examined by Quantimet 720 from BESI and by other techniques. *Geoderma* 30:285-302.

Christenson, H.K., and R.G. Horn. 1985. Solvation forces measured in non-aqueous liquids. *Chem. Scr.* 25:37-41.

Churchman, G.J., J.O. Skjemstad, and J.M. Oades. 1993. Influence of clay minerals and organic matter on effects of sodicity on soils. *Aust. J. Soil Res.* 31:779-800.

Cochrane, H.R., G. Scholz, and A.M.E. van Vreeswyk. 1994. Sodic soils in Western Australia. *Aust. J. Soil Res. 32:359-388.*

Cockroft, B., and F.M. Martin. 1981. Irrigation. In: Oades, J.M., D.G. Lewis, and K. Norrish (Eds.). *Red-Brown Earths of Australia.* pp. 133-147. University of Adelaide and CSIRO Div. of Soils, Adelaide, Australia.

Cockroft, B., and W.K. Mason. 1987. Irrigated agriculture. In: Conner, D.J. and D.F. Smith (Eds.). *Agriculture in Victoria.* pp. 159-177. Australian Institute of Agricultural Science, Victoria Branch, Melbourne, Australia.

Cole, B.A., C. Ratansen, P. Maiklad, T.B. Liggins, and S. Chirapuntu. 1977. Dispersive clay in irrigation dams in Thailand. In: Sherard, J.L., and R.S. Decker (Eds.). *Dispersive Clays, Related Piping, and Erosion in Geotechnical Projects.* pp. 25-41. American Society for Testing and Materials, ASTM STP 623.

Coleman, N.T., and G.W. Thomas. 1967. The basic chemistry of soil acidity. In: Pearson, R.W., and F. Adams (Eds.). *Soil Acidity and Liming.* pp. 1-42. American Society of Agronomy, Madison, WI.

Colen, H. 1995. Personal communication.

Collins, T., and D.G. Allen. 1991. A bench scale aerated lagoon for studying the removal of chlorinated organics. *Tappi. J.* 74:231-234.

Collis-George, N., and D.E. Smiles. 1963. An examination of cation balance and moisture characteristic methods of determining the stability of soil aggregates. *J. Soil Sci.* 14:21-32.

Cook, G.D. 1988. Degradation of Soil Structure under Continuous Cultivation on the Darling Downs, Queensland. Ph.D. Thesis. University of Queensland, Australia.

Cooper, B. 1994. *Central and North West Regions Water Quality Program. 1993/1994 Report on Pesticide Monitoring.* New South Wales Department of Water Resources TS94.087, Sydney, Australia.

Cooper, B., J. Graice, and J. Kimber. 1994. *Central and North West Regions Water Quality Program.* New South Wales Department of Water Resources TS94.090, Sydney, Australia.

Coughlan, K.J. 1984. The structure of Vertisols. In: McGarity, J.W., E.H. Houltz, and H.B. So (Eds.). *Properties and Utilisation of Cracking Clay Soils.* Proceedings of a Symposium. pp 87-96. Rev. Rural Sci., University of New England, Armidale, Australia.

Coughlan, K.J., and R.J. Loch. 1984. The relationship between aggregation and other soil properties in cracking clay soils. *Aust. J. Soil Res.* 22:59-88.

Coughlan, K.J., R.J. Loch, and W.E. Fox. 1978. Binary packing theory and the physical properties of aggregates. *Aust. J. Soil Res.* 16:283-289.

Cramer, G.R., and A. Läuchli. 1986. Ion activities in solution in relation to Na^+-Ca^{2+} interactions at the plasmalemma. *J. Exp. Bot.* 37:320-330.

Cresswell, H.P., and J.A. Kirkegaard. 1995. Subsoil amelioration by plant roots: The process and the evidence. *Aust. J. Soil Res.* 33:221-239.

Crouch, R.J. 1977. Tunnel-gully erosion and urban development: A case study. In: Sherard, J.L., and R.S. Decker (Eds.). *Dispersive Soils, Related Piping, and Erosion in Geotechnical Projects.* pp. 58-68. American Society for Testing and Materials, ASTM STP 623.

Crouch, R.J., J.W. McGarity, and R.R. Storrier. 1986. Tunnel formation processes in the Riverina area of NSW, Australia. *Earth Surf. Processes Landforms* 11:157-168.

Crouch, R.J., K.C. Reynolds, and R.W. Hicks. 1991. Soils and their use for earthworks. In: Chapman, P.E.V., and B.W. Murphy (Eds.). *Soil: Their Properties and Management.* pp. 268-297. Sydney University Press, Sydney, Australia.

Cuker, B.E., P.T. Gama, and J.M. Burkholder. 1990. Type of suspended clay influences lake productivity and phytoplankton community response to phosphorus loading. *Limn. Oceanography* 37:822-830.

Cuker, B.E., and J. Hudson. 1992. Type of suspended clay influences zooplankton response to phosphorus loading. *Limnol. Oceanogr.* 35:830-839.

Curtin, D., F. Selles, and H. Steppuhn. 1992. Influence of salt concentration and sodicity on the solubility of phosphate in soils. *Soil Sci.* 153:409-416.

Curtin, D., F. Selles, and H. Steppuhn. 1995a. Sodium-calcium exchange selectivity as influenced by soil properties and method of determination. *Soil Sci.* 159:176-184.

Curtin, D., H. Steppuhn, A.R. Mermut, and F. Selles. 1995b. Sodicity in irrigated soils in Saskatchewan: Chemistry and structural stability. *Can. J. Soil Sci.* 75:177-185.

Curtin, D., H. Steppuhn, and F. Selles. 1993a. Plant responses to sulfate and chloride salinity: Growth and ionic relations. *Soil Sci. Soc. Am. J.* 57:1304-1310.

Curtin, D., H. Steppuhn, and F. Selles. 1994. Effects of magnesium on cation selectivity and structural stability of sodic soils. *Soil Sci. Soc. Am. J.* 58:730-737.

Curtin, D., and J.K. Syers. 1990. Extractability and adsorption of sulphate in soils. *J. Soil Sci.* 41:305-312.

Curtin, D., J.K. Syers, and N.S. Bolan. 1993b. Phosphate sorption by soil in relation to exchangeable cation composition and pH. *Aust. J. Soil Res.* 31:137-149.

Curtin, D., J.K. Syers, and G.W. Smillie. 1987. The importance of exchangeable cations and resin sink characteristics in the release of soil phosphorus. *J. Soil Sci.* 38:711-716.

Dahiya, I.S., and R. Anlauf. 1990. Sodic soils in India: Their reclamation and management. *Z. Kulturtech. Landentwick.* 31:26-34.

Dalang, F., J. Buffle, and W. Haerdl. 1984. Study of the influence of fulvic substances on the adsorption of copper (II) ions at kaolinite surfaces. *Environ. Sci. Tech.* 18:135-141.

Dalrymple, J.B., and C.Y. Jim. 1984. Experimental study of soil microfabrics induced by isotropic stresses of wetting and drying. *Geoderma* 34:43-68.

Daly, H. 1994. *Central and North West Regions Water Quality Monitoring Program: 1993/1994 Report on Nutrients and General Water Quality Monitoring.* New South Wales Department of Water Resources TS94.088, Sydney, Australia.

Davidson, J.L., and J.P. Quirk. 1961. The influence of dissolved gypsum on pasture establishment on irrigated sodic clays. *Aust. J. Agric. Res.* 12:100-110.

Davidson, N.J., R. Galloway, and G. Lazeresen. 1993. Limits to the productivity of *Atriplex* in salt-affected duplex soils. In: Davidson, N., and R. Galloway (Eds.). *Productive Use of Saline Land.* pp. 108-111. Australian Council for International Agricultural Research, Canberra, Australia.

Dean's Committee, University of California. 1968. *Agricultural Development of New Lands on the West Side of the San Joaquin Valley: Land, Crops, and Economics.* Report 1. College of Agriculture and Environmental Science, Davis, CA.

Dementyeva, T.G., and V.Y. Motuzov. 1988. Pochvenniye faktori urazhayev risa y kormovih kultur na orositelnih sistemah Prikaspiyskoy nizhmennosti. In: *Pochvi solontzovih territoriy i metodi ih izucheniya.* pp. 51-58. Pochvennyiant I.N., Moscow, Russia.

Dent, F.J. 1992. Salt-affected soils in the Asia and Pacific Region. An overview. In: Monthareon, L. (Ed.). *Strategies for Utilizing Salt-Affected Lands.* pp. 259-268. Proc. Int. Workshop Salt Affected Soils Bangkok, Thailand.

Derjaguin, B.V., and N.V. Churaev. 1974. Structural component of disjoining pressure. *J. Colloid Interface Sci.* 49:249-255.

Desh-Raj, and D. Raj. 1990. Experiences in waste land development: A case study. In: Mathur, A.N., and N.S. Rathore (Eds.). *Renewable Energy and Environment.* Himanshu Publications, Udaipur, India.

de'Sigmond, A.A.J. 1926. Contribution to the theory of the origin of alkali soils. *Soil Sci.* 21:455-475.

de'Sigmond, A.A.J. 1927a. *Hungarian Alkali Soils and Methods of Their Reclamation.* University of California Printing Office, Berkeley, CA.

de'Sigmond, A.A.J. 1927b. The classification of alkali and salty soils. *Trans. 1st Int. Congr. Soil Sci.* 1:320-344.

de'Sigmond, A.A.J., A. Arany, and A. Herke. 1927. The effect of calcium and aluminum salts in alkali soil reclamation. *First Int. Congr. Soil Sci.* 2:512-517.

de Villiers, J.M. 1962. A Study of Soil Formation in Natal. Ph.D. Diss. University of Natal, Pietermaritzburg, South Africa.

Devitt, D., W.M. Jarrell, and K.L. Stevens. 1981. Sodium-potassium ratios in soil solution and plant response under saline conditions. *Soil Sci. Soc. Am. J.* 45:80-86.

Dexter, A.R. 1988a. Advances in characterization of soil structure. *Soil Tillage Res.* 11:199-238.

Dexter, A.R. 1988b. Strength of soil aggregates and of aggregate beds. *Catena Suppl.* 11:35-52.

Doneen, L.D. 1948. The quality of irrigation water and soil permeability. *Soil Sci. Soc. Am. Proc.* 13:523-526.

Downes, R.G. 1946. Tunnelling erosion in North-eastern Victoria. *J. Counc. Sci. Ind. Res.* 19:283-292.

Doyle, R.B., and F.M. Habraken. 1993. The distribution of sodic soils in Tasmania. *Aust. J. Soil Res.* 31:931-948.

Draycott, A.P. 1993. Nutrition. In: Cooke, D.A., and R.K. Scott (Eds.). *The Sugar Beet Crop.* pp. 238-277. Chapman and Hall, London, UK.

Drew, M.C., and E.J. Sisworo. 1979. The development of waterlogging damage in young barley plants in relation to plant nutrition status and changes in soil properties. *New Phytol.* 82:301-314.

Droubi, A., C. Cheverry, B. Fritz, and Y. Tardy. 1976. Géochimie des eaux et des sels dans les sols des polders du lac Tchad: application d'un modéle thermodynamique de simulation de l'évaporation. *Chem. Geol.* 17:165-177.

Ducellier, J. 1963. Contribution a l'étude des formations cristallines et métamorphiques du Centre et du Nord de la Haute Volta. *Mém. BRGM.* 10 Paris, France.

Duley, F.L. 1939. Surface factors affecting the rate of intake of water by soils. *Soil Sci. Soc. Am. Proc.* 4:60-64.

Dunnivant, F.M., P.M. Jardine, D.L. Taylor, and J.F. McCarthy. 1992. Co-transport of cadmium and hexachlorobiphenyl by dissolved organic carbon through columns containing aquifer material. *Environ. Sci. Tech.* 26:360-368.

Durgin, P.B., and J.G. Chaney. 1984. Dispersion of kaolinite by dissolved organic matter from Douglas-fir roots. *Can. J. Soil Sci.* 64:445-455.

Dutt, G.R., T.W. Terkeltoub, and R.S. Rauschkolb. 1972. Prediction of gypsum and leaching requirements for sodium-affected soils. *Soil Sci.* 114:93-103.

Eaton, F.M. 1940. Effect of exchange sodium on the moisture equivalent and the wilting coefficient of soils. *J. Agric. Res.* 61:401-425.

Edwards, A.P., and J.M. Bremner. 1967. Microaggregates in soils. *J. Soil Sci.* 18:64-73.

Egorov, V.V., V.M. Fridland, E.N. Ivanova, N.N. Rozov, V.A. Nosin, and T.A. Friev. 1986. *Classification and Diagnostics of Soils of the USSR.* Oxonian Press, New Delhi, India.

Elimelech, M., C.R. O'Melia. 1990. Kinetics of deposition of colloidal particles in porous media. *Environ. Sci. Tech.* 24:1528-1536.

Ellis, J.H., and O.G. Caldwell. 1935. Magnesium clay solonetz. *Trans. 3rd Int. Congr. Soil Sci.* I:348-350.

El-Morsy, E.A., M. Malik, and J. Letey. 1991a. Polymer effects on the hydraulic conductivity of saline and sodic soil conditions. *Soil Sci.* 151:430-435.

El-Morsy, E.A., M. Malik, and J. Letey. 1991b. Interactions between water quality and polymer treatment on infiltration rate and clay migration. *Soil Technol.* 4:221-231.

Elrashid, M.A., and G.A. O'Connor. 1982. Influence of solution composition on sorption of zinc by soils. *Soil Sci. Soc. Am. J.* 46:1153-1158.

El-Swaify, S.A. 1973. Structural changes in tropical soils due to anions in irrigation water. *Soil Sci.* 24:137-144.

El-Swaify, S.A., and L.D. Swindale. 1969. Hydraulic conductivity of some tropical soils as a guide to irrigation water quality. *Trans. 9th Int. Congr. Soil Sci.* 1:381-389.

EMBRAPA. 1971. Levantamento de reconhecimento dos solos do sul do estado de Mato Grosso. *EMBRAPA Boletim Téchnico* 18. Rio de Janeiro, Brazil.

EMBRAPA. 1977. Levantamento exploratório-reconhecimento de solos da margem direita do Rio São Francisco, estado da Bahia. *EMBRAPA Boletim Téchnico* 52. Recife, Brazil.

Emerson, W.W. 1954. The determination of the stability of soil crumbs. *J. Soil Sci.* 5:235-250.

Emerson, W.W. 1967. A classification of soil aggregates based on their coherence in water. *Aust. J. Soil Sci.* 5:47-57.

Emerson, W.W. 1977. Bonding of aggregates by soil organic matter. In: Russell, J.S., and E.L. Greacen (Eds.). *Soil Factors in Crop Production in a Semi-arid Environment.* University of Queensland Press/Australian Society of Soil Science, St. Lucia, Australia.

Emerson, W.W. 1983. Interparticle bonding. *In Soils: An Australian Viewpoint.* pp. 477-498. Commonwealth Scientific and Industrial Research Organization Publications, Melbourne, Australia/Academic Press, London, UK.

Emerson, W.W., and C.L. Chi. 1977. Exchangeable calcium, magnesium and sodium and the dispersion of illites in water. II. Dispersion of illites in water. *Aust. J. Soil Res.* 15:255-262.

Emerson, W.W., R.C. Foster, and J.M. Oades. 1986. Organomineral complexes in relation to soil aggregation and structure. In: Huang, P.M., and M. Schnitzer (Eds.). *Interactions of Soil Minerals with Natural Organics and Microbes.* pp. 521-548. Soil Science Society of America, Madison, WI.

Emerson, W.W., and B.H. Smith. 1970. Magnesium, organic matter and soil structure. *Nature* 228:453-454.

Enfield, C.F., G. Bengsston, and R. Lindquist. 1989. Influence of macromolecules on chemical transport. *Environ. Sci. Tech.* 23:1278-1286.

Enfield, C.G., and S.R. Yates. 1987. *Chemical Transport to Groundwater.* American Society of Agronomy, Madison, WI.

Epstein, E., and W.J. Grant. 1973. Soil crust formation as affected by raindrop impact. In: Hadas, A., D. Swartzendruber, P.E. Rijtema, M. Fuchs, and B. Yaron (Eds.). *Physical Aspects of Soil Water and Salts in Ecosystems.* pp. 195-201. Springer Verlag, New York, NY.

Erdélyi, M. 1979. Hydrodynamics of the Hungarian Basin. *VITUKI Proc.* 18.

Eriksson, E. 1952. Cation exchange equilibria on clay minerals. *Soil Sci.* 74:103-113.

Eskov, A. I. 1991. The evolution of trench steppe solonetz. In: *Genesis and Control of Fertility of Salt-Affected Soils.* pp. 257-260. Dokuchaev Soil Institute, Moscow, Russia.

Evangelou, V.P., J. Wang, and R.E. Phillips. 1994. New developments and perspectives on soil potassium quantity/intensity relationships. *Adv. Agron.* 52:173-227.

Evans, L.T., and E.W. Russell. 1959. The adsorption of humic and fulvic acids by clays. *J. Soil Sci.* 10:119-132.

Evans, R.G., C.J. Smith, J.D. Oster, and B.A. Myers. 1990. Saline water application effects on furrow infiltration of Red-Brown Earths. *Trans. ASAE* 33:1563-1572.

FAO. 1964. *Report on the Soils of Paraguay.* 2nd ed. Food and Agriculture Organization, Rome, Italy/UNESCO, Paris, France.

FAO. 1974. *Soil Map of the World. I. Legend.* UNESCO, Paris, France.

FAO. 1978. *Soil Map of the World. VIII. North and Central Asia.* UNESCO, Paris, France.

FAO. 1988. *Soil Map of the World. Revised Legend.* World Soil Resources Report 60. Food and Agriculture Organization, Rome, Italy.

FAO. 1991. *World Soil Resources. An Explanatory Note on the FAO World Soil Resources Map at 1:25 000 000 Scale.* World Soil Resources Rep. 66. Food and Agriculture Organization, Rome, Italy.

FAO. 1993. *Global and National Soils and Terrain Digital Databases (SOTER).* World Soil Resources Rep. 74. Food and Agriculture Organization, Rome, Italy.

Felhendler, R., I. Shainberg, and H. Frenkel. 1974. Dispersion and hydraulic conductivity of soils in mixed solution. *Trans. 10th Int. Congr. Soil Sci.* 1:103-112.

Ferguson, B.K. 1994. *Stormwater Infiltration.* Lewis Publishers, Boca Raton, FL.

Ferguson, B.K., and P.R. Gonnsen. 1993. Stream rehabilitation in a disturbed industrial watershed. In: Hatcher, K.J. (Ed.). *Proc. Georgia Water Resources Conf.* University of Georgia Institute of Natural Resources, Athens, GA.

Ferreiro, E.A., and A.K. Helmy. 1974. Flocculation of Na-montmorillonite by electrolytes. *Clay Miner.* 10:203-213.

Ferris, I.G., R. Fawcett, P.R. Stork, B.M. Haigh, N. Pederson, and A. Rovira. 1995. Persistence and leaching of sulfonylurea herbicides. In: Naidu, R., M.E. Sumner, and P. Rengasamy (Eds.). *Australian Sodic Soils: Distribution, Properties, and Management.* pp. 229-238. Commonwealth Scientific and Industrial Research Organization Publications, Melbourne, Australia.

Fireman, M. 1944. Permeability measurements on disturbed soil samples. *Soil Sci.* 58:337-353.

Fireman, M., and G.B. Bodman. 1939. The effect of saline irrigation water upon the permeability and base status of soils. *Soil Sci. Soc. Am. Proc.* 4:71-77.

Fitzpatrick, R.W., S.C. Boucher, R. Naidu, and E. Fritsch. 1994. Environmental consequences of soil sodicity. *Aust. J. Soil Res.* 32:1069-1093.

Fitzpatrick, R.W., S.C. Boucher, R. Naidu, and E. Fritsch. 1995. Environmental consequences of soil sodicity. In: Naidu, R., M.E. Sumner, and P. Rengasamy (Eds.). *Australian Sodic Soils: Distribution, Properties, and Management.* pp. 163-176. Commonwealth Scientific and Industrial Research Organization Publications, Melbourne, Australia.

Ford, G.W., J.J. Martin, P. Rengasamy, S.C. Boucher, and A. Ellington. 1993. Soil sodicity in Victoria. *Aust. J. Soil Res.* 31:869-910.

Foree, E.G., and J.S. Tapp. 1979. Design of chemical treatment system for removal of suspended solids from surface mining runoff from a 1000-acre watershed. *Symp. on Surface Mining Hydrology, Sedimentology and Reclamation.* University of Kentucky, Lexington, KY.

Foster, R.C. 1981. Polysaccharides in soil fabrics. *Science* 214:665 667.

Foster, R.C. 1988. Microenvironments of soil microorganisms. *Biol. Fertil. Soils* 6:189-203.

Frenkel, H. 1984. Reassessment of water quality criteria for irrigation. In: Shainberg, I., and J. Shalhevet (Eds.). *Soil Salinity under Irrigation.* pp.143-172. Springer-Verlag, Berlin, Germany.

Frenkel, H., and N. Alperovitch. 1983. Factors affecting the estimation of exchangeable sodium percentage in soils from Israel. *Hassadeh* 63:1291-1296.

Frenkel, H., M.V. Fey, and G.J. Levy. 1992a. Organic and inorganic anion effects on reference and soil clay critical flocculation concentration. *Soil Sci. Soc. Am. J.* 56:1762-1766.

Frenkel, H., Z. Gerstl, and N. Alperovitch. 1989. Exchange-induced dissolution of gypsum and the reclamation of sodic soils. *J. Soil Sci.* 40:599-611.

Frenkel, H., J.O. Goertzen, and J.D. Rhoades. 1978. Effects of clay type and content, exchangeable sodium percentage, and electrolyte concentration on clay dispersion and soil hydraulic conductivity. *Soil Sci. Soc. Am. J.* 48:32-39.

Frenkel, H., and A. Hadas. 1981. Effect of tillage and gypsum incorporation on rain runoff and crust strength in field soils irrigated with saline-sodic water. *Soil Sci. Soc. Am. J.* 45:156-158.

Frenkel, H., G.J. Levy, and M.V. Fey. 1992b. Clay dispersion and hydraulic conductivity of clay-sand mixtures as affected by the addition of various anions. *Clays Clay Min.* 40:515-521.

Frenkel, H., and I. Shainberg. 1975. Chemical and hydraulic changes in soils irrigated with brackish water. In: *Proc. Int. Symp. Brackish Water Factor Dev.* pp. 175-198. University of Beersheva, Beersheva, Israel.

Fuentes Godo, P.C., C. Raffa, and G. Otamendi. 1980. Biological approach to reclaiming saline-sodic soils in irrigation areas (in Spanish). *Actas IX Reunion Argentina de la Cienca del Suelo.* 2:807-815.

Gambrell, R.P., J.B. Wiesepape, W.H.J. Patrick, and M.C. Duff. 1991. The effects of pH, redox, and salinity on metal release from a contaminated sediment. *Water Air Soil Pollut.* 57-58:359-367.

Gardner, E.A., and K.J. Coughlan. 1982. Physical factors determining soil suitability for irrigated crop production in the Burdekin-Elliot River area. Agricultural Chemistry Branch Technical Rep. 20, Queensland Department of Primary Industries, Indooroopilly, Queensland, Australia.

Gardner, E.A., R.J. Shaw, G.D. Smith, and K.J. Coughlan. 1984. Plant available water capacity: Concept, measurement and prediction. In: McGarity, J.W., E.H. Houltz, and H.B. So (Eds.). *Properties and Utilisation of Cracking Clay Soils.* pp. 164-175. Proceedings of a Symposium, Rev. Rural Sci. 5, University of New England, Armidale, NSW, Australia.

Gardner, W.R., M.S. Mayhugh, J.O. Goertzen, and C.A. Bower. 1959. Effect of electrolyte concentration and ESP on diffusivity of water in soils. *Soil Sci.* 88:270-274.

Gaynor J.D., and V.V. Volk. 1981. Runoff losses of atrazine and terbutryn from unlimed and limed soil. *Environ. Sci. Tech.* 15:440-443.

Ghosh, K., and M. Schnitzer. 1980. Macromolecular structures of humic substances. *Soil Sci.* 129:266-276.

Gill, H.S., and I.P. Abrol. 1990. Tree based land use systems for utilization of salt affected soils. *Trans. 14th Int. Congr. Soil Sci.* 7:174-179.

Gill, H.S., and I.P. Abrol. 1993. Afforestation and amelioration of salt-affected soils in India. In: Davidson, N., and R. Galloway (Eds.). *Productive Use of Saline Land.* pp. 23-28. Australian Council for International Agricultural Research, Canberra, Australia.

Gilliam, J.W., and G.D. Bubenzer. 1987. Soil erosion and productivity. *Southern Cooperative Ser. Bul.* 360. University of Wisconsin, Madison, WI.

Gillman, G.P., and E.A. Sumpter. 1986. Modification to the compulsive exchange method for the measurement of exchange characteristics of soils. *Aust. J. Soil Res.* 24:61-66.

Glazovskaya, M.A. 1984. *Soils of the World.* A. A. Balkema, Rotterdam, Netherlands.

Godagnone, R., J.C. Salazar, and R.M. Di Giacomo. 1991. Propuesta de una nueva clasificación de algunos molisoles con horizonte nátrico en la Pampa Deprimida (P'cia. Buenos Aires). *Proc. XIII Cong. Argent. Cien. Suelo* I:206-207.

Goertzen, J.O., and C.A. Bower. 1958. Carbon dioxide from plant roots as a factor in the replacement of adsorbed sodium in calcareous soils. *Soil Sci. Soc. Am. Proc.* 22:36-37.

Golchin, A., J.M. Oades, J.O. Skjemstad, and P. Clarke. 1994. Soil structure and carbon cycling. *Aust. J. Soil Res.* 32:1043-1068.

Goldberg, S., and H.S. Forster. 1990. Flocculation of reference clays and arid-zone soil clays. *Soil Sci. Soc. Am. J.* 54:714-718.

Goldberg, S., B.S. Kapoor, and J.D. Rhoades. 1990. Effect of aluminum and iron oxides and organic matter on flocculation and dispersion of arid zone soils. *Soil Sci.* 150:588-593.

Goldberg, S., D.L. Suarez, and R.A. Glaubig. 1988. Factors affecting clay dispersion and aggregate stability of arid zone soils. *Soil Sci.* 146:317-325.

Gotoh, S., and W.H. Patrick. 1972. Transformation of manganese in a waterlogged soil as affected by redox potential and pH. *Soil Sci. Soc. Am. Proc.* 36:728-742.

Govers, G. 1987. Spatial and temporal variability in rill development processes at the Huldenberg Experimental Site. *Catena Suppl.* 8:17-34.

Graber, E.R., and U. Mingelgrin. 1994. Clay swelling and regular solution theory. *Environ. Sci. Technol.* 28:2360-2365.

Grattan, S.R., and C.M. Grieve. 1992. Mineral element acquisition and growth response of plants in saline environments. *Agric. Ecosys. Environ.* 38:275-300.

Green, R.E., and S.W. Karickhoff. 1990. Sorption estimates for modeling. In: Cheng, H.H. (Ed.). *Pesticides in the Soil Environment: Processes, Impacts and Modeling.* pp. 79-101. Soil Science Society of America, Madison, WI.

Greene, H., and O.W. Snow. 1939. Soil improvement in the Sudan Gezira. *J. Agric. Sci.* 29:1-34.

Greene, R.S.B., A.M. Posner, and J.P. Quirk. 1978. A study of the coagulation of montmorillonite and illite suspensions by CaCl$_2$ using the electron microscope. In: Emerson,W.W., R.D. Bond, and A.R. Dexter (Eds.). *Modification of Soil Structure*. pp. 35-40. John Wiley, New York, NY.

Greene, R.S.B., and I.B. Wilson. 1989. Amelioration of some physical properties and nutrient availability of an exposed B horizon of a Red-Brown Earth. *Soil Use Manage.* 5:66-71.

Greenland, D.J. 1963. Adsorption of polyvinyl alcohols by montmorillonite. *J. Colloid Sci.* 18:647-664.

Greenland, D.J. 1965a. Interaction between clays and organic compounds in soils. Part I. Mechanisms of interaction between clays and defined organic compounds. *Soils Fert.* 28:415-425.

Greenland, D.J. 1965b. Interaction between clays and organic compounds in soils. Part II. Adsorption of soil organic compounds and its effect on soil properties. *Soils Fert.* 28:521-532.

Greenland, D.J. 1979. Structural organization of soils and crop production In: Lal, R., and D.J. Greenland (Eds.). *Soil Physical Properties and Crop Production in the Tropics*. pp. 47-56. John Wiley, New York, NY.

Greenland, D.J., and M.H.B. Hayes. 1978. *The Chemistry of Soil Constituents*. p. 19. John Wiley, Chichester, UK.

Gregory, J. 1989. Fundamentals of flocculation. *Crit. Rev. Environ. Control* 19:185-230.

Grewal, S.S., and I.P. Abrol. 1989. Amelioration of sodic soil by rainwater conservation and Karnal grass sown in the interspace of trees. *J. Indian Soc. Soil Sci.* 37:371-376.

Grierson, I.T. 1975. The Effect of Gypsum on the Chemical and Physical Properties of a Range of Red-Brown Earths. M. Agric. Sci. Thesis, University of Adelaide, Australia.

Grierson, I.T. 1978. Gypsum and Red Brown Earths. In: Emerson, W. W., R. D. Bond, and A.R. Dexter (Eds.). *Modifications of Soil Structure*. pp. 315-324. John Wiley, New York, NY.

Grieve, C.M. 1993. Personal communication.

Grieve, C. M., and Fujyama, H. (1987). The response of two rice cultivars to external Na/Ca ratio. *Plant Soil* 103:245-250.

Grieve, C.M., and E.V. Maas. 1988. Differential effects of sodium/calcium ratio on sorghum genotypes. *Crop Sci.* 28:659-665.

Grove, J.H., C.S. Fowler, and M.E. Sumner. 1982. Determination of the charge character of selected acid soils. *Soil Sci. Soc. Am. J.* 46:32-38.

Gschwend, P.M., D. Backhus, and J.K. MacFarlane. 1988. *Mobilization of Colloids in Groundwater due to Infiltration of Water Near an Electric Generating Station*. Massachusetts Institute of Technology Energy Laboratory Rep. MIT-EL 88-004, Cambridge, MA.

Gschwend, P.M., and M.D. Reynold. 1987. Monodisperse ferrous phosphate colloids in an anoxic groundwater plume. *J. Contam. Hydrol.* 1:309-327.

Gu, B., and H.E. Doner. 1993. Dispersion and aggregation of soils as influenced by organic and inorganic polymers. *Soil Sci. Soc. Am. J.* 57:709-716.

Guggenberger, G., B. Glaser, and W. Zech. 1994. Heavy metal binding by hydrophobic and hydrophilic dissolved organic carbon fractions in a Spodosol A and B horizon. *Water Air Soil Pollut.* 72:111-127.

Gunn, R.H. 1967. A soil catena on denuded laterite profiles in Queensland. *Aust. J. Soil Res.* 5:117-132.

Gupta, A.P., S.S. Khanna, and N.K. Tomar. 1985. Effect of sodicity on the utilization of phosphatic fertilizers by wheat. *Soil Sci.* 139:47-52.

Gupta, R.K., and I.P. Abrol. 1990a. Salt-affected soils: Their reclamation and management for crop production. *Adv. Soil Sci.* 11:223-288.

Gupta, R.K., and I.P. Abrol. 1990b. Reclamation and management of alkali soils. *Indian J. Agric. Sci.* 60:1-16.

Gupta, R.K., D.K. Bhumbla, and I.P. Abrol. 1984a. Effect of sodicity, pH, organic matter, and calcium carbonate on the dispersion behaviour of soils. *Soil Sci.* 137:245-251.

Gupta, R.K., D.K. Bhumbla, and I.P. Abrol. 1984b. Sodium-calcium exchange equilibria in soils as affected by calcium carbonate and organic matter. *Soil Sci.* 138:109-114.

Gupta, R.K., and H. Chandra. 1972. Effect of gypsum in reducing boron hazard of saline irrigation waters and irrigated soil. *Ann. Arid. Zone* 11:228-230.

Gupta, R.K., R. Chhabra, and I.P. Abrol. 1981. The relationship between pH and exchangeable sodium in a sodic soil. *Soil Sci.* 131:215-219.

Gupta, R.K., R.R. Singh, and I.P. Abrol. 1989. Influence of simultaneous changes in sodicity and pH on the hydraulic conductivity of an alkali soil under rice culture. *Soil Sci.* 147:28-33.

Gupta, R.K., R.R. Singh, and K.K. Tanji. 1990. Phosphorus release in sodium ion dominated soils. *Soil Sci. Soc. Am. J.* 54:1254-1260.

Hadas, A., and J. Hagin. 1972. Boron adsorption by soils as influenced by potassium. *Soil Sci.* 113:189-193.

Hall, A.D. 1904. The effect of long-continued use of sodium nitrate on the constitution of the soil. *J. Chem. Soc. Trans.* 85:964-971.

Hall, D.J.M. 1990. An assessment of techniques for improving the structure of degraded Red-Brown Earths. Final report to the Wheat Industry Research Committee of New South Wales, NSW Agriculture and Fisheries, Trangie, NSW, Australia.

Hallsworth, E.G., and H.D. Waring. 1964. Studies in pedogenesis in New South Wales. VIII. An alternative hypothesis for the formation of the Solodized-Solonetz of the Pilliga district. *J. Soil Sci.* 15:159-177.

Halvorson, A.D. 1990. Management of dryland seeps. In: Tanji, K.K. (Ed.). *Agricultural Salinity Assessment and Management*. pp. 372-392. American Society of Civil Engineers, New York, NY.

Hamaker, J.W., and J.M. Thompson. 1972. Adsorption. In: Goring, C.A.I., and J.W. Hamaker (Eds.). *Organic Chemicals in the Soil Environment, Vol 1*. pp. 49-143. Marcel Dekker, Inc., New York.

Hamblin, A.P. 1985. The influence of soil structure on water movement, crop root growth, and water uptake. *Adv. Agron.* 38:95-158.

Harker, D.B., G.R. Webster, and R.R. Cairns. 1977. Factors contributing to crop response on a deep-plowed Solonetz soil. *Can. J. Soil Sci.* 57:279-287.

Harris, S.A. 1958. The gilgaied and bad-structured soils of central Iraq. *J. Soil Sci.* 9:169-185.

Harter, R.D., and R. Naidu. 1995. Role of metal-organic complexation in metal sorption by soils. *Adv. Agron.* 55:219-264.

Hayes, M.H.B. 1980. The role of natural and synthetic polymers in stabilising soil aggregates. In: Berkeley, R.C.W., J.M. Lynch, J. Melling , P.R. Rutter, and V.B. Ellis (Eds.). *Microbial Adhesion to Surfaces.* pp. 263-296. Horwood Ltd., Chichester, UK.

Hayes, M.H.B., M. Stacy, and J.M. Thompson. 1972. Adsorption of s-triazine herbicides by soil organic matter: Preparations. In: *Isotopes and Radiation in Soil Organic Matter Studies.* pp 75-90. International Atomic Energy Agency, Vienna, Austria.

Hayes, M.H.B., and R.S. Swift. 1978. The chemistry of soil organic colloids. In: Greenland, D.J., and M.H.B. Hayes (Eds.). *The Chemistry of Soil Constituents.* John Wiley, New York, NY.

Haynes, R.J. 1986. *Mineral Nitrogen in the Plant-Soil System.* Academic Press, Orlando, FL.

Hayward, H.E., and C.H. Wadleigh. 1949. Plant growth on saline and alkali soils. *Adv. Agron.* 1:1-38.

Hazelton, P.A. 1992. *Soil Landscapes of the Kiama 1:100 000 Sheet.* NSW Department of Conservation and Land Management, Sydney, Australia.

Hearn, A.B., and G.A. Constable. 1984. Irrigation for crops in a sub-humid environment. VII. Evaluation of irrigation strategies for cotton. *Irrig. Sci.* 5:75-94.

Heck, R.J., and A.R. Mermut. 1992a. The chemistry of saturation extracts of Solonetzic and associated soils. *Can. J. Soil Sci.* 72:43-56

Heck, R.J., and A.R. Mermut. 1992b. Genesis of Natriborolls (Solonetzic) in a closed lake basin in Saskatchewan, Canada. *Soil Sci. Soc. Am. J.* 56:842-848

Hedley, M.J., P.H. Nye, and R.E. White. 1982. Plant-induced changes in the rhizosphere of rape (*Brassica napus* var. Emerald) seedlings II. Origin of the pH change. *New Phytol.* 91:31-44.

Heede, B.H. 1971. Characteristics and processes of soil piping in gullies. *USDA Rocky Mt.For.Range Exp. Station Res. Pap.* RM 68.

Heinzen, R.T. and K. Arulanandan. 1976. Factors influencing dispersive clays and methods of identification. In: Sherard, J.L., and R.S. Decker (Eds.). *Dispersive Clays, Related Piping, and Erosion in Geotechnical Projects.* pp. 202-217. American Society for Testing and Materials, ASTM STP 623.

Helalia, A.M. and J. Letey. 1988a. Polymer type and water quality effects on soil dispersion. *Soil Sci. Soc. Am. J.* 52:243-246.

Helalia, A.M., and J. Letey. 1988b. Cationic polymer effects on infiltration rates with a rainfall simulator. *Soil Sci. Soc. Am. J.* 52:247-250.

Helfferich, F.G. 1962. *Ion Exchange.* McGraw-Hill, New York, NY.

Helmers, S. J., and S. Miyamota. 1990. Mechanical and chemical practices for reducing salinity in pecan orchards. In: *Visons of the Future.* pp. 374-377. Proc. 3rd Nat. Irrig. Symp. ASAE Publ. 04-90. St. Joseph, MI.

Helmke, P.A., and R. Naidu. 1996. Fate of contaminants in the soil environment: Metal contaminants. In: R.Naidu, R.S. Kookana, D.P. Oliver, S. Rogers, and M.J. McLaughlin (Ed.). *Contaminants and the Soil Environment in the Australasia-Pacific Region.* pp. 69-94. Kluwer Academic Publishers, Dordrecht, Netherlands.

Hilgard, E.W. 1877. *Report to the President of the University.* Experiment Station Rep., Univ. of California, Oakland, CA.

Hilgard, E.W. 1889. The rise of the alkali in the San Joaquin Valley. *Calif. Agric. Exp. Sta. Bull.* 83.

Hillel, D. 1980. *Applications of Soil Physics.* Academic Press, New York, NY.

Hingston, F.J. 1964. Reactions between boron and clays. *Aust. J. Soil Res.* 2:83-95.

Hitrov, N.B. 1988. Nabuhanie slitnih solontzevatih chernozemov Stavropolya. In: *Pochvi solontzovih territoriy i metodi ih izucheniya.* pp. 142-149. Otvetstvennyui Redaktor I.N., Moscow, Russia.

Hodgson, A.S. 1982. The effects of duration, timing and chemical amelioration of short-term waterlogging during furrow irrigation of cotton in a cracking grey clay. *Aust. J. Agric. Res.* 33:1019-1028.

Hodgson, A.S., and K.Y. Chan. 1984. Deep moisture extraction and crack formation by wheat and safflower in a Vertisol following irrigated cotton rotations. In: McGarity, J.W., E.H. Hoult, and H.B. So (Eds.). *The Properties and Utilization of Cracking Clay Soils.* pp. 299-304. University of New England, Armidale, Australia.

Hodgson, A.S., J.F. Holland, and P. Rayner. 1989. Effects of field slope and duration of furrow irrigation on growth and yield of six grain legumes on a waterlogging-prone Vertisol. *Field Crops Res.* 22:165-180.

Hodgson, J.F., W.L. Lindsay, and J.F. Trierweiler. 1966. Micronutrient cation complexing in soil solution: II. Complexing of zinc and copper in displaced solution from calcareous soils. *Soil Sci. Soc. Am. Proc.* 30:723-726.

Hoffman, G.J. 1986. Guidelines for reclamation of salt-affected soils. *App. Agric. Res.* 1:65-72.

Hoffman, G.J., and M. van Genuchten. 1983. Soil properties and efficient water use: Water management for salinity control. In: H.M. Taylor, W.R. Jordan, and T.R. Sinclair (Eds.). *Limitations to Efficient Water Use in Crop Production.* pp. 73-85. American Society of Agronomy, Madison, WI.

Hoffman, M.R., E.C. Yost, S.J. Eisenreich, and W.J. Maier. 1981. Characterizations of soluble and colloidal phase metal complexes in river water by ultrafiltration: A mass balance approach. *Environ. Sci. Tech.* 15:655-661.

Holburt, M.B. 1984. Colorado River salinity: The user's perspective. In: French, R.H. (Ed.). *Salinity in Watercourses and Reservoirs.* pp. 13-22. Butterworth Publishers, Boston, MA.

Holloway, R.E., and A.M. Alston. 1992. The effects of salt and boron on growth of wheat. *Aust. J. Agric. Res.* 43:987-1001.

Holmes, J.W. and H.C.T. Stace. 1968. On the domed structure and anisotropy of the B horizon of a Solodized Solonetz. *Aust. J. Soil Res.* 6:149-157.

Horn, C.P. 1983. The Effect of Cations on Soil Structure. Dip. Agr. Sci. Thesis. University of New England, Armidale, Australia.

Hosking, P.L. 1967. Tunneling erosion in New Zealand. *J. Soil Water Conserv.* 22:149-151.

Hottin, G., and O.F. Ouedraogo. 1975. Notice explicative de la carte géologique à 1/1 000 000 de la république de la Haute-Volta. Direction de la Géologie et des Mines. Ouagadougou, Upper Volta.

Hoyle, B.J., H. Yamada, and T.D. Hoyle. 1972. Aggresizing to eliminate objectionable clods. *Calif. Agric.* 26:3-5.

Hrasko, J. 1971. Salt-affected soils in Czechoslovakia and the problems of their utilization. In: Szabolcs, I. (Ed.). *European Solonetz Soils and Their Reclamation.* pp. 49-60. Akadémiai Kiadó, Budapest, Hungary.

Hubble, G.D., R.F. Isbell, and K.H. Northcote. 1983. Features of Australian Soils. In: *Soils: An Australian Viewpoint.* pp. 17-47. Commonwealth Scientific and Industrial Research Organization Publications, Melbourne, Australia/Academic Press, London, UK.

Huberty, M.R., and A.U. Pillsbury. 1941. Factors influencing infiltration rates into some California soils. *Trans. Am. Geophys. Union* 22:686-697.

Huheey, J.E., E.A. Keiter and R.L. Keiter. 1994. *Inorganic Chemistry.* Harper Collins, New York, NY.

Hulme, P.J., D.C. McKenzie, T.S. Abbott, and D.A. MacLeod. 1991. Changes in physical properties of a Vertisol following an irrigation of cotton as influenced by the previous crop. *Aust. J. Soil Res.* 29:425-442.

Hunsaker, V.E., and P.F. Pratt. 1971. Calcium-magnesium exchange equilibria in soils. *Soil Sci. Soc. Am. Proc.* 35:151-152.

Hunter, M.N., P.L.M. de Jabrun, and D.E. Byth. 1980. Response of nine soybean lines to soil moisture conditions close to saturation. *Aust. J. Exp. Agric. Anim. Husb.* 20:339-345.

Hussain, A., and P. Gul. 1993. Selection of tree species suitable for saline and waterlogged areas in Pakistan. In: Davidson, N., and R. Galloway (Eds.). *Productive Use of Saline Land.* pp. 53-55. Australian Council for International Agricultural Research, Canberra, Australia.

Hutchins, S.R., M.B. Tomson, and C.H. Ward. 1983. Trace organic contamination of groundwater from a rapid infiltration site: Laboratory-field coordination study. *Environ. Toxicol. Chem.* 2:195-216.

Ilyas, M., R.W. Miller, and R.H. Qureshi. 1993. Hydraulic conductivity of saline-sodic soil after gypsum application and cropping. *Soil Sci. Soc. Am. J.* 57:1580-1585.

Inanaga, S. 1991. Improvement of sloping beds of row for crop establishment in saline low land in Gangzhou Prefecture, China. In: *Impacts of Salinization and Acidification on Terrestrial Ecosystem and Its Rehabilitation.* pp. 143-148. Proc. Int. Symp. September 26-26, 1991. Tokyo, Japan.

Inskeep, W.P., and P.R. Bloom. 1986. Kinetics of calcite precipitation in the presence of water soluble organic ligands. *Soil Sci. Soc. Am. J.* 50:1167-1172.

INTA. 1990a. *Atlas de Suelos de la Republica Argentina. I. Buenos Aires.* Instituto Nacional de Tecnología Agropecuaria, Buenos Aires, Argentina.

INTA. 1990b. *Atlas de Suelos de la Republica Argentina. II. Santa Fé.* Instituto Nacional de Tecnología Agropecuaria, Buenos Aires, Argentina.

Isbell, R.F. 1958. The occurence of a highly alkaline Solonetz soil in Southern Queensland. *Queensl. J. Agric. Sci.* 15:15-23.

Isbell, R.F. 1994. *A Classification System for Australian Soils.* 3rd approximation. CSIRO Division of Soils. Tech. Rep. 2/1993 (unpublished).

Isbell, R.F. 1995a. Sealing, crusting and hardsetting conditions in Australian soils. In: So H.B., G.D. Smith, S.R. Raine, B.M. Schafer, and R.J. Loch (Eds.). *Sealing, Crusting and Hardsetting Soils: Productivity and Conservation.* pp. 15-30. Australian Society of Soil Science, Queensland Branch, Brisbane, Australia.

Isbell, R.F. 1995b. The use of sodicity in Australian soil classification. In: Naidu, R., M.E. Sumner, and P. Rengasamy (Eds.). *Australian Sodic Soils: Distribution, Properties and Management.* pp. 47-52. Commonwealth Scientific and Industrial Research Organization Publications, Melbourne, Australia.

Isbell, R.F., R. Reeve and J.T. Hutton. 1983. Salt and sodicity. In: *Soils: An Australian Viewpoint.* pp. 107-117. Commonwealth Scientific and Industrial Research Organization Publications, Melbourne/Academic Press, London.

Israelachvili, J.N. 1991. *Intermolecular and Surface Forces.* 2nd ed. Academic Press, New York, NY.

Israelachvili, J.N., and P.M. McGuiggan. 1988. Forces between surfaces in liquids. *Science* 241:795-800.

Iu, K.L., I.D. Pulford, and H.J. Duncan. 1981. Influence of waterlogging and lime or organic matter additions on the distribution of trace metals in an acid soil. II. Zinc and copper. *Plant Soil* 59:327-333.

Iwata, S., T. Tabuchi, and B.P. Warkentin. 1995. *Soil-water Interactions: Mechanisms and Applications.* pp. 440. Marcel Dekker, New York, NY.

Jackson, M.B., and M.C. Drew. 1984. Effects of flooding on growth and metabolism of herbaceous plants. In: Kozlowski, T.T. (Ed.). *Flooding and Plant Growth.* pp. 47-128. Academic Press, London, UK.

Jackura, K.A. 1980. Infiltration drainage of highway surface water. California Department of Transportation Rep. FHWA/CA/TL-80-04.

Jacquin, F., N. Mallouhi, and T. Gallali. 1979. Etude sur l'intensité des transfers de matière organiques sous l'influence de la salinité. *C. R. Seánces Acad. Sci. Ser. D (Paris)* 289:1229-1232.

Jardine, P.M., G.V. Wilson, R.J. Luxmoore, and J.F. McCarthy. 1989. Transport of inorganic and natural organic tracers through an isolated pedon in a forest watershed. *Soil Sci. Soc. Am. J.* 53:317-323.

Jayawardane, N.S. 1979. An equivalent salt solutions method for predicting hydraulic conductivities of soils for different salt solutions. *Aust. J. Soil Res.* 17:423-428.

Jayawardane, N.S. 1983. Further examination of the equivalent salt solution method for predicting hydraulic conductivities of soils for different salt solutions. *Aust. J. Soil Res.* 21:105-108.

Jayawardane, N.S. 1992. Prediction of unsaturated hydraulic conductivity changes of a loamy soil in different salt solutions by using the equivalent salt solutions concept. *Aust. J. Soil Res.* 30:565-571.

Jayawardane, N.S., and J. Blackwell. 1985. Effect of gypsum enriched slots on moisture movement and aeration in an irrigated swelling clay soil. *Aust. J. Soil. Res.* 23:481-492.

Jayawardane, N.S., and J. Blackwell. 1986. Effects of gypsum-enriched slotting on infiltration rates and moisture storage in a swelling clay soil. *Soil Use Manage.* 2:114-118.

Jayawardane, N.S., and J. Blackwell. 1990. Use of the neutron method in assessing changes in soil strength of undisturbed and ameliorated transitional Red-Brown Earths during soil drying cycles. *Aust. J. Soil Res.* 28:167-176.

Jayawardane, N.S., and P.S. Blackwell. 1991. Relationship between equivalent salt solution series of different soils. *J. Soil Sci.* 42:95-102.

Jayawardane, N.S., J. Blackwell, G. Kirchof, and W.A. Muirhead. 1994. Slotting: A deep tillage technique for ameliorating sodic, acid and other degraded soils and for land treatment of waste. In: Jayawardane, N.S., and B.A. Stewart (Eds.). *Subsoil Management Techniques.* pp. 109-146. Lewis Publishers, Boca Raton, FL.

Jayawardane, N.S., J. Blackwell, and W.A. Muirhead. 1987a. Research strategies in planning and evaluating soil ameliorative techniques on clay soils. In: K.J. Coughlan, and P.N. Troung (Eds.). *Effects of Management Practices on Soil Physical Properties.* pp. 213-217. Queensland Department of Primary Industries Conference and Workshop Series, Brisbane, Australia.

Jayawardane, N.S., J. Blackwell, and M. Stapper. 1987b. Effects of changes in moisture profiles of a transitional Red-Brown Earth due to surface and slotted gypsum applications. *Aust. J. Agric. Res.* 38:239-251.

Jayawardane, N.S., and K.Y. Chan. 1994. The management of soil physical properties limiting crop production in Australian sodic soils: A review. *Aust. J. Soil Res.* 32:13-44.

Jayawardane, N.S., and W.S. Meyer. 1985. Measuring air-filled porosity changes in an irrigated swelling clay soil. *Aust. J. Soil Res.* 23:15-22.

Jayawardane, N.S., and S.A. Prathapar. 1992. Effect of soil loosening on hydraulic properties of a duplex clay soil. *Aust. J. Soil Res.* 30:959-975.

Jaynes, D., and D.J. Hundsaker. 1989. Spatial and temporal variability of water content and infiltration on a flood- irrigated field. *Trans. ASAE* 26:1422-1429.

Jeffrey, J.J., and N.C. Uren. 1983. Copper and zinc species in the soil solution and the effects of soil pH. *Aust. J. Soil Res.* 21:479-488.

Jekel, M.R. 1986. The stabilization of dispersed mineral particles by adsorption of humic substances. *Water Res.* 20:1543-1554.

Jenkin, J.J. 1986. Western civilization. In: Russell, J.S., and R. Isbell (Eds.). *Australian Soils: The Human Impact.* pp. 134-156. University of Queensland Press, St. Lucia, Australia.

Jenny, H. 1941. *Factors of Soil Formation.* McGraw-Hill, New York, NY.

Jones, A.R. 1984. Controlling salinity in the Colorado River Basin, the arid west. In: French, R.H. (Ed.). *Salinity in Watercourses and Reservoirs.* pp. 337-347. Butterworth Publishers, Boston, MA.

Joshi, Y.C., A. Qadar, A.R. Bal, and R.S. Rana. 1980. Sodium/potassium index of wheat seedlings in relation to sodicity tolerance. *Proc. Int. Symp. on Salt Affected Soils: Principles and Practices for Reclamation and Management.* pp. 457-460. Central Soil Salinity Research Institute, Karnal, India.

Jugsujinda, A., and W.H. Patrick. 1977. Growth and nutrient uptake by rice in a flooded soil under controlled aerobic-anaerobic and pH conditions. *Agron. J.* 69:705-710.

Jury, W.A., W.W. Jarrell, and D. Devitt. 1979. Reclamation of saline-sodic soils by leaching. *Soil Sci. Soc. Am. J.* 43:1100-1106.

Juste, C., and J. Delas. 1970. Comparaison, par une méthode respirométrique, des stabilités biologiques d'un humate de calcium et d'un humate de sodium. *C. R. Seánces Acad. Sci. Ser. D (Paris)* 270:1127-1129.

Juste, C., J. Delas, and M. Langon. 1975. Comparaison de la stabilité biologique de différents humates métalliques. *C. R. Seánces Acad. Sci. Ser. D* (Paris) 281:1685-1688.

Kabata-Pendias, A., and H. Pendias. 1991. *Trace Elements in Soils and Plants.* 2nd ed. CRC Press, Boca Raton, FL.

Kadry, L.T. 1969. A note on the alluvial soils of Iraq, their salinity, sodicity and reclaimability. *Agrokém. Talajtan* 18:159.

Kalbasi, M., G.J. Racz, and L.A. Lewen-Rudgers. 1978. Reaction products and solubility of applied Zn compounds in some Manitoba soils. *Soil Sci.* 125:55-64.

Kan, A.T., and M.B. Tomson. 1990. Ground water transport of hydrophobic organic compounds in the presence of dissolved organic matter. *Environ. Toxicol. Chem.* 9:253-263.

Kanwar, J.S. 1969. Salt-affected soils in India, their nature and distribution. *Agrokém. Talajtan* 18:79-86.

Kanwar, J.S., D.R. Bhumabla, and N. Singh. 1965. Studies on the reclamation of saline and sodic soils in the Punjab. *Indian J. Agric. Sci.* 35:43-51.

Kaplan, D.I., P.M. Bertsch, D.C. Adriano, and W.P. Miller. 1993. Soil-borne mobile colloids as influenced by water flow and organic carbon. *Environ. Sci. Tech.* 27:1193-1200.

Kaushik, B.D. 1990. Cyanobacteria: Some aspects related to their role as ameliorating agent of salt-affected soils. In: Rajarao, V.N. (Ed.). *Perspectives in Phycology.* Proc. Int. Symp. Phycology. pp. 405-410. Madras, India.

Kay, B.D. 1990. Rates of change of soil structure under different cropping systems. *Adv. Soil Sci.* 12:1-52.

Kazman, Z., I. Shainberg, and M. Gal. 1983. Effect of low levels of exchangeable Na and applied phosphogypsum on the infiltration rate of various soils. *Soil Sci.* 35:184-192.

Kelley, W.P. 1937. *The Reclamation of Alkali Soils.* Citrus Exp. Sta. Pap. 385. University of California, Riverside, CA.

Kelley, W.P. 1948. *Cation Exchange in Soils.* Reinhold Publishing, New York, NY.

Kelley, W.P. 1951. *Alkali Soils: Their Formation, Properties and Reclamation.* Reinhold Publishing, New York, NY.

Kemper, W.D., and E.J. Koch. 1966. Aggregate stability of soils from western United States and Canada. *USDA Tech. Bull.* 1355.

Keren, R. 1990. Water-drop kinetic energy effect on infiltration in sodium-calcium-magnesium soils. *Soil Sci. Soc. Am. J.* 54:983-987.

Keren, R. 1991. Specific effect of magnesium on soil erosion and water infiltration. *Soil Sci. Soc. Am. J.* 55:783-787.

Keren, R., and R.G. Gast. 1981. Effects of wetting and drying and of exchangeable cations on boron adsorption by Na montmorillonite. *Soil Sci. Soc. Am. J.* 45:45-48.

Keren, R., D. Sador, H. Frenkel, I. Shainberg, and A Meiri. 1990. Irrigation with sodic and brackish water and its effects on the soil and on cotton yields. *Hassadeh* 70:1822-1829.

Keren, R., I. Shainberg, H. Frenkel, and Y. Kalo. 1983. A field study of the effect of exchangeable sodium on surface runoff from Loess soil. *Soil Sci. Soc. Am. J.* 47:1001-1007.

Keren, R., and M.J. Singer. 1988. Effect of low electrolyte concentration on hydraulic conductivity of sodium/calcium montmorillonite system. *Soil Sci. Soc. Am. J.* 52:368-373.

Kertész, M., and T. Tóth. 1994. Mapping of soil salinization status of solonetzic landscapes based on seminatural vegetation. *Agrokém. Talajtan* 43:113-132.

Khanduja, S.D. 1987. Short-rotation firewood forestry on sodic soils in northern India: Imperatives. *Indian J. For.* 10:75-79.

Kile, D.E., and C.T. Chiou. 1989. Water solubility enhancement of nonionic organic contaminants. *Adv. Chem. Ser.* 219:131-158.

Kim, J.I., G. Buckau, F. Baumgartner, H.C. Moon, and D. Lux. 1984. Colloid generation and the actinide migration in the Gorleben groundwaters. In: McVay, G.L. (Ed.). *Scientific Basis for Nuclear Waste Management.* 7: 31-40. Elsevier Publishers, New York, NY.

Kim, K., and W.P. Miller. 1996. Effect of ionic strength and slope gradient on infiltration and interrill erosion of two Georgia soils. *Soil Sci. Soc. Am. J.*

Kingsbury, R.W., and E. Epstein. 1986. Salt sensitivity in wheat. A case for specific ion toxicity. *Plant Physiol.* 80:651-654.

Kinniburgh, D.G., and M.L. Jackson. 1982. Concentration and pH dependence of calcium and zinc adsorption by iron hydrous oxide gel. *Soil Sci. Soc. Am. J.* 46:56-61.

Klute, A. 1982. Tillage effects on hydraulic properties of a soil: A review. In: Unger, P.W., D.M. van Doren, F.D. Whisler, and E.L. Skidmore (Eds.). *Predicting Tillage Effects on Soil Physical Properties and Processes.* pp. 29-43. American Society of Agronomy, Madison, WI.

Knisel, W.G. 1980. CREAMS: A field-scale model for chemical, runoff and erosion for agricultural management systems. *USDA Soil Conservation Service Rep.* 26, Washington, DC.

Kohut, C.K., and M.J. Dudas. 1994. Characteristics of clay minerals in saline alkaline soils in Alberta, Canada. *Soil Sci. Soc. Am. J.* 58:1260-1269.

Kookana, R.S. 1989. Equilibrium and Kinetic Aspects of Sorption, Desorption and Mobility of Pesticides in Soils. Ph.D. Thesis, University of Western Australia, Nedlands, Australia.

Kookana, R.S., and L.A.G. Aylmore. 1994. Retention and release of diquat and paraquat herbicides in soils. *Aust. J. Soil Res.* 31:97-109.

Kookana, R.S., R. Naidu, and K.G. Tiller. 1994. Sorption nonequilibrium during cadmium transport through soils. *Aust. J. Soil Res.* 32:635-651.

Korentajer, L., P.C. Henry, and B.E. Eisenberg. 1991. A simple effects analysis of the effects of phosphogypsum application on phosphorus transport to runoff water and eroded soil sediments. *J. Environ. Qual.* 20 596-603.

Korentajer, L., R. Stern, and M. Aggasi. 1993. Slope effects on cadmium load of eroded sediments and runoff water. *J. Environ. Qual.* 22:639-645.

Kosmas, C., and N. Moustakas. 1990. Hydraulic conductivity and leaching of an organic saline-sodic soil. *Geoderma* 46:363-370.

Kovda, V. A., and C. van den Berg. 1973. Reclamation of saline and alkali soils. In: Kovda, V.A., and C. van den Berg (Eds.). *Irrigation, Drainage and Salinity: An International Source Book.* pp. 430-480. Hutchinson and Co., Paris, France.

Kreit, J.E., I. Shainberg, and A.J. Herbillon. 1982. Hydrolysis and decomposition of hectorites in dilute solutions. *Clays Clay Min.* 30:223-231.

Kretzschmar, R., W.P. Robarge, and S.B. Weed. 1993. Flocculation of kaolinitic soil clays: Effects of humic substances and iron oxides. *Soil Sci. Soc. Am. J.* 57:1277-1283.

Kreybig, L., and A. Endrédy. 1935. Uber die Abhängigkeit des Vorkommens von Alkaliböden im oberen Tisza-Gebiete Ungarns von der absoluten Höhenlage. *Trans. 3rd Int. Congr. Soil. Sci.* I:357-360.

Kumar, A. 1985. Karnal grass for reclaiming alkali soils. Tech. Bull. 5. Central Soil Salinity Research Institute, Karnal, India.

Kumar, A. 1990. Effect of gypsum compared with that of grasses on the yield of forage crops on highly sodic soil. *Exp. Agric.* 26:185-188.

Kumar, A., and I.P. Abrol. 1986. Grasses in alkali soils. Tech. Bull. 11, Central Soil Salinity Research Institute, Karnal, India.

Kumar, V., V.S. Ahlawat, and R.S Antil. 1985. Interactions of nitrogen and zinc in pearl millet: I. Effect of nitrogen and zinc levels on dry matter yield and concentration and uptake of nitrogen and zinc in pearl millet. *Soil Sci.* 139:351-356.

Kumar, V., B.K. Bhatia, and U.C. Shukla. 1981. Magnesium and zinc relationships in relation to dry matter and the concentration and uptake of nutrients in wheat. *Soil Sci.* 131:151-155.

Kumar, V., and M. Singh. 1980. Sulphur and zinc interactions in relation to yield, uptake and utilization of sulphur in soybean. *Soil Sci.* 130:18-25.

Kundell, J.E., and T.C. Rasmussen. 1995. *Erosion and Sedimentation: Scientific and Regulatory Issues.* Georgia Board of Natural Resources, Atlanta, GA.

Kunishi, H.M., and D.E. Glotfelty. 1985. Sediment, season, and salinity effects on phosphorus concentrations in an estuary. *J. Environ. Qual.* 14:292-296.

Ladd, J.N., R.C. Foster, P. Nannipieri, and J.M. Oades. 1996. Soil structure and biological activity. *Soil Biochem.* 9: 23-78.

Ladd, J.N., R.C. Foster, and J.O. Skjemstad. 1993. Soil structure: Carbon and nitrogen metabolism. *Geoderma* 56:401-434.

Lal, R., B.J. Bridge, and N. Collis-George. 1970. The effect of column diameter on the infiltration rate into a swelling soil. *Aust. J. Soil Res.* 8:185-193.

Läuchli, A., and E. Epstein. 1990. Plant responses to saline and sodic conditions. In: Tanji, K.K. (Ed.). *Agricultural Salinity Assessment and Management.* pp. 112-137. ASCE Manuals Rep. Eng. Practice 71. American Society of Civil Engineers, New York, NY.

Laura, R.D. 1973. Effects of sodium carbonate on carbon and nitrogen mineralisation of organic matter added to soil. *Geoderma* 9:15-26.

Laura, R.D. 1976. Effects of alkali salts on carbon and nitrogen mineralization of organic matter in soil. *Plant Soil* 44:587-596.

Laura, R.D. 1991. Effect of sodicity on the mineralization of nitrogen in soil. *Int. J. Trop. Agric.* 9:174-181.

Lavado, R.S. 1983. Evaluación de la relación entre la composición química del agua de lluvia y el distinto grado de salinidad y sodicidad de distintos suelos. *Rev. Fac. Agron.* 4:135-139.

Lavado, R.S., and M. Alconada. 1994. Soil properties on grazed and ungrazed plots of a grassland sodic soil. *Soil Technol.* 7:75-81.

Lavado, R.S., and R.R. Cairns. 1980. Solonetzic soil properties and yield of wheat, oats and barley as affected by deep plowing and ripping. *Soil Tillage Res.* 1:69-79.

Lavado, R.S., G. Hevia, A.A. Parodi, and L. Mormeneo. 1983. Liberación de iones en suelos en función de la concentración y tipo de sales. Mem. Reun. Int. Suelos Afect. Sales en América Latina. Maracay, Venezuela.

Lavado, R.S., and N.B. Reinaudi. 1988. Wind blown dust from salty areas as a source of fluoride for plants. *Fluoride* 19:14-18.

Lavado, R.S., and M.A. Taboada. 1988. Water, salt, and sodium dynamics in a Natraquoll in Argentina. *Catena* 15:577-594.

Lax, A., E. Diaz, V. Castillo, and J. Albaladejo. 1994. Reclamation of physical and chemical properties of a salinized soil by organic amendment. *Arid Soil Res. Rehabil.* 8:9-17.

Leal, J.E., M.E. Sumner, and L.T. West. 1994. Evaluation of available P with different extractants on Guatemalan soils. *Commun. Soil Sci. Plan Anal.* 25:1161-1196.

Lebron, I., and D.L. Suarez. 1992. Electrophoretic mobility of illite and micaceous soil clays. *Soil Sci. Soc. Am. J.* 56:1106-1115.

Lehrsch, G.A., and C.W. Robbins. 1994. Cottage cheese (acid) whey effects on sodic soil aggregate stability. *Arid Soil Res. Rehabil.* 8:19-31.

Leinweber, P., G. Reuter, and K. Brozio. 1993. Cation exchange capacity of organo-mineral particle size fractions in soil for long term experiments. *J. Soil Sci.* 44:111-119.

Leonard, R.A. 1990. Movement of pesticides into surface waters. In: Cheng, H.H. (Ed.). *Pesticides in the Soil Environment: Processes, Impacts and Modelling.* pp. 305-349. Soil Science Society of America, Madison, WI.

Leprun, J-C. 1977. Esquisse pédologique à 1/50 000 des alentours da la Mare d'Oursi avec notice et analyses des sols. ORSTOM, Paris, France.

Letey, J. 1985. Relationship between soil physical properties and crop production. *Adv. Soil Sci.* 1:277-291.

Lettenmaier, D.P., E.R. Hooper, C. Wagoner, and K.B. Faris. 1991. Trends in stream quality in the continental United States, 1978-1987. *Water Resour. Res.* 27:327-339.

Levy, G.J., M. Agassi, H.J.C. Smith, and R. Stern. 1993a. Microaggregate stability of kaolinitic and illitic soils determined by ultrasonic energy. *Soil Sci. Soc. Am. J.* 57:803-808.

Levy, G.J., N. Alperovitch, A.J. van der Merwe, and I. Shainberg. 1989. The hydrolysis of kaolinitic soils as affected by the type of the exchangeable cation. *J. Soil Sci.* 40:613-620.

Levy, G.J., M. Ben-Hur, and M. Agassi. 1991. The effect of polyacrylamide on runoff, erosion and cotton yield from fields irrigated with moving sprinkler systems. *Irrig. Sci.* 12:55-60.

Levy, G.J., H. Eisenberg, and I. Shainberg. 1993b. Clay dispersion as related to soil properties and water permeability. *Soil Sci.* 155:15-22.

Levy, G.J., J. Levin, M. Gal, M. Ben-Hur, and I. Shainberg. 1992. Polymer effects on infiltration and soil erosion during consecutive simulated sprinkler irrigations. *Soil Sci. Soc. Am. J.* 56:902-907.

Levy, G.J., J. Levin, and I. Shainberg. 1994. Seal formation and interrill soil erosion. *Soil Sci. Soc. Am. J.* 58:203-209.

Levy, G.J., J. Levin, and I. Shainberg. 1995. Polymer effects on runoff and soil erosion from sodic soils. *Irrig. Sci.* 6:9-14.

Levy, G.J., I. Shainberg, N. Alperovitch, and A.J. van der Merwe. 1991. Effect of Na-hexametaphosphate on the hydraulic conductivity of kaolinite-sand mixture. *Clays Clay Min.* 39:131-136.

Levy, G.J., and H.v.H. van der Watt. 1988. Effects of clay mineralogy and soil sodicity on soil infiltration rate. *S. Afr. J. Plant Soil* 5:92-96.

Levy, G.J., and H.v.H. van der Watt. 1990. Effect of exchangeable potassium on the hydraulic conductivity and infiltration rate of some South African soils. *Soil Sci.* 149:69-77.

Levy, G.J., H.v.H. van der Watt, and H.M. du Plessis. 1988. Effect of Na/Mg and Na/Ca systems on soil hydraulic conductivity and infiltration. *Soil Sci.* 146:303-310.

Levy, R., and D. Hillel. 1968. Thermodynamic equilibrium constants of Na/Ca exchange in some Israeli soils. *Soil Sci.* 106:393-398.

Lewis, D.T., and J.V. Drew. 1973. Slick spots in Southern Nebraska: Patterns and Genesis. *Soil Sci. Soc. Am. Proc.* 37:600-606.

Lickacz, J. 1993. Management of Solonetzic soils. Soils and Water Agri~fax, Agdex 518-8. Print Media Branch, Alberta Agriculture, Edmonton, Alberta, Canada.

Lickacz, J. 1994. Personal communication.

Lickacz, J., and G. Coy. 1991. An overview of deep tillage research in Alberta. Saskatchewan Soils and Crops Workshop, Saskatoon, Saskatchewan. Alberta Agriculture, Food and Rural Development, Edmonton, Alberta, Canada.

Lind, W.T., D.S. Doyle, D.S. Vodopich, B.G. Trotter, J.G. Limon, and L. Davalos-Lind. 1992. Clay turbidity: Regulation of phytoplankton production in a large, nutrient-rich tropical lake. *Limnol. Oceanogr.* 37:549-565.

Lindsay, W.L. 1979. *Chemical Equilibria in Soils*. John Wiley, New York, NY.

Lion, L.W., R.S. Altman, and J.O. Leckie. 1982. Trace metal adsorption characteristics of esturine particulate matter: Evaluation of Fe/Mn oxide and organic surface coatings. *Environ. Sci. Tech.* 16:660-666.

Liu, H., and G. Amy. 1993. Modeling partitioning and transport interactions between natural organic matter and polynuclear aromatic hydrocarbons in groundwater. *Environ. Sci. Tech.* 27:1553-1562.

Livingston, E.H., H.H. Harper, and J.L. Herr. 1994. The use of alum injection to treat stormwater. pp. 31-53. *Proc. 2nd. Ann. Conf. Soil Water Manag. Urb. Devel.*, Sydney, Australia.

Lobartini, J.C., K.H. Tan, L.E. Asmussen, R.A. Leonard, D. Himmelsbach, and A.R. Gingle. 1991. Chemical and spectral differences in humic matter from swamps, streams and soils in the southeastern United States. *Geoderma* 49:241-254.

Loch, R.J., and T.E. Donnollan. 1983. Field rainfall simulator studies on two clay soils of the Darling Downs, Queensland. II. Aggregate breakdown, sediment properties and soil erodibility. *Aust. J. Soil Res.* 21:47-58.

Loganathan, P., R.G. Burau, and D.W. Fuerstenau. 1977. Influence of pH on the sorption of Co^{2+}, Zn^{2+}, and Ca^{2+} by a hydrous managanese oxide. *Soil Sci. Soc. Am. J.* 41:57-62.

Loneragan, J.F. 1975. The availability and adsorption of trace elements in soil-plant systems and their relation to movement and concentrations of trace elements in plants. In: Nicholas, D.J.D. and A. R. Egan (Eds.). *Trace Elements in Soil-Plant Systems*. pp. 100-133. Academic Press, New York, NY.

Loveday, J. 1984. Amendments for reclaiming sodic soils. In: Shainberg, I., and J. Shalhevet (Eds.). *Soil Salinity under Irrigation: Processes and Management*. pp. 220-237. Springer-Verlag, Berlin, Germany.

Loveday, J., and R. Ayers. 1994. Personal communication.

Loveday, J., and B.J. Bridge. 1983. Management of salt-affected soils. In: *Soils: An Australian Viewpoint*. pp. 841-855. Commonwealth Scientific and Industrial Research Organization Publications, Melbourne, Australia/Academic Press, London.

Loveday, J., and J.C. Pyle. 1973. The Emerson dispersion test and its relationship to hydraulic conductivity. *CSIRO Division of Soils Tech. Pap.* 15.

Loveday, J., and D.R. Scotter. 1966. Emergence response of subterannean clover to dissolved gypsum in relation to soil properties and evaporative conditions. *Aust. J. Soil Res.* 4:55-68.

Loveland, P.J., J. Hazelden, and R.G. Sturdy. 1987. Chemical properties of salt-affected soils in north Kent and their relationship to soil instability. *J. Agric. Sci. (Camb.)* 109:1-6.

Low, P.F. 1979. Nature and properties of water in montmorillonite. *Soil Sci. Soc. Am. J.* 43:651-658.

Low, P.F. 1980. The swelling of clay: II. Montmorillonites. *Soil Sci. Soc. Am. J.* 44:667-676.

Low, P.F. 1991. Structural and other forces involved in the swelling of clays. *NATO Advanced Workshop on Clay Swelling and Expansive Soils*. Cornell University, Ithaca, NY.

Lyle, C.W., A.H. Mehanni, and A.P. Repsys. 1986. Leaching rates under a perennial pasture irrigated with saline water. *Irrig. Sci.* 7:277-286.

Lytle, D. T. 1993. Digital soil databases for the United States. In: Goodchild, M.F., B.O. Parks, and L.T. Steyaert (Eds.). *Environmental Modeling with GIS*. pp. 386-391. Oxford University Press, New York, NY.

Maas, E.V., and C.M. Grieve. 1987. Sodium-induced calcium deficiency in salt-stressed corn. *Plant Cell Environ.* 10:559-564.

Maas, E.V., and G.J. Hoffman. 1977. Crop salt tolerance: Current assessment. *J. Irrig. Drain. Div. Am. Soc. Civ. Eng.* 103:115-130.

MacEwan, D.M.C. 1948. Adsorption by montmorillonite and its relation to surface adsorption. *Nature* 162:935-936.

MacEwan, R.J., W.K. Gardner, A. Ellington, D.G. Hopkins, and A.C. Bakker. 1992. Tile and mole drainage for control of waterlogging in duplex soils of south-eastern Australia. *Aust. J. Exp. Agric.* 32:865-878.

Macvicar, C.N., J.M. de Villiers, R.F. Loxton, E. Verster, J.J.N. Lambrechts, F.R. Merryweather, J. le Roux, T.H. van Rooyen, and H.J. von M. Harmse. 1977. *Soil Classification. A Binomial System for South Africa*. Soil and Irrigation Research Institute. Dept. of Agricultural Technical Services, Pretoria, South Africa.

Madak, D.P., K.P. Singh, H. Chandra, and P.K. Ray. 1992. Mobile and bound forms of trace metals in the sediments of the Lower Ganges. Water Res. 26:1541-1548.

Madhun, Y.A., J.L. Young, and V.H. Freed. 1986. Binding of herbicides by water soluble organic materials from soil. *J. Envir. Qual.* 15:64-68.

Magistad, O.C., and J.E. Cristiansen. 1944. Saline soils, their nature and management. *U.S. Dept. of Agriculture Circ.* 707.

Magyar, P. 1928. Beiträge zu den Pflanzensoziologischen und Geobotanischen Verhältnissen der Hortobágy-Steppe. *Erdészeti Kisérletek* 30:217-226.

Mahanty, J., and B.W. Ninham. 1976. *Dispersion Forces*. Academic Press, London, UK.

Malhi, S.S., D.W. McAndrew, and M.R. Carter. 1992a. Effect of tillage and N fertilization of a Solonetzic soil on barley production and some soil properties. *Soil Tillage Res.* 2: 95-107.

Malhi, S.S., D.W. McAndrew, and M.R. Carter. 1992b. Effect of surface-applied Ca amendments and N on Solonetzic soil properties and composition of barley. *Arid Land Res. Rehabil.* 6:71-81.

Malik, K.A., Z. Aslam, and M. Naqvi. 1986. Kallar grass: A plant for saline land. Nuclear Institute for Agriculture and Biology, Faisalabad, Pakistan.

Malik, K.A. and K. Haider. 1977. Decomposition of carbon-14-labelled plant material in saline-sodic soils. In: *Soil Organic Matter Studies*. IAEA, FAO Proc. Symp. International Atomic Energy Agency, Vienna, Austria.

Marcar, N.E., D.F. Crawford, and P.M. Leppert. 1993. The potential of trees for the utilisation and management of salt-affected land. In: Davidson, N., and R. Galloway (Eds.). *Productive Use of Saline Land*. pp. 17-22. Australian Council for International Agricultural Research, Canberra, Australia.

Marion, G.M., and K.L. Babcock. 1976. Predicting specific conductance and salt concentration in dilute aqueous solutions. *Soil Sci.* 122:181-187.

Marschner, H. 1986. *Mineral Nutrition of Higher Plants.* Academic Press, London.

Marshall, C.E. 1964. *The Physical Chemistry and Mineralogy of Soils.1. Soil Materials.* John Wiley, New York, NY.

Marshall, T.J. 1958. A relation between permeability and size distribution of pores. *J. Soil Sci.* 9:1-8.

Marshall, T.J., and J.W. Holmes. 1988. *Soil Physics.* 2nd ed. Cambridge University Press, Cambridge, England.

Martell, A.E. 1964. Stability constants of metal-ion complexes. Section II: Organic ligands. In: Sillen, L.G., and A.E. Martell (Eds.). *Stability Constants of Metal-Ion Complexes.* The Chemical Society, London, UK.

Martens, D.C. 1968. Plant availability of extractable boron, copper and zinc as related to selected soil properties. *Soil Sci.* 106:23-28.

Martin, J.P., and S.J. Richards. 1959. Influence of exchangeable hydrogen and calcium, and of sodium, potassium and ammonium at different hydrogen levels on certain physical properties of soils. *Soil Sci. Soc. Am. Proc.* 23:335-338.

Mason, W.K., K.E. Pritchard, and D.R. Small. 1987. Effects of early season waterlogging on maize growth and yield. *Aust. J. Agric. Res.* 38:27-35.

Masshady, A.S., and D.L. Rowell. 1978. Soil alkalinity: Alkalinity and alkalinity development. *J. Soil Sci.* 29:65-75.

Massoud, F.I. 1977. Basic principles for prognosis and monitoring of salinity and sodicity. *Proc. Int. Conf. on Management of Saline Water for Irrigation.* pp. 432-454. Texas Technical University, Lubbock, TX.

Materechera, S.A., A.R. Dexter, and A.M. Alston. 1991. Penetration of very strong soils by seedling roots of different plant species. *Plant Soil* 135:31-41.

Mathur, N.K., and A.K. Sharma. 1984. Eucalyptus in reclamation of saline and alkaline soils in India. *Indian For.* 110:9-15.

Mayer, U., and V. Gutmann. 1975. The functional approach to ionization phenomena in solutions. *Adv. Inorg. Chem. Radiochem.* 17:189-230.

McAndrew, D.W., and S.S. Malhi. 1990. Long-term effect of deep plowing Solonetzic soil on chemical characteristics and crop yield. *Can. J. Soil Sci.* 70:565-570.

McAndrew, D.W., and S.S. Malhi. 1992. Long-term N fertilization of a Solonetzic soil: Effects on chemical and biological properties. *Soil Biol. Biochem.* 24:619-623.

McAndrew, D.W., S.S. Malhi, H. Steppuhn, and D. Curtin. 1992. Various methods of amelioration of Solonetzic soils. *Proc. 29th Alberta Soil Science Workshop,* pp. 278-282, Lethbridge, Alberta, Canada.

McBride, M.B. 1989. Surface chemistry of soil minerals. In: Dixon, J.B., and S.B. Weed. (Eds.). *Minerals in Soil Environments.* 2nd ed. pp. 35-88. Soil Science Society of America, Madison, WI.

McBride, M.B. 1994. *Environmental Chemistry of Soils.* Oxford University Press, Oxford, U.K.

McBride, M.B., and J.J. Blasiak. 1979. Zinc and copper solubility as a function of pH in an acid soil. *Soil Sci. Soc. Am. J.* 43:866-870.

McCabe, G.D., and W.J. O'Brien. 1983. The effects of suspended silt on feeding and reproduction of *Daphnia pulex. Amer. Midl. Nat.* 110:324-337.

McCarthy, J.F., T.M. Williams, L. Liand, P.M. Jardine, L.W. Jolly, D.L. Taylor, A.V. Palumbo, and L.W. Cooper. 1993. Mobility of natural organic matter in a sandy aquifer. *Environ. Sci. Tech.* 27:667-676.

McCarthy, J.F., and J.M. Zachara. 1989. Subsurface transport of contaminants. *Environ. Sci. Tech.* 23:496-502.

McCown, R.L., G.G. Murtha, and G.D. Smith. 1976. Assessment of available water storage capacity of soils with restricted permeability. *Water Resour. Res.* 12:1255-1259.

McCutcheon, S.C., J.L. Martin, and T.O. Barnwell. 1993. Water quality. In: Maidment, D.R. *(Ed.). Handbook of Hydrology.* McGraw-Hill, New York, NY.

McDowell, L.L., G.H. Willis, C.E. Murphree, L.M. Southwick, and S. Smith. 1981. Toxaphene and sediment yields in runoff from a Mississippi Delta watershed. *J. Environ. Qual.* 10:120-125.

McDowell-Boyer, L.M. 1992. Chemical mobilization of micronsized particles in saturated porous media under steady flow conditions. *Environ. Sci. Tech.* 26:586-593.

McGarity, J.W., and R.J.K. Myers. 1968. Denitrifying activity in Solodized Solonetz of eastern Australia. *Soil Sci. Soc. Am. Proc.* 32:812-817.

McGarry, D., and K.Y. Chan. 1984. Preliminary investigation of clay soil's behaviour under furrow irrigated cotton. *Aust. J. Soil Res.* 22:99-108.

McIntyre, D.S. 1958. Permeability measurement of soil crusts formed by raindrop impact. *Soil Sci.* 85:185-189.

McIntyre, D.S. 1979. Exchangeable sodium, subplasticity and hydraulic conductivity of some Australian soils. *Aust. J. Soil Res.* 17:115-120.

McIntyre, D.S., J. Loveday, and C.L. Watson. 1982a. Field studies of water and salt movement in an irrigated swelling clay soil. I. Infiltration during ponding. *Aust. J. Soil Res.* 20:81-90.

McIntyre, D.S., J. Loveday, and C.L. Watson. 1982b. Field studies of water and salt movement in an irrigated swelling clay soil. II. Profile hydrology during ponding. *Aust. J. Soil Res.* 20:91-99.

McIntyre, D. S., J. Loveday, and C.L. Watson. 1982c. Field studies of water and salt movement in an irrigated swelling clay soil. III. Salt movement during ponding. *Aust. J. Soil Res.* 20:101-105.

McKenzie, D.C., D.J.M. Hall., T.S. Abbott, A.M. Kay, and J.D. Sykes. 1992. Soil management for irrigated cotton. *NSW Agriculture Agfact* 6:53.

McKenzie, D.C., T.S. Abbott, K.Y. Chan, P.G. Slavich, and D.J.M. Hall. 1993. The nature, distribution and management of sodic soils in New South Wales. *Aust. J. Soil Res.* 31:839-868.

McLaughlin, M.J., and K.G. Tiller. 1994. Chloro-complexation of cadmium in soil solutions of saline/sodic soils increases phytoavailability of cadmium. *Trans. 15th Int. Congr. Soil Sci.* 3b:195-196.

McLaughlin, M.J., K.G. Tiller, T.A. Beech, and M.K. Smart. 1994. Soil salinity causes elevated cadmium concentrations in field-grown potato tubers. *J. Environ. Qual.* 34:1013-1018.

McNeal, B.L. 1968. Prediction of the effect of mixed-salt solutions on soil hydraulic conductivity. *Soil Sci. Soc. Am. Proc.* 32:190-193.

McNeal, B.L. 1974. Soil salts and their affects on water movement. In: van Schilfgaarde, J. (Ed.). *Drainage for Agriculture.* American Society of Agronomy, Madison, WI.

McNeal, B.L., and N.T. Coleman. 1966. Effect of solution composition on soil hydraulic conductivity. *Soil Sci. Soc. Am. Proc.* 30:308-312.

McNeal, B.L., D.A. Layfield, W.A. Norvell, and J.D. Rhoades. 1968. Factors influencing hydraulic conductivity of soils in the presence of mixed-salt solutions. *Soil Sci. Soc. Am. Proc.* 32:187-190.

McNeal, B.L., W.A. Norvell, and N.T. Coleman. 1966a. Effect of solution composition on soil hydraulic conductivity and on the swelling of extracted soil clays. *Soil Sci. Soc. Am. Proc.* 30:308-315.

McNeal, B.L., G.A. Pearson, J.T. Hatcher, and C.A. Bower. 1966b. Effect of rice culture on the reclamation of sodic soils. *Agron. J.* 58:238-240.

Mead, J.A., and K.Y. Chan. 1988. Effect of deep tillage and seedbed preparation on the growth and yield of wheat on a hardsetting soil. *Aust. J. Exp. Agr.* 28:491-498.

Mech, S.J., G.M. Horner, L.M. Cox, and E.E. Cory. 1967. Soil profile modification by backhoe mixing and deep plowing. *Trans. ASAE* 10:775-779.

Mehrotra, N.K., V.K. Khanna, and S.C. Agarwal. 1986. Soil-sodicity-induced zinc deficiency in maize. *Plant Soil* 92:63-71.

Mehta, K.K. 1985. Alkali soils: Steps for reclamation. Better Farming in Salt Affected Soils, Brochure 2, Central Soil Salinity Research Institute, Karnal, India.

Mehta, S.C., S.R. Poonia, and R. Pal. 1984. Adsorption and immobilization of zinc in calcium- and sodium-saturated soils from a semiarid region, India. *Soil Sci.* 137:108-115.

Merrill, S.D., and E.J. Doering. 1985. Predicting unsaturated hydraulic conductivity of sodic soils. Paper 85-2007, Summer Meeting, American Society of Agricultural Engineers, St. Joseph, MI.

Meyer, L.D., and W.C. Harmon. 1984. Susceptibility of agricultural soils to interrill erosion. *Soil Sci. Soc. Am. J.* 48:1152-1157.

Meyer, L.D., W.H. Wischmeier, and G.R. Foster. 1970. Mulch rates for erosion control on steep slopes. *Soil Sci. Soc. Am. Proc.* 34:929-931.

Meyer, W.S., H.D. Barrs, R.C.G. Smith, N.S. White, A.D. Heritage, and D.L. Short. 1985. The effect of irrigation on soil oxygen status and root and shoot growth of wheat in clay soil. *Aust. J. Agric. Res.* 36:171-185.

Mezumen, U., and R. Keren. 1981. Boron adsorption by soils using a phenomenological adsorption equation. *Soil Sci. Soc. Am. J.* 45:722-726.

Middleton, H.E. 1930. Properties of soils which influence soil rrosion. *U.S. Dept. of Agriculture Tech. Bull.* 178.

Mikkelsen, D.S., and S. Kuo. 1977. Zinc fertilization and behavior in flooded soils. CAB Special Publ. 5., Farnham Royal, UK.

Miller, J.J., and S. Pawluk. 1994. Genesis of Solonetzic soils as a function of topography and seasonal dynamics. *Can. J. Soil Sci.* 74:207-217.

Miller, W.P. 1987. Infiltration and soil loss of three gypsum-amended Ultisols under simulated rainfall. *Soil Sci. Soc. Am. J.* 51:1314-1320.

Miller, W.P., and M.K. Baharuddin. 1986. Relationship of soil dispersibility to infiltration and erosion of Southeastern soils. *Soil Sci.* 142:235-240.

Miller, W.P., and M.K. Baharuddin. 1987a. Interrill erodibility of highly weathered soils. *Comm. Soil Sci. Plant Anal.* 18:933-946.

Miller, W.P., and M.K. Baharuddin. 1987b. Particle size of interrill-eroded sediments from highly weathered soils. *Soil Sci. Soc. Am. J.* 51:1610-1615.

Miller, W.P., and D.E. Radcliffe. 1992. Soil crusting in the southeastern United States. In: Sumner, M.E., and B.A. Stewart (Eds.). *Soil Crusting: Chemical and Physical Processes.* pp. 233-266. Lewis Publishers, Boca Raton, FL.

Miller, W.P., and J. Scifres. 1988. Effect of sodium nitrate and gypsum on infiltration and erosion of a highly weathered soil. *Soil Sci.* 145:304-309.

Mishra, B., and R.K. Bhattacharya. 1980. Limits of varietal tolerances to sodicity in rice. *Proc. Int. Symp. on Salt Affected Soils: Principles and Practices for Reclamation and Management.* pp. 502-507. Central Soil Salinity Research Institute, Karnal, India.

Misono, M., E. Ochiai, Y. Saito, and Y. Yoneda. 1967. A new dual parameter scale for the strength of Lewis acids and bases with the evaluation of their softness. *J. Inorg. Nucl. Chem.* 29:2685-2691.

Mitchell, A.R., and T.J. Donovan, 1991. Field infiltration of a salt-loaded soil: Evidence of a permeability hysteresis. *Soil Sci. Soc. Am. J.* 55:706-710.

Miyamota, S., and C. Enriquez. 1990. Comparative effects of chemical amendments on salt and Na leaching. *Irrig. Sci.* 11:83-92.

Miyamoto, S., R.J. Prather, and J.L. Stroehlein. 1975a. Sulphuric acid and leaching requirements for reclaiming sodium-affected calcareous soils. *Plant Soil* 43:573-585.

Miyamoto, S., J. Ryan, and J.L. Stroehlein. 1975b. Potentially beneficial uses of sulfuric acid in southwestern agriculture. *J. Environ. Qual.* 4:431-437.

Mohammed, E.T.Y., J. Letey, and R. Branson. 1979. Sulphur compounds in water treatment. *Sulphur Agric.* 3:7-11.

Molope, M.B., I.C. Grieve, and E.R. Page. 1987. Contributions by fungi and bacteria to aggregate stability of cultivated soils. *J. Soil Sci.* 38:71-77.

Moore, D.C., and M.J. Singer. 1990. Crust formation effects on soil erosion processes. *Soil Sci. Soc. Am. J.* 54:1117-1123.

Moore, D.C., M.J. Singer, and W.H. Olson. 1989. Improving orchard soil structure and water penetration. *Calif. Agric.* 43:7-9.

Moraghan, J.T., and H.J. Mascagni. 1991. Environmental and soil factors affecting micronutrient deficiencies and toxicities. In: Mortvedt, J.J., F.R. Cox, L.M. Shuman, and R.M. Welch (Eds.). *Micronutrients in Agriculture.* 2nd ed. Soil Science Society of America, Madison, WI.

More, S.D. 1994. Effect of farmwastes and organic manures on soil properties, nutrient availability and yield of rice-wheat grown on sodic Vertisol. *J. Indian Soc. Soil Sci.* 42:253-256.

Morin, J., and Y. Benyamini. 1977. Rainfall infiltration into bare soils. *Water Resour. Res.* 13:813-817.

Morras, H., and L. Candioti. 1982. Relación entre permeabilidad, ciertos caracteres analíticos y situación topográfica de algunos suelos de los bajos submeridionales (Santa Fé). *Rev. Invest. Agropec.* 26:23-32.

Morras, H., and S. Perman. 1977. Características pedológicas y utilización de algunos suelos halomórficos de los bajos submeridionales, Argentina. *Turrialba* 27:387-403.

Mortensen, J.L. 1962. Adsorption of hydrolysed polyacrylonitrile on kaolinite. *Clays Clay Min.* 9:530-545.

Mortland, M.M. 1970. Clay organic complexes and interactions. *Adv. Agron.* 22:75-114.

Mortvedt, J. 1972. *Micronutrients in Agriculture.* American Society of Agronomy, Madison, WI.

Moulin, V.M., M.S. Caceci, and M.J. Theyssier. 1989. Complexation behavior of humic substances from granitic groundwater towards Am(III). In: Allard, B., H. Boren, and A. Grimvall (Eds.). *Humic Substances in the Aquatic Environment.* pp. 305-313. Springer-Verlag, New York, NY.

Muirhead, W.A., E. Christen, and J. Moll. 1994. Mole drains under slots reduce waterlogging and salinity in an irrigated clay soil of semi-arid Australia. *XII World Congr. Agric. Eng. Rep. 94-A-018.*

Muirhead, W.A., E. Humphreys, N.S. Jayawardane, J. Moll, and D. Foley. 1992. Water table control using shallow subsurface drains in irrigated clay soil. National Soil Conf. p. 15. Australian Society of Soil Science, Glen Osmond, Australia.

Mullins, C.E., D.A. MacLeod, K.H. Northcote, J.M. Tisdall, and I.M. Young. 1990. Hard-setting soils: Behavior, occurrence, and management. *Adv. Soil Sci.* 11:37-108.

Muneer, M., and J.M. Oades. 1989a. The role of Ca-organic interactions in soil aggregate stability. I. Laboratory studies with ^{14}C-glucose, $CaCO_3$ and $CaSO_4 \cdot 2H_2O$. *Aust. J. Soil Res.* 27:389-399.

Muneer, M., and J.M. Oades. 1989b. The role of Ca-organic interactions in soil aggregate stability. II. Field studies with ^{14}C-labelled straw, $CaCO_3$ and $CaSO_4 \cdot 2H_2O$. *Aust. J. Soil Res.* 27 401-409.

Munn, L.C., and M.M. Boehm. 1983. Soil genesis in a Natrargid-Haplargid complex in northern Montana. *Soil Sci. Soc. Am. J.* 47:1186-1192.

Murdoch, G. 1964. Soil survey and soil classification in Swaziland 1955-63. *Sols Afr.* 9:117-135.

Murdoch, G., and J.P. Andriesse. 1964. *A Soil Irrigability Survey of the Lower Usutu Basin (south) in the Swaziland Lowveld.* Department of Technical Co-operation Overseas Research Publ.3, Her Majesty's Stationery Office, London, UK.

Murphree, C.E., C.K. Mutchler, and K.C. McGregor. 1985. Sediment yield from a 259 ha flatlands watershed. *Trans. ASAE* 28:1120-1123.

Murphy, B.W. 1981. Is this soil going to be a headache?. Upper and Lower Macquarie Areas Tech. Pap. 13/80, Soil Conservation Service, Wellington, NSW, Australia.

Murray, R.S., and J.P. Quirk. 1982. The physical swelling of clays in solvents. *Soil Sci. Soc. Am. J.* 46:865-868.

Murthy, R.S., L.R. Hirekerur, S.B. Deshpande, and B.V. Venkata Rao. 1982. *Benchmark Soils of India.* National Bureau of Soil Survey and Land Use Planning (ICAR), New Delhi, India.

Myers, R.J.K., and J.W. McGarity. 1971. Factors influencing high denitrifying activity in the subsoil of Solodized Solonetz. *Plant Soil* 35:145-160.

Myers, R.J.K., and J.W. McGarity. 1972. Denitrification in undisturbed cores from a Solodized Solonetz B horizon. *Plant Soil* 37:81-89.

Naidu, R., N.S. Bolan, R.S. Kookana, and K.G. Tiller. 1994. Ionic strength and pH effects on the sorption of cadmium and the surface charge of soils. *Eur. J. Soil Sci.* 45:419-429.

Naidu, R., N.J. de Lacy, I.D. Hollingsworth, and R.W. Fitzpatrick. 1995a. Seasonal changes or iron and dissolved organic carbon concentrations in streams in the Warren Catchment, South Australia. In: Naidu, R., M.E. Sumner, and P. Rengasamy (Eds.). *Australian Sodic Soils: Distribution, Properties and Management.* pp. 185-189. Commonwealth Scientific and Industrial Research Organization Publications, Melbourne, Australia.

Naidu, R., M.E. Sumner, and P. Rengasamy. 1995b. National Conference on Sodic Soils: Summary and conclusions. In: Naidu, R., M.E. Sumner, and P. Rengasamy (Eds.). *Australian Sodic Soils: Distribution, Properties and Management.* pp. 343-346. Commonwealth Scientific and Industrial Research Organization Publications, Melbourne, Australia.

Naidu, R., R.W. Fitzpatrick, I.O. Hollingsworth, and D.R. Williamson. 1993a. Effect of landuse on the composition of throughflow water immediately above clayey B horizons in the Warren Catchment, South Australia. *Aust. J. Exp. Agric.* 33:239-244.

Naidu, R., N.J. McKenzie, S. McClure, and R.W. Fitzpatrick. 1996. Soil solution composition and aggregate stability changes caused by long-term farming at four contrasting sites in South Australia. *Aust. J. Soil Res.* 34:511-527.

Naidu, R., H. Merry, G.J. Churchman, M.J. Wright, R.S. Murray, R.W. Fitzpatrick, and B.A. Zarcinas. 1993b. Sodicity in South Australia: A review. *Aust. J. Soil Res.* 31:911-930.

Naidu, R., and P. Rengasamy. 1993. Ion interactions and constraints to plant nutrition in Australian sodic soils. *Aust. J. Soil Res.* 31:801-819.

Naidu, R., P. Rengasamy, N.J. deLacy, and B.A.Z. Zarcinas. 1995c. Soil solution composition of some sodic soils. In: Naidu, R., M.E. Sumner, and P. Rengasamy. (Eds.). *Australian Sodic Soils: Distribution, Properties and Management.* Commonwealth Scientific and Industrial Research Organization Publications, Melbourne, Australia.

Naidu, R., M.E. Sumner, and P. Rengasamy. 1995d. *Australian Sodic Soils: Distribution, Properties and Management.* Commonwealth Scientific and Industrial Research Organization Publications, Melbourne, Australia.

Naidu, R., R.K. Syers, R.W. Tillman, and J.H. Kirkman. 1990. Effect of liming on phosphate sorption by acid soils. *J. Soil Sci.* 41:157-164.

Naidu, R., K.G. Tiller, and D. Oliver. 1995e. Influence of soil solution composition on aqueous chemistry of cadmium in soils. Int. Conf. Geochem. Trace Elem., Paris, France.

Nakayama, F.S. 1970. Hydrolysis of $CaCO_3$, Na_2CO_3 and $NaHCO_3$ and their combinations in the presence and absence of external CO_2 source. *Soil Sci.* 109:391-398.

Narkis, N., M. Rebhun, and H. Sperber. 1968. Flocculation of clay suspensions in the presence of humic and fulvic acids. *Isr. J. Chem.* 6 295-305.

National Research Council. 1989. *Irrigation-Induced Water Quality Problems.* National Academy Press, Washington, DC.

Naylor, D.V., and R. Overstreet. 1969. Sodium-calcium exchange behaviour in organic soils. *Soil Sci. Soc. Am. Proc.* 33:848-851.

Nelson, D.M., W.R. Penrose, J.O. Karttuhen, and P. Mehlhaff. 1985. Effects of dissolved organic carbon on the adsorption properties of plutonium in natural waters. *Environ. Sci. Tech.* 19:127-131.

Nelson, J.L., and S.W. Melsted. 1955. The chemistry of added Zn to soils and clays. *Soil Sci. Soc. Am. Proc.* 19:323-326.

Nelson, P.N., J.A. Baldock, and J.M. Oades. 1993. Concentration and composition of dissolved organic carbon in streams in relation to catchment soil properties. *Biogeochem.* 19:27-50.

Nelson, P.N., E. Cotsaris, J.M. Oades, and D.B. Bursill. 1990. Influence of soil clay content on dissolved organic matter in stream waters. *Aust. J. Mar. Freshwater Res.* 41:761-774.

Nelson, P.N., J.M. Oades, and J.N. Ladd. 1996. Decomposition of ^{14}C-labelled plant material in a salt-affected soil. *Soil Biol. Biochem.* 28:433-441.

Newell, A.D., and J.G. Sanders. 1986. Relative copper binding capacities of dissolved organic compounds in a coastal plain estuary. *Environ. Sci. Tech.* 20:817-821.

Nickel, S.H. 1977. A rheological approach to dispersive clays. In: Sherard, J.L., and R.S. Decker (Eds.). *Dispersive Clays, Related Piping and Erosion in Geotechnical projects.* pp. 303-312. American Society for Testing and Materials, ASTM STP 623.

Nightingale, H.I., and W.C. Bianchi. 1977. Groundwater turbidity resulting from artificial recharge. *Ground Water* 15:146-152.

Ninham, B.W. 1985. The background to hydration forces. *Chem. Scr.* 25:3-6.

Noble, J.C., and C.R. Kleinig. 1971. Response of irrigated grain sorghum to broadcast gypsum and phosphorus on a heavy clay soil. *Aust. J. Expt. Agric. Anim. Husb.* 11:53-58.

Norlyn, J.D., and E. Epstein. 1984. Variability in salt tolerance of four triticale lines at germination and emergence. *Crop Sci.* 24:1090-1092.

Norrish, K. 1954. The swelling of montmorillonite. *Trans. Faraday Soc.* 18:120-134.

Norrish, K., and K.G. Tiller. 1976. Subplasticity in Australian soils. V. Factors involved and techniques of dispersion. *Aust. J. Soil Res.* 14:273-289.

Northcote, K.H. 1979. *A Factual Key for the Recognition of Australian Soils.* 4th ed. Rellim Technical Publications, Adelaide, South Australia.

Northcote, K.H. 1988. Soils and Australian viticulture. In: Coombe, B.G., and P.R. Dry (Eds.). *Viticulture: Resources in Australia.* pp. 61-90. Australian Industrial Publications, Adelaide, Australia.

Northcote, K.H., G.G. Beckmann, E. Bettanay, H.M. Churchward, D.C. van Dijk, G.M. Dimmock, G.D. Hubble, R.F. Isbell, W.M. McArthur, G.G. Murtha, K.D. Nicolls, T.R. Paton, C.H. Thompson, A.A. Webb, and M.J. Wright. 1960-1968. *Atlas of Australian Soils. Sheets 1-10, with Explanatory Booklets.* CSIRO and Melbourne University Press, Melbourne.

Northcote, K.H., and J.K.M. Skene. 1972. Australian soils with saline and sodic properties. Soil Publ. 27, CSIRO Publications, Melbourne, Australia.

Novotny, V., and G. Chesters. 1989. Delivery of sediment and pollutants from nonpoint sources: A water quality perspective. *J. Soil Water Conserv.* 44:568-576.

Nyborg, M. 1978. Special fertility research topics. In: Toogood, J.A., and R.R. Cairns (Eds.). *Solonetzic Soils Technology and Management in Alberta.* Bull. B-78-1, 2nd ed. pp. 46. University of Alberta, Edmonton, Alberta, Canada.

Nye, P.H. 1986. Acid-base changes in the rhizosphere. *Adv. Plant Nutr.* 3:129-153.

Nyhan, J.W., B.J. Drennon, W.V. Abeele, M.L. Wheeler, W.D. Purtymun, G. Trujillo, W.J. Herrera, and J.W. Booth. 1985. Distribution of plutonium and americium beneath a 33-year old liquid waste disposal site. *J. Environ. Qual.* 14:501-509.

Oades, J.M. 1978. Mucilages at the root surface. *J. Soil Sci.* 29:1-16.

Oades, J.M. 1984. Soil organic matter and structural stability: Mechanisms and implications for management. *Plant Soil* 76:319-337.

Oades, J.M. 1988. The retention of organic matter in soils. *Biogeochem.* 5:35-70.

Oades, J.M. 1989. An introduction to organic matter in mineral soils. In: Dixon, J.B., and S.B. Weed (Eds.). *Minerals in Soil Environments.* pp. 89-160. Soil Science Society of America, Madison, WI.

Oades, J.M. 1993. The role of biology in the formation, stabilization and degradation of soil structure. *Geoderma* 56:377-400.

Oades, J.M. 1995. Recent advances in organomineral interactions: Implications for carbon cycling and soil. In: Huang, P.M., J. Berthelin, J.M. Bollag, W.B. McGill, and A.L. Page (Eds.). *Environmental Impact of Soil Component Interactions: Natural and Anthropogenic Organics.* Lewis Publishers, Boca Raton, FL.

Oades, J.M., G.P. Gillman, and G. Uehara. 1989. Interactions of soil organic matter and variable-charge clays. In: Coleman, D.C., J.M. Oades, and G. Uehara (Eds.). *Dynamics of Soil Organic Matter in Tropical Ecosystems*. pp. 69-95. Niftal Project, University of Hawaii, HI.

Oades, J.M., and A.G. Waters. 1991. Aggregate hierarchy in soils. *Aust. J. Soil Res.* 29:815-828.

Obrejanu, G., and G. Sandu. 1971. Amelioration of Solonetz and solonetzized soils in the Socialist Republic of Romania. In: Szabolcs, I.(Ed.). *European Solonetz Soils and Their Reclamation*. pp. 99-130. Akadémiai Kiadó. Budapest, Hungary.

Oertel, A.C. 1961. Pedogenesis of some Red-Brown Earths based on trace-element profiles. *J. Soil Sci.* 12:242-258.

Oertel, A.C., and G. Blackburn. 1970. Pedogenesis of a Solodized Solonetz, based on duplicate soil profiles. *Aust. J. Soil Res.* 8:59-70.

Oertli, J.J., and E. Grgurevic. 1975. Effect of pH on the adsorption of B by excised barley roots. *Agron. J.* 67:278-280.

Ogner, G., and M. Schnitzer. 1971. Chemistry of fulvic acid, soil humic fraction, and its relation to lignin. *Can. J. Chem.* 49:101-106.

Olsen, R.E., and G. Mesri. 1971. Mechanisms controlling compressibility of clays. *J. Soil Mech. Found. Div. Proc. Am. Soc. Civ. Eng.* 96:1863-1978.

Olsen, S.R., C.V. Cole, F.S. Watanabe, and L.A. Dean. 1954. Estimation of available phosphorus in soils by extraction with sodium bicarbonate. *U.S. Dept. Of Agriculture Circ.* 939.

Olsson, K.A., B. Cockroft, and P. Rengasamy. 1994. Improving and managing subsoil structure for high productivity from temperate crops on beds. In: Jayawardane, N.S., and B.S. Stewart (Eds.). *Subsoil Management Techniques*. pp. 1-31. Lewis Publishers, Boca Raton, FL.

Ong, H.L., and R.E. Bisque. 1968. Coagulation of humic colloids by metal ions. *Soil Sci.* 106:220-224.

Oprea, C.V., E. Stepanescu, and I. Vlas. 1971. *Solurile Saline si Alcaline*. Ceres, Bucharest, Rumania.

ORSTOM. 1968. Etude pédologique de la Haute-Volta région: Centre Nord. ORSTOM, Dakar, Senegal.

Oster, J.D. 1982. Gypsum use in irrigated agriculture: A review. *Fert. Res.* 3:73-89.

Oster, J.D. 1993. Sodic soil reclamation. In: Leith H., and A. Al Massom (Eds.). *Towards the Rational Use of High Salinity Tolerant Plants*. pp. 485-490. Kluwer Academic Publishers, Dordrecht, Netherlands.

Oster, J.D. 1994a. Management of irrigation water and its ecological impact. *Trans. 15th World Congr. Soil Sci.* 3a:332-345.

Oster, J.D. 1994b. Irrigation with poor quality water. *Agric. Water Manage.* 25:271-297.

Oster, J.D., and H. Frenkel. 1980. The chemistry of the reclamation of sodic soils with gypsum and lime. *Soil Sci. Soc. Am. J.* 44:41-45.

Oster, J.D., and J.D. Rhoades. 1975. Calculated drainage water compositions and salt burdens resulting from irrigation with western US river waters. *J. Environ. Qual.* 4:73-79.

Oster, J.D., and F.W. Schroer. 1979. Infiltration as influenced by irrigation water quality. *Soil Sci. Soc. Am. J.* 43:444-447.

Oster, J.D., and I. Shainberg. 1979. Exchangeable cation hydrolysis and soil weathering as affected by exchangeable sodium. *Soil Sci. Soc. Am. J.* 43:70-75.

Oster, J.D., I. Shainberg, and I.P. Abrol. 1995. Reclamation of salt-affected soils. In: Agassi, M. (Ed.). *Soil Erosion, Conservation and Rehabilitation*. pp. 315-352. Marcel Dekker, New York, NY.

Oster, J.D., I. Shainberg, and J.D. Wood. 1980. Flocculation value and gel structure of Na/Ca montmorillonite and illite suspensions. *Soil Sci. Soc. Am. J.* 44:955-959.

Oster, J.D., and M.J. Singer. 1984. Water penetration problems in California soils. LAWR Tech. Rep. 10011. University of California, Davis, CA.

Oster, J.D., M.J. Singer, A. Fulton, W. Richardson, and T. Prichard. 1992. Water penetration problems in California soils: Prevention, diagnoses and solutions. Kearney Foundation of Soil Science. DANR, University of California. Riverside, CA.

Ostrikova, K.T. 1991. *Soils of Volgograd Region*. Guide Book. Nauka, Moscow, Russia.

Overstreet, R., J.C. Martin, and H.M. King. 1951. Gypsum, sulfur, and sulfuric acid for reclaiming an alkali soil of the Fresno series. *Hilgardia* 21:113-127.

Overstreet, R., J.C. Martin, R.K. Schulz, and O.D. McCutcheon. 1955. Reclamation of an alkali soil of the Hacienda series. *Hilgardia* 24:53-68.

Palaveyev, T.D., and M.D. Penkov. 1990. Properties of surface waterlogged clay soils containing exchangeable magnesium. *Sov. Soil Sci.* 22:87-96.

Palmer, B., and P. Hazelton. 1994. Sydney Water Board St. George water area main failure analysis. In: Peterson, D.R., A.J. Weatherley, and R.E. White (Eds.). *Soil in the City*. University of Melbourne, Melbourne, Australia.

Panayiotopoulos, K.P. 1989. Packing of sands: A review. *Soil Tillage Res.* 13:101-121.

Parfitt, R.L., and D.J. Greenland. 1970. Adsorption of polysaccharides by montmorillonite. *Soil Sci. Soc. Am. Proc.* 34:862-865.

Paruelo, J.M., and O.E. Sala. 1990. Caracterización de las inundaciones en la Depresión del Salado (Buenos Aires, Argentina): Dinámica de la napa freática. *Turrialba* 40:5-11.

Pashley, R.M. 1985. The effects of hydrated cation adsorption on surface forces between mica crystals and its relevance to colloidal systems. *Chem. Scr.* 25:22-27.

Pashley, R.M., and J.P. Quirk. 1984. The effect of cation valency on DLVO and hydration forces between macroscopic sheets of muscovite mica in relation to clay swelling. *Colloids Surf.* 9:1-15.

Pearson, G.A. 1960. Tolerance of crops to exchangeable sodium. *U.S. Dept. of Agriculture Inf. Bull.* 216.

Pearson, G.A., and L. Bernstein. 1958. Influence of exchangeable sodium on yield and chemical composition of plants: II. Wheat, barley, oats, rice, tall fescue, and wheatgrass. *Soil Sci.* 86:254-261.

Penrose, W.R., W.L. Polzer, E.H. Essington, D.M. Nelson, and K.A. Orlandini. 1990. Mobility of plutonium and americium through a shallow aquifer in a semiarid region. *Environ. Sci. Tech.* 24:228-234.

Peters, T.W. 1978. Solonetzic soils technology and management. In: Toogood, J.A., and R.R. Cairns (Eds.). *Solonetzic Soils Technology and Management in Alberta.* Bull. B-78-1, 2nd ed. pp. 8-25. University of Alberta, Edmonton, Alberta, Canada.

Petersen, W., and M. Böttger. 1991. Contribution of organic acids to the acidification of the rhizosphere of maize seedlings. *Plant Soil* 132:159-163.

Peterson, S.M., and G.E. Batley. 1989. The role of non-settling particles in pollutant transport: Endosulfan. In: Moore, I.D. (Ed.). *Modeling the Fate of Chemicals in the Environment.* pp. 103-108. CRES, Australian National University, Canberra, Australia.

Pettit, L.D., and H.K.J. Powell. 1993. *Stability Constants Database.* International Union of Pure and Applied Chemistry/ Academic Software, Timble, UK.

Poirrie, M.A., B.R. Bordelon, and J.L. Laseter. 1972. Adsorption and concentration of dissolved carbon-14 DDT by coloring colloids in surface waters. *Environ. Sci. Tech.* 6:1033-1035.

Pojasok, T., and B.D. Kay. 1990. Effect of root exudates from corn and bromegrass on soil structural stability. *Can. J. Soil Sci.* 70:351-362.

Ponnamperuma, F.N. 1972. The chemistry of submerged soils. *Adv. Agron.* 24:29-96.

Poonia, S.R., S.C. Mehta, and Raj Pal. 1984. Sodification of soil in relation to organic matter, total electrolyte concentration and nature of cations and anions. *J. Indian Soc. Soil Sci.* 32:663-668.

Poonia, S.R., and O. Talibudeen. 1977. Sodium-calcium exchange equilibria in salt-affected and normal soils. *J. Soil Sci.* 28:276-288.

Prather, R.J. 1977. Sulfuric acid as an amendment for reclaiming soils high in boron. *Soil Sci. Soc. Am. J.* 41:1098-1101.

Prather, R.J., J.O. Goertzen, J.D. Rhoades, and H. Frenkel. 1978. Efficient amendment use in sodic soil reclamation. *Soil Sci. Soc. Am. J.* 42: 782-786.

Pratt, P.F., and B.L. Grover. 1964. Monovalent-divalent cation exchange equlibria in soils in relation to organic matter and type of clay. *Soil Sci. Soc. Am. Proc.* 28:32-35.

Pratt, P.F., and D.L. Suarez. 1990. Irrigation water quality assessments. In: Tanji, K.K. (Ed.). *Agricultural Salinity Assessment and Management.* pp. 220-236. ASCE Manuals and Reports on Engineering Practice 71. American Society of Civil Engineering, New York, NY.

Pratt, P.F., and D.W. Thorne. 1948. Solubility and physiological availability of phosphate in sodium and calcium systems. *Soil Sci. Soc. Am. Proc.* 13:213-217.

Pratt, P.F., L.D. Whitting, and B.L. Grover. 1962. Effect of pH on sodium-calcium exchange equilibria in soils. *Soil Sci. Soc. Am. Proc.* 26:227-230.

Puls, R.W., and R.M. Powell. 1992. Transport of inorganic colloids through natural aquifer material: Implications for contaminant transport. *Environ. Sci. Tech.* 26:614-621.

Pupisky, H., and I. Shainberg. 1979. Salt effects on the hydraulic conductivity of a sandy soil. *Soil Sci. Soc. Am. J.* 43:429-433.

Purves, W.D., and W.D. Blyth. 1969. A study of associated hydromorphic and sodic soils on redistributed Karroo sediments. *Rhod. J. Agric. Res.* 7:99-109.

Puttaswamygowda, B.S., and P.F. Pratt. 1973. Effect of straw, $CaCl_2$ and submergence on a sodic soil. *Soil Sci. Soc. Am. Proc.* 37:208-212.

Quirk, J.P. 1978. Some physico-chemical aspects of soil structural stability: A review. In: Emerson, W.W., R.D. Bond, and A.R. Dexter (Eds.). *Modification of Soil Structure.* pp. 3-16. John Wiley, New York, NY.

Quirk, J.P. 1986. Soil permeability in relation to sodicity and salinity. *Phil. Trans. R. Soc. London* A 316:297-317.

Quirk, J.P. 1994. Interparticle forces: A basis for the interpretation of soil physical behaviour. *Adv. Agron.* 53:121-183.

Quirk, J.P., and L.A.G. Aylmore. 1971. Domains and quasi-crystalline regions in clay systems. *Soil Sci. Soc. Am. Proc.* 35:652-654.

Quirk, J.P., and R.S. Murray. 1991. Towards a model for soil structural behaviour. *Aust. J. Soil Res.* 29:829-867.

Quirk, J.P., and R.K. Schofield. 1955. The effect of electrolyte concentration on soil permeability. *J. Soil Sci.* 6:163-178.

Rahman, A.W., and D.L. Rowell. 1979. The influence of magnesium in saline and sodic soils: A specific effect or a problem of cation exchange? *J. Soil Sci.* 30:535-546.

Rajkai, K., and E. Molnár. 1981. Soil-water-plant relationships in a salt-affected area of the Great Hungarian Plain. *Agrokém. Talajtan* 30:97-104.

Rao, D.L.N. 1987. Slow-release urea fertilizers: Effect on floodwater chemistry, ammonia volatilization and rice growth in an alkali soil. *Fert. Res.* 13:209-221.

Rao, D.L.N., and L. Batra. 1983. Ammonia volatilization from applied nitrogen in alkali soils. *Plant Soil* 70:219-228.

Rao, D.L.N., and R.G. Burns. 1991. The influence of blue-green algae on the biological amelioration of alkali soils. *Biol. Fertil. Soils* 11:306-312.

Rao, D.L.N., and S.K. Ghai. 1986. Effect of phenylphosphorodiamidate on urea hydrolysis, ammonia volatilization and rice growth in an alkali soil. *Plant Soil* 94:313-320.

Rao, P.S.C., P. Nkedi-Kizza, J.M. Davidson, and L.T. Ou. 1983. Retention and transformation of pesticides in relation to nonpoint source pollution from croplands. In: Schaller, F.W., and G.W.Bailey (Eds.). *Agricultural Management and Water Quality.* pp. 126-140. Iowa State University Press, Ames, IA.

Rashid, A., J.K. Khattak, M.Z. Khan, M.J. Iqbal, F. Akbar, and P. Khan. 1993. Selection of halophytic forage shrubs for the Peshawar valley, Pakistan. In: Davidson, N., and R. Galloway (Eds.). *Productive Use of Saline Land.* pp. 56-61. Australian Council for International Agricultural Research, Canberra, Australia.

Rasmussen, W.W., and B.L. McNeal. 1973. Predicting optimum depth of profile modification by deep plowing for improving saline-sodic soils. *Soil Sci. Soc. Am. Proc.* 37:432-437.

Rasmussen, W.W., D.P. Moore, and L.A. Alban. 1972. Improvement of a Solonetzic (slick spot) soil by deep plowing, subsoiling, and amendments. *Soil Sci. Soc. Am. Proc.* 36:137-142.

Ravina, I., and Z. Markus. 1975. The effect of high exchangeable potassium percentage on soil properties and plant growth. *Plant Soil* 42:661-672.

Rawitz, E., W.B. Hoogmoed, and J. Morin. 1986. The effects of tillage practices on crust properties, infiltration and crop response under semi-arid condtions. In: F. Callebaut, D. Gabriels, and M. de Boodt (Eds.). *Assessment of Soil Surface Sealing and Crusting*. pp. 278-284. Flanders Research Center for Soil Erosion and Soil Conservation, Gent, Belgium.

Rayment, G.E., and F.R. Higginson. 1992. *Australian Laboratory Handbook of Soil and Water Chemical Methods*. Inkata Press, Sydney, Australia.

Reeve, R.C. 1953. A method for determining the stability of soil structure based upon air and water permeability measurements. *Soil Sci. Soc. Am. Proc.* 17:324-329.

Reeve, R.C. 1958. The transmission of water by soils as influenced by chemical and physical properties. *Trans. 5th Int. Congr. Agric. Eng.* pp. 21-32.

Reeve, R.C., L.E. Allison, and D.F. Peterson. 1948. Reclamation of saline-alkali soils by leaching. Utah Agric. Exp. Sta. Bull. 335.

Reeve, R.C., C.A. Bower, R.H. Brooks, and F.B. Gschwend. 1954. A comparison of the effects of exchangeable Na and K upon the physical condition of soils. *Soil Sci. Soc. Am. Proc.* 18:140-132.

Reeve, R.C., J.N. Luthin, and W.W. Donnan. 1957. Drainage investigation methods. In: Luthin, J.N. (Ed.). *Drainage of Agricultural Lands*. pp. 395-445. American Society of Agronomy, Madison, WI.

Reeve, R.C., and G.H. Tamaddoni. 1965. Effect of electrolyte concentration on laboratory permeability and field intake rate of a sodic soil. *Soil Sci.* 99:261-266.

Reichert, J.M., and L.D. Norton. 1994. Aggregate stability and rain-impacted sheet erosion of air-dried and prewetted clayey surface soils under intense rain. *Soil Sci.* 158:159-169.

Rengasamy, P. 1983. Clay dispersion in relation to changes in the electrolyte composition of dialysed Red-Brown Earths. *J. Soil Sci.* 34:723-732.

Rengasamy, P. 1987. Importance of calcium in irrigation with saline-sodic water: A viewpoint. *Agric. Water Manage*. 12:207-219.

Rengasamy, P., R.S.B. Greene, and G.W. Ford. 1984a. The role of clay fraction in the particle arrangement and stability of soil aggregates: A Review. *Clay Res.* 3:53-67.

Rengasamy, P., R.S.B. Greene, and G.W. Ford. 1986. Influence of magnesium on aggregate stability in sodic Red-Brown Earths. *Aust. J. Soil Res.* 24:229-237.

Rengasamy, P., R.S.B. Greene, G.W. Ford, and A.H. Mehanni. 1984b. Identification of dispersive behaviour and the management of Red-Brown Earths. *Aust. J. Soil Res.* 22:413-431.

Rengasamy, P., J.A. Kempers, and K.A. Olsson. 1991. Dispersive potential of Natrixeralfs and their crusting strength. *Clay Res.* 10:6-10.

Rengasamy, P., and R. Naidu. 1994. Dispersive potential of sodic soils as influenced by the charge on their clay fractions. In: Churchman, G.J., R.W. Fitzpatrick, and R.A. Eggleton (Eds.). *Clays: Controlling the Environment*. pp. 467-470. Commonwealth Scientific and Industrial Resrarch Organization Publications, Melbourne, Australia.

Rengasamy, P., R. Naidu, T.A. Beech, K.Y. Chan, and C. Chartres. 1992. Crust formation as related to the dispersive potential in Australian soils. *Catena* 26:65-76.

Rengasamy, P., and K.A. Olsson. 1991. Sodicity and soil structure. *Aust. J. Soil Res.* 29:935-952.

Rengasamy, P., and K.A. Olsson. 1993. Irrigation and sodicity. *Aust. J. Soil Res.* 31:821-837.

Rhoades, J.D. 1972. Quality of irrigation water. *Soil Sci.* 113:277-284.

Rhoades, J.D., and R.D. Ingvalson. 1969. Macroscopic swelling and hydraulic conductivity properties of four vermiculite soils. *Soil Sci. Soc. Am. Proc.* 33:364-369.

Rhoades, J.D., D.B. Kruger, and M.J. Reed. 1968. The effect of soil mineral weathering on the sodium hazard of irrigation waters. *Soil Sci. Soc. Am. Proc.* 32:643-647.

Rhoades, J.D., and J. Loveday. 1990. Salinity in irrigated agriculture. In: Stewart, B.A., and D.R. Nielson (Eds.). *Irrigation of Agricultural Crops*. pp. 1089-1142. American Society of Agronomy, Madison, WI.

Ribaudo, M.O., and C.E. Young. 1989. Estimating the water quality benefits from soil erosion control. *Water Res. Bul.* 25:71-78.

Rice, W.A. 1978. Microbiological relationships of Solonetzic soils. In: Toogood, J.A., and R.R. Cairns (Eds.). *Solonetzic Soils Technology and Management in Alberta*. Bull. B-78-1, 2nd ed. pp. 59-62. University of Alberta, Edmonton, AB.

Richards, L. A. 1954. *Diagnosis and Improvement of Saline and Alkali Soils*. USDA Handbook 60. US Govt. Printing Office, Washington, DC.

Richardson, J.L., L.P. Wilding, and R.B. Daniels. 1992. Recharge and discharge of groundwater in aquic conditions illustrated with flownet analysis. *Geoderma* 53:65-78.

Richie, J.A. 1963. Earthwork tunnelling and the application of soil testing procedure. *J. Soil Conser. Serv. NSW.* 19:111-129.

Rimmer, D.L., and D.J. Greenland 1976. Effect of $CaCO_3$ on the swelling of a soil clay. *J. Soil Sci.* 27:129-139.

Rimmer, D.L., M.E. Hamad, and J.K. Syers. 1992. Effects of pH and sodium ions on $NaHCO_3$-extractable phosphate in calcareous sodic soils. *J. Sci. Food Agric.* 60:383-385.

Ritchie, J.T., D.E. Kissel, and E. Burnett. 1972. Water movement in undisturbed swelling clay soil. *Soil Sci. Soc. Am. Proc.* 36:874-879.

Road Transport Authority. 1992. Report on Kiama Bends. Road Transport Authority, Wollongong, NSW, Australia.

Robbins, C.W. 1986. Sodic calcareous soil reclamation as affected by different amendments and crops. *Agron. J.* 78:916-920.

Robbins, C.W., R.J. Wagenet, and J.J. Jurinak. 1980. A combined transport chemical equilibrium model for calcareous and gypsiferous soils. *Soil Sci. Soc. Am. J.* 44:1191-1194.

Robertson, W.D., J.F. Barker, Y. LeBeau, and S. Marcoux. 1984. Contamination of an unconfined sand aquifer by waste pulp liquor: A case study. *Ground Water* 22:191-197.

Roehl, J.E. 1962. Sediment source areas, and delivery ratios influencing morphological factors. *Int. Assoc. Hydrol. Sci.* 59:202-213.

Rose, C.W., and H. Ghadiri. 1991. Transport and enrichment of soil sorbed chemicals. In: Moore, I.D. (Ed.). *Modelling the Fate of Chemicals in the Environment.* pp. 90-102. CRES, Australian National University, Canberra, Australia.

Rovira, A.D., and Ridge, E.H. 1983. Soil-borne root diseases and wheat. In: *Soils: An Australian Viewpoint.* pp. 721-734. Commonwealth Scientific and Industrial Research Organization Publications, Melbourne, Australia/Academic Press, London, UK.

Rowell, D.L., D. Payne, and N. Ahmad. 1969. The effect of the concentration and movement of solutions on the swelling, dispersion and movement of clay in saline and alkali soils. *J. Soil Sci.* 20:176-188.

Rozanov, A. 1961. *The Serozems of Central Asia.* Israel Program for Scientific Translations, Jerusalem, Israel.

Ruehrwein, R.A., and D.W. Ward. 1952. Mechanism of clay aggregation by polyelectrolytes. *Soil Sci.* 73:485-492.

Russell, J.S., E.J. Damprath, and C.S. Andrew. 1988. Phosphorus sorption of subtropical acid soils as influenced by the nature of the cation suite. *Soil Sci. Soc. Am. J.* 52:1407-1410.

Russo, D., and E. Bresler. 1977. Analysis of the saturated and unsaturated hydraulic conductivity in mixed sodium and calcium soil systems. *Soil Sci. Soc. Am. J.* 41:706-712.

Ryan, J., S. Miyamoto, and J.L. Stroehlein. 1974. Solubility of manganese, iron, and zinc as affected by application of acid to calcareous soils. *Plant Soil* 40:421-427.

Ryan, J., J.L. Stroehlein, and S. Miyamoto. 1975. Effect of surface-applied sulfuric acid on growth and nutrient availability of five range grasses in calcareous soils. *J. Range Manage.* 28:411-414.

Sadana, U.S., and M.S. Bajwa. 1985. Manganese equilibrium in submerged sodic soils as influenced by application of gypsum and green manuring. *J. Agric. Sci. Camb.* 104:257-261.

Sadana, U.S., and P.N. Takkar. 1988. Effect of sodicity and zinc on soil solution chemistry of manganese under submerged conditions. *J. Agric. Sci. Camb.* 111:51-55.

Saeed, M., and R.L. Fox. 1977. Relationship between suspension pH and Zn solubility in acid and calcareous soils. *Soil Sci.* 124:129-203.

Sajwan, K.S., and W.L. Lindsay. 1986. Effect of redox on zinc deficiency in paddy rice. *Soil Sci. Soc. Am. J.* 50:1264-1269.

Sandhu, G.R., and Z. Aslam. 1980. Economic utilization of salt-affected soils. In: *Salt-Affected Soils.* Proc. Int. Symp. On Salt Affected Soils. pp. 142-148. Central Soil Salinity Research Institute, Karnal, India.

Sandoval, F.M. 1978. Deep plowing improves sodic claypan soils. *ND Agric. Exp. Sta. Farm Res.* 35:15-18.

Schachtschabel, P. 1940. Untersuchungen über die sorption der tonmineralien und organischen boden-kolloide. *Kolloid-Beihefte* 51:199.

Schloms B.H.A., F. Ellis, and J.J.N. Lambrechts. 1983. Soils of the Cape Coastal Platform. Fynbos Palaeoecology: A Preliminary Synthesis. In: Deacon, H.J., Q.J. Hendy, and J.J.N. Lambrechts (Eds.). *South African National Scientific Programs Report 75.* pp70-99. Council for Scientific and Industrial Research, Pretoria, South Africa.

Schnitzer, M. 1978. Humic substances: Chemistry and reactions. In: Schnitzer, M., and S.U. Khan (Eds.). *Soil Organic Matter.* pp. 1-64. Elsevier, Amsterdam, Netherlands.

Schnitzer, M., and S.I.M. Skinner. 1963. Organo-metallic interactions in soils: I. Reactions between a number of metal ions and organic matter of a Podzol Bh horizon. *Soil Sci.* 96:86-93.

Schofield, R.K., and H.R. Samson. 1954. Flocculation of kaolinite due to the attraction of oppositely charged crystal faces. *Discuss. Faraday Soc.* 18:138-145.

Schulz, R.K., R. Overstreet, and I. Barshad. 1964. Some unusual ionic exchange properties of sodium in certain salt-affected soils. *Soil Sci.* 99:161-165.

Scifres, J.L. 1989. Runoff and Erosion in a Wheat/soybean Cropping System Amended with Gypsum. M.S. Thesis University of Georgia, Athens, GA.

Seelig, B.D., J.L. Richardson, and W.T. Barker. 1990. Characteristics and taxonomy of sodic soils as a function of landform position. *Soil Sci. Soc. Am. J.* 54:1690-1697.

Seeliger, M.T. 1968. Fertilizer for cereal soils. *J. Dept. Agric. South Aust.* 72:62-67.

Sekhon, B.S., and M.S. Bajwa. 1993. Effect of organic matter and gypsum in controlling soil sodicity in rice-wheat-maize systems irrigated with sodic waters. *Agric. Water Man.* 24:15-25.

Senesi, N., and Y. Chen. 1989. Interaction of toxic organic chemicals with humic substances. In: Gerstl, Z., Y. Chen, U. Mingelgrin, and B. Yaron (Eds.). *Toxic Organic Chemicals in Porous Media.* pp. 37-90. Springer-Verlag, Berlin, Germany.

Shafer, G.J. 1978. Pinhole test for dispersive soil: Suggested change. *J. Geotech. Eng. Div. ASCE.* 104:760-765.

Shainberg, I., N. Alperovitch, and R. Keren. 1987a. Charge density and Na/K/Ca exchange on smectites. *Clays Clay Min.* 35:68-73.

Shainberg, I., N. Alperovitch, and R. Keren. 1988. Effect of magnesium on the hydraulic conductivity of sodic smectite-sand mixtures. *Clays Clay Min.* 36:432-438.

Shainberg, I., E. Bresler, and Y. Klausner. 1971. Studies on Na/Ca montmorillonite systems. I. The swelling pressure. *Soil Sci.* 111:214-219.

Shainberg, I., and M. Gal. 1982. The effect of lime on the response of soils to sodic conditions. *J. Soil Sci.* 33:489-498.

Shainberg, I., R. Keren, N. Alperovitch, and D. Goldstein. 1987b. Effect of exchangeable potassium on the hydraulic conductivity of smectite-sand mixtures. *Clays Clay Min.* 35:305-310.

Shainberg, I., and J. Letey. 1984. Response of soils to sodic and saline conditions. *Hilgardia* 52:1-57.

Shainberg, I., and G.J. Levy. 1992. Physico-chemical effects of salts upon infiltration and water movement in soils. In: Wagenet, R.J., P. Baveye, and B.A. Stewart (Eds.). *Interacting Processes in Soil Science.* pp. 37-94. Lewis Publishers, Boca Raton, FL.

Shainberg, I., and G.J. Levy. 1994. Organic polymers and soil sealing in cultivated soils. *Soil Sci.* 158:267-273.

Shainberg, I., G.J. Levy, P. Rengasamy, and H. Frenkel. 1992a. Aggregate stability and seal formation as affected by drops' impact energy and soil amendments. *Soil Sci.* 154:113-119.

.Shainberg, I., J.D. Rhoades, and R.J. Prather. 1981a. Effect of low electrolyte concentration on clay dispersion and hydraulic conductivity of a sodic soil. *Soil Sci. Soc. Am. J.* 45:273-277.

Shainberg, I., J.D. Rhoades, D.L. Suarez, and R.J. Prather. 1981b. Effect of mineral weathering on clay dispersion and hydraulic conductivity of sodic soils. *Soil Sci. Soc. Am. J.* 45:287-291.

Shainberg, I., and J. Shalhevet. 1984. *Soil Salinity under Irrigation: Processes and Management.* Springer-Verlag, Berlin, Germany.

Shainberg, I., and J. J. Singer. 1986. Suspension concentration effects on deposition crusts and soil hydraulic conductivity. *Soil Sci. Soc. Am. J.* 50:1537-1540.

Shainberg, I., M.E. Sumner, W.P. Miller, M.P.W. Farina, M.A. Pavan, and M.V. Fey. 1989. Use of gypsum on soils: A review. *Adv. Soil Sci.* 9:1-112.

Shainberg, I., D. Warrington, and J.M. Laflen. 1992b. Soil dispersibility, rain properties, and slope interaction in rill formation and erosion. *Soil Sci. Soc. Am. J.* 56:278-283.

Shainberg, I., D. Warrington, and P. Rengasamy. 1990. Effect of PAM and gypsum application on rain infiltration and runoff. *Soil Sci.* 149:301-307.

Shanmuganathan, R.T., and J.M. Oades. 1983a. Influence of anions on dispersion and physical properties of the A horizon of a Red-Brown Earth. *Geoderma* 29:257-277.

Shanmuganathan, R.T., and J.M. Oades. 1983b. Modification of soil physical properties by addition of calcium compounds. *Aust. J. Soil Res.* 21:285-300.

Sharma, S.K. 1986. Mechanism of tolerance in rice varieties differing in sodicity tolerance. *Plant Soil* 93:141-145.

Sharma, S.K. 1988. Recent advances in afforestation of salt-affected soils in India. *Adv. For. Res. India* 2:17-31.

Sharma, D.P., K.K. Mehta, and G. Singh. 1983. Use of dairy waste for the reclamation of sodic soils. *Curr. Agric.* 7:13-17.

Sharma, B.D., P.S. Sidhu, and V.K. Nayyar. 1992. Distribution of micronutrients in arid zone soils of Punjab and their relation with soil properties. *Arid Soil Res. Rehabil.* 6:233-242.

Sharma, M.L., and D.R. Williamson. 1984. Secondary salinization of water resources in Southern Australia. In: French, R.H. (Ed.). *Salinity in Watercourses and Reservoirs.* pp. 571-582. Butterworth Publishers, Boston, MA.

Sharma, B.M., and J.S.P. Yadav. 1986. Leaching losses of iron and manganese during reclamation of alkali soil. *Soil Sci.* 142:149-163.

Sharpley, A.N. 1985. Selective erosion of plant nutrients in runoff. *Soil Sci. Soc. Am. J.* 49:1527-1534.

Sharpley, A.N. 1995. Identifying sites vunerable to phosphorus loss in agricultural runoff. *J. Environ. Qual.* 24:947-951.

Sharpley, A.N., S.C. Chapra, R. Wedepohl, J.T. Sims, T.C. Daniel, and K.R. Reddy. 1994. Managing agricultural phosphorus for protection of surface waters: Issues and options. *J. Environ. Qual.* 23:437-451.

Sharpley, A.N., D. Curtin, and J.K. Syers. 1988. Changes in water-extractability of soil inorganic phosphate induced by sodium saturation. *Soil Sci. Soc. Am. J.* 52:637-640.

Sharpley, A.N., and P.J.A. Withers. 1994. The environmentally sound management of agricultural phosphorus. *Fert. Res.* 39:133-146.

Shaw, R.L. 1988. Soil salinity and sodicity. In: Fergus, I.F. (Ed.). *Understanding Soils and Soil Data.* pp. 109-134. Queensland Branch, Australian Society of Soil Science, Brisbane, Australia.

Shaw, R.L. 1994. Estimation of the electrical conductivity of saturation extracts from the electrical conductivity of 1:5 soil:water suspensions and various soil properties. Mimeo, Queensland Department of Primary Industries, Brisbane, Australia.

Shaw, R.J. 1996. A Unified Soil Property and Sodicity Model of Salt Leaching and Water Movement. Ph.D. thesis, University of Queensland, Brisbane, Australia.

Shaw, R.J., L.J. Brebber, C.R. Ahern, and M. Weinand. 1994. A review of sodicity and sodic soil behaviour in Queensland. *Aust. J. Soil Res.* 32:143-172.

Shaw, R.J., and I. Gordon. 1994. Salinity in cotton areas. Proc. 7th Austr. Cotton .Conf., pp. 279-288. Australian Cotton Growers Research Association, Gold Coast, Queensland, Australia.

Shaw, R.J., and P.J. Thorburn. 1985. Prediction of leaching fraction from soil properties, irrigation water and rainfall. *Irrig. Sci.* 6:73-83.

Shaw, R.J., and D.F. Yule. 1978. Assessment of soils for irrigation, Emerald, Queensland. Agricultural Chemistry Branch Tech. Rep. 13, Queensland Department of Primary Industries, Indooroopilly, Australia.

Sherard, J.L. 1976. Pinhole test for identifying dispersive soils. *J. Geotech. Eng. Div. ASCE.* 102:69-85.

Sherard, J.L., L.P. Dunnigan, and R.S. Decker. 1976. Identification and nature of dispersive soils. *J. Geotech. Div. ASCE* 102:287-301.

Sherard, J.L., R.S. Decker, and N.L. Rykes. 1972. Piping in earth dams of dispersive clay. *Proc. Amer. Soc. Civ. Eng. Spec. Conf. On Performance on Earth and Earth-Supported Structures* 1:587-626. St. Joseph, MI.

Sherard, J.L., L.P. Dunnigan, and R.S. Decker. 1977. Some engineering problems with dispersive clays. In: Sherard, J.L. and R.S. Decker (Eds.). *Dispersive Clays, Related Piping, and Erosion in Geotechnical Projects.* pp. 3-12. American Society for Testing and Materials, ASTM STP 623.

Sheridan, J.M., and R.K. Hubbard. 1987. Transport of solids in streamflow from Coastal Plain watersheds. *J. Environ. Qual.* 16:131-136.

Sholkovitz, E.R., and D. Copland. 1981. The coagulation, solubility and adsorption properties of Fe, Mn, Cu, Ni, Co and humic acids in a river water. *Geochim. Cosmochim. Acta* 45:181-189.

Short, S.A., R.T. Lawson, and J. Ellis. 1988. $^{234}U/^{238}U$ and $^{230}Th/^{234}U$ activity ratios in the colloidal phases of aquifers in lateritic weathered zones. *Geochim. Cosmochim. Acta.* 52:2555-2563.

Shukla, U.C., S.B. Mittal, and R.K. Gupta. 1980. Zinc adsorption in some soils as affected by exchangeable cations. *Soil Sci.* 129:67-371.

Shuman, M.S. 1992. Dissociation pathways and species distribution of Al bound to an aquatic fulvic acid. *Environ. Sci. Tech.* 6:1033-1035.

Sims, J.T. 1986. Soil pH effects on the distribution and plant availability of manganese, copper and zinc. *Soil Sci. Soc. Am. J.* 45:367-373.

Simms, H.J., and D.R. Rooney. 1965. Gypsum for difficult clay wheatgrowing soils. *J. Agric. Victoria* 63:401-409.

Simunek, J., and D.L. Suarez. 1994. Two-dimensional transport model for variable saturated porous media with major ion chemistry. *Water Res. Res.* 30:1115-1133.

Singer, M.J., P. Janitzky, and J. Blackard. 1982. The influence of exchangeable sodium percentage on soil erodibility. *Soil Sci. Soc. Am. J.* 46:117-121.

Singh, G., N.T. Singh, and I.P. Abrol. 1994. Agroforestry techniques for the rehabilitation of degraded salt-affected lands in India. *Land Degrad. Rehabil.* 5:223-242.

Singh, J.P., S.P.S. Karwasra, and M. Singh. 1988. Distribution and forms of copper, iron, manganese and zinc in calcareous soils of India. *Soil Sci.* 146:359-366.

Singh, K.N. 1985. Cultural practices for growing rice in alkali soils. Better Farming in Salt Affected Soils, Brochure 9, Central Soil Salinity Research Institute, Karnal, India.

Singh, M., and S.P. Singh. 1980. Zn-P interaction in submerged paddy soils. *Soil Sci.* 129:282-289.

Singh, M., and S.P. Singh. 1983. Effects of zinc and phosphorus on adsorption of iron and nitrogen by submerged paddy. *Soil Sci.* 135:71-78.

Singh, M.V., and I.P. Abrol. 1985. Solubility and adsorption of Zn in a sodic soil. *Soil Sci.* 140:5407-5411.

Singh, M.V., R. Chabra, and I.P. Abrol. 1982. Effect of zinc levels on the growth and yield of rice and wheat in a sodic soil. Ann. Rep. Central Soil Salinity Research Institute, Karnal, India.

Singh, M.V., R. Chhabra, and I.P. Abrol. 1983. Factors affecting DTPA-extractable zinc in sodic soils. *Soil Sci.* 136:359-366.

Singh, M.V., and R.E. Franklin. 1974. Availability of native and applied zinc (Zn^{65}) as affected by nitrogen carriers in normal and saline soils. *Plant Soil* 40:699-702.

Singh, N.T. 1974. Physico-chemical changes in sodic soils incubated at saturation. *Plant Soil* 40:303-311.

Singh, R., R.G. Gerritse, and L.A.G. Aylmore. 1990. Adsorption-desorption behaviour of selected pesticides in some Western Australian soils. *Aust. J. Soil Res.* 28:227-243.

Singh, R., N.T. Singh, and Y. Arora. 1980. The use of spent wash for the reclamation of sodic soils. *J. Indian Soc. Soil Sci.* 28:38-41.

Singh, R.B., P.S. Minhas, C.P.S. Chauhan, and R.K. Gupta. 1992. Effect of high salinity and SAR waters on salinisation, sodification and yields of pearl-millet and wheat. *Agric. Water Manage.* 21:93-106.

Singh, S.B., R. Chhabra, and I.P. Abrol. 1979. Effect of exchangeable sodium on the yield and chemical composition of raya (*Brassica juncea* L.). *Agron. J.* 71:767-770.

Singh, Y., C.S. Khind, Bijay-Singh, and B. Singh. 1991. Efficient management of leguminous green manures in wetland rice. *Adv. Agron.* 45:135-189.

Sinha, N.P,. and S.N. Jha. 1984. Studies on pyrite as amendment for sodic soil and as nutrients. *Indian J. Agric. Chem.* 17:75-101.

Skene, T.M., and J.M. Oades. 1995. The effects of sodium adsorption ratio and electrolyte concentration on water quality: laboratory studies. *Soil Sci.* 159:65-73.

Slade, P.G., and J.P. Quirk. 1991. The limited crystalline swelling of smectite in $CaCl_2$, $MgCl_2$ and $LaCl_3$ solutions. *J. Colloid Interface Sci.* 144:18-26.

Slade, P.G., J.P. Quirk, and K. Norrish. 1991. Crystalline swelling of smectite samples in concentrated NaCl solutions in relation to layer charge. *Clays Clay Min.* 39:234-238.

Sleeman, J.R. 1964. Structure variation within two Red-Brown Earth profiles. *Aust. J. Soil Res.* 2:146-161

Smettam, K.R.J., D.J. Chittleborough, B.G. Richards, and F.W. Leany. 1991. The influence of macropores on runoff generation from a hillslope soil with a contrasting textural class. *J. Hydrol.* 122:235-252.

Smillie, G.W., D. Curtin, and J.K. Syers. 1987. Influence of exchangeable Ca on phosphate retention in weakly acid soils. *Soil Sci. Soc. Am. J.* 51:1169-1172.

Smith, G.D., K.J. Coughlan, and W.E. Fox. 1978. The role of texture in soil structure. In: Emerson, W.W., R.D. Bond, and A.R. Dexter (Eds.). *Modification of Soil Structure.* pp. 79-86. Wiley, Chichester, UK.

Smolders, E., R.M. Lambrechts, M.J. McLaughlin, and K.G. Tiller. 1995. The influence of chloride on the cadmium uptake by Swiss chard (*Beta vulgaris* L. cv Fordhook Giant). I. Soil experiments. Proc. 3rd Int. Conf. Biogeochem. Trace Elements Symposium B1, Insyaprint S.A., Tours, France.

So, H.B., and L.A.G. Aylmore. 1993. How do sodic soils behave? The effects of sodicity on soil physical behaviour. *Aust. J. Soil Res.* 31:761-777.

So, H.B., D.W. Tayler, W.J. Yates, and J.W. McGarity. 1978. Amelioration of structurally unstable grey and brown clays. In: Emerson, W.W., R.D. Bond, and A.R. Dexter (Eds.). *Modifications of Soil Structure.* pp. 325-334. John Wiley and Sons, New York, NY.

Soil Classification Working Group. 1991. *Soil Classification: A Taxonomic System for South Africa.* Department of Agricultural Development, Pretoria, South Africa.

Sokoloff, V.P. 1938. Effect of neutral salts of sodium and calcium on carbon and nitrogen of soils. *J. Agric. Res.* 57:201-216.

Sonar, K.R., and R.V. Ghugare. 1982. Release of Fe, Mn and P in a calcareous Vertisol and yield of upland rice as influenced by presowing soil water treatments. *Plant Soil* 68:11-18.

Soriano, A. 1991. Rio de la Plata grasslands. In: Coupland, R.T. (Ed.). *Ecosystems of the World: Natural Grasslands.* pp. 387-407. Elsevier, Amsterdam, Netherlands.

Spaargaren, O. 1994. World Reference Base for Soil Resources (Draft). International Society of Soil Science/International Soil Reference and Information Centre, Wageningen, Netherlands/ FAO, Rome, Italy.

Spain, A.V., R.F. Isbell, and M.E. Probert. 1983. Soil organic matter. In: *Soils: An Australian Viewpoint.* pp. 551-563. Commonwealth Scientific and Industrial Research Organization Publications, Melbourne, Australia/Academic Press, London, U.K.

Sparks, D.L. 1986. *Soil Physical Chemistry*. CRC Press, Boca Raton, FL.

Spoor, G. 1995. Application of mole drainage in the solution of subsoil management problems. In: Jayawardane, N.S., and B.A. Stewart (Eds.). *Subsoil Management Techniques*. pp. 67-108. Lewis Publishers, Boca Raton, FL.

Spoor, G., and R.J. Godwin. 1978. An experimental investigation into deep loosening of soil by rigid tines. *J. Agric. Eng. Res.* 23:243-258.

Sposito, G. 1984. *The Surface Chemistry of Soils*. Oxford University Press, New York, NY.

Sposito, G. 1989. *The Chemistry of Soils*. Oxford University Press, New York, NY.

Sposito, G., and C.S. Le Vesque. 1985. Sodium-calcium-magnesium exchange on Wyoming bentonite in the presence of adsorbed sodium. *Soil Sci. Soc. Am. J.* 49:1153-1159.

Sposito, G., and S.V. Mattigod. 1977. On the chemical foundation of the sodium adsorption ratio. *Soil Sci. Soc. Am. J.* 41:323-329.

Sposito, G., and S.V. Mattigod. 1979. *Geochem: A Computer Program for the Calculation of Chemical Equilibria in Soil Solutions and Other Natural Water Systems*. Kearney Foundation of Soil Science, University of California, Riverside, CA.

Srivastava, A.K., and O.P. Srivastava. 1993. Cation exchange capacity in relation to soil sodicity in amended saline-sodic soil. *J. Indian Soc. Soil Sci.* 41:155-157.

Stace, H.C.T., G.D. Hubble, R. Brewer, K.H. Northcote, J.R. Sleeman, M.J. Mulcahy, and E.G. Hallsworth. 1968. *A Handbook of Australian Soils*. Rellim Technical Publications, Adelaide, Australia.

Stall, J.B. 1972. Effects of sediment on water quality. *J. Environ. Qual.* 1:353-360.

Stannard, M.E., and I.D. Kelly. 1977. The irrigation potential of the lower Namoi Valley. Water Resource Commission, New South Wales, Australia.

Statton, C.T., and J.K. Mitchell. 1977. Influence of eroding solution composition on dispersive behavior. In: Sherard, J.L. and R.S. Decker (Eds.). *Dispersive Clays, Related Piping and Erosion in Geotechnical Projects*. pp. 398-497. American Society for Testing and Materials, ASTM STP 623.

Stephens, C.G., and C.M. Donald. 1958. Australian soils and their responses to fertilizers. *Adv. Agron.* 10:167-256.

Stern, R., M. Ben-Hur, and I. Shainberg. 1991a. Clay mineralogy effect on rain infiltration, seal formation and soil losses. *Soil Sci.* 152:455-462.

Stern, R., M.C. Laker, and A.J. van der Merwe. 1991b. Field studies on the effect of soil conditioners and mulch on runoff from kaolinitic soils. *Aust. J. Soil Res.* 29:249-251.

Stevenson, F.J. 1982. Organic matter reactions involving pesticides in soil. In: Stevenson, F.J. (Ed.). *Humus Chemistry: Genesis, Composition and Reactions*. pp. 403-419. Wiley-Interscience Publishers, New York, NY.

Stevenson, F.J. 1991. Organic matter-micronutrient reactions in soil. In: Mortvedt, J.J., F.R. Cox, L.M. Shuman, and R.M. Welch (Eds.). *Micronutrients in Agriculture*. 2nd ed. pp. 145-186. Soil Science Society of America, Madison, WI.

Stevenson, F.J. 1992. *Humus Chemistry: Genesis, Composition, Reactions*. 2nd ed. John Wiley, New York, NY.

Stevenson, F.J., and A. Fitch. 1981. Reactions with organic matter. In: Loneregan, J.F., A.D. Robson, and R.D. Graham (Eds.). *Copper in Soils and Plants*. pp. 69-96. Academic Press, Sydney, Australia.

Stevenson, F.J., and A. Fitch. 1986. Chemistry of complexation of metal ions with soil solution organics. In: Huang, P.M. and M. Schnitzer (Eds.). *Interactions of Soil Minerals with Natural Organics and Microbes*. pp. 29-58. Soil Science Society of America, Madison, WI.

Stocking, M.A. 1979. Catena of sodium-rich soil in Rhodesia. *J. Soil Sci.* 30:139-146.

Stroehlein, J.L., S. Miyamoto, and J. Ryan. 1978. Sulfuric acid for improving irrigation waters and reclaiming sodic soils. Rep. 78-5, Dept. of Soils, Water and Engineering, University of Arizona, Tucson, AZ.

Stuber, R.J., G. Gebhart, and O.E. Maughan. 1982. Habitat suitability index models: Largemouth bass. *U.S. Fish and Wildlife Service*. FWS/OBS-82-10-16.

Stumm, W. 1992. *Chemistry of the Solid-Water Interface*. John Wiley, New York, NY.

Stumm, W., R. Kummert, and L. Sigg. 1980. A ligand exchange model for the adsorption of inorganic and organic ligands at hydrous oxide interfaces. *Croat. Chem. Acta*. 53:291-312.

Suarez, D.L. 1981. Relationship between pH_c and sodium adsorption ratio (SAR) and an alternative method of estimating SAR of soil or drainage water. *Soil Sci. Soc. Am. J.* 45:469-475.

Suarez, D.L., and C.M. Grieve. 1988. Predicting cation ratios in corn from saline solution composition. *J. Exp. Bot.* 39:605-612.

Suarez, D.L., J.D. Rhoades, R. Savado, and C.M. Grieve. 1984. Effect of pH on saturated hydraulic conductivity and soil dispersion. *Soil Sci. Soc. Am. J.* 48:50-55.

Subhashini, D., and B.D. Kaushik. 1981. Amelioration of sodic soils with blue-green algae. *Aust. J. Soil Res.* 19:361-366.

Suhayda, C.G., R.E. Redmann, B.L. Harvey, and A.L. Cipywnyk. 1992. Comparative response of cultivated and wild barley species to salinity stress and calcium supply. *Crop Sci.* 32:154-163.

Sumner, M.E. 1992. The electrical double layer and clay dispersion. In: Sumner, M.E., and B.A. Stewart (Eds.). *Soil Crusting: Chemical and Physical Processes*. pp. 1-32. Lewis Publishers, Boca Raton, FL.

Sumner, M.E. 1993a. Sodic soils: New perspectives. *Aust. J. Soil Res.* 31:683-750.

Sumner, M.E. 1993b. Gypsum and acid soils: The world scene. *Adv. Agron.* 51: 1-31.

Sumner, M.E., and M.P.W. Farina. 1986. Phosphorus interactions with other nutrients and lime in field cropping systems. *Adv. Soil Sci.* 5:201-236.

Sumner, M.E., and W.P. Miller. 1992. Soil crusting in relation to global soil degradation. *Am. J. Alter. Agric.* 7:56-62.

Sumner, M.E., and W.P. Miller. 1996. Cation-exchange capacity and exchange coefficients. In: Sparks, D.L. (Ed.). *Methods of Soil Analysis*. American Society of Agronomy, Madison, WI.

Swanson, R.A., and G.R. Dutt. 1973. Chemical and physical processes that affect atrazine distribution in soils. *Soil Sci. Soc. Am. J.* 37:872-876.

Swarup, A. 1980. Effect of submergence and farmyard manure application on the yield and nutrition of rice and sodic soil reclamation. *J. Indian Soc. Soil Sci.* 28:532-534.

Swatzen-Allen, S.L., and E. Matijevic. 1974. Surface and colloid chemistry of clays. *Chem. Rev.* 74:385-400.

Syers, J.K., and D. Curtin. 1989. Inorganic reactions controlling phosphorus cycling. In: Tiessen, H. (Ed.). *Phosphorus Cycles in Terrestrial and Aquatic Ecosystems.* pp. 17-29. SCOPE/UNEP Regional Workshop 1: Europe. University of Saskatchewan, Saskatoon, Canada.

Sylvester, B.A., L.S. Garton, and R.L. Autenriethm. 1994. Aquatic sediments. *Water Envir. Res.* 66:496-531.

Szabolcs, I. 1979. *Review of Research on Salt Affected Soils.* UNESCO, Paris, France.

Szabolcs, I. 1989. *Salt-affected Soils.* CRC Press, Boca Raton, FL.

Szabolcs, I. 1991. Soil classification related properties of salt-affected soils. In: Kimbal, J. (Ed.). *Characterization, Classification, and Utilization of Cold Aridisols and Vertisols.* Proc. 6th Int. Soil Correlation Meeting (VIISCOM)-USDA Soil Conservation Service, National Soil Survey Center, Lincoln, NE.

Tanji, K.K., L.D. Doneen, G.V. Ferry, and R.S. Ayers. 1972. Computer simulation analysis on reclamation of salt-affected soils in San Joaquin Valley, California. *Soil Sci. Soc. Am. Proc.* 36:127-133.

Tanji, K.K., and F.F. Karajeh. 1993. Saline drain water reuse in agroforestry systems. *J. Irrig. Drain. Eng.* 119:841-849.

Tarchitzky, J., Y. Chen, and A. Banin. 1993. Humic substances and pH effects on sodium- and calcium-montmorillonite flocculation and dispersion. *Soil Sci. Soc. Am. J.* 57:367-372.

Tardy, Y., C. Cheverry, and B. Fritz. 1974. Néoformation d'une argile magnésienne dans les dépressions interdunaires du lac Tchad. *C. R. Acad. Sci. Paris Ser. D* 278:1999-2002.

Taylor, A.J., and K.A. Olsson. 1987. Effect of gypsum and deep ripping on lucerne (*Medicago sativa* L.) yields on a Red-Brown Earth under flood and spray irrigation. *Aust. J. Exp. Agr.* 27:841-849.

Taylor, H.M. 1971. Effects of soil strength on seedling emergence, root growth and crop yield. In: Barnes K.K, W.M. Carleton, H.M. Taylor, R.I. Throckmorton, and G.E. van den Berg (Eds.). *Compaction of Agricultural Soils.* pp. 292-305. American Society of Agricultural Engineers Monograph, St. Joseph, MI.

Tennant, D., G. Scholtz, J. Dixon, and B. Purdie. 1992. Physical and chemical characteristics of duplex soils and their distribution in the south-west of Western Australia. *Aust. J. Exp. Agric.* 32:827-844.

Theisen, M.S. 1994. *The Expanding Role of Geosynthetics in Erosion and Sediment Control.* Synthetic Industries, Chattanooga, TN.

Theng, B.K.G. 1982. Clay-polymer interactions: Summary and perspectives. *Clays Clay Min.* 30:1-10.

Theng, B.K.G., and H.W. Scharpenseel. 1976. The adsorption of ^{14}C-labelled humic acid by montmorillonite. In: Bailey, S.W. (Ed.). *Proc. Int. Clay Conf.* pp. 643-653. Applied Publishing, Wilmette, IL.

Thomas, D.S.G., and P.A. Shaw. 1991. *The Kalahari Environment.* Cambridge University Press, Cambridge, UK.

Thomas, G.W. 1977. Historical developments in soil chemistry: Ion exchange. *Soil Sci. Soc. Am. J.* 41:230-238.

Thompson J.G., and W.D. Purves. 1978. *A Guide to the Soils of Rhodesia.* Rhod. Agric. J. Tech. Handbook 3.

Thorburn, P.J., E.A. Gardner, A.F. Geritz, and K.J. Coughlan. 1989. The effect of wetting pre-treatment on the desorption moisture characteristic of Vertisols. *Aust. J. Soil Res.* 27:27-38.

Thorburn, P.J., C.W. Rose, R.J. Shaw, and D.F. Yule. 1990. Interpretation of solute profile dynamics in irrigated soils. I. Mass balance approaches. *Irrig. Sci.* 11:199-207.

Thorup, J.T. 1969. pH effect on root growth and water uptake by plants. *Agron. J.* 61: 25-227.

Tiller, K.G. 1983. Micronutrients. In: *Soils: An Australian Viewpoint.* pp. 365-387. Commonwealth Scientific and Industrial Research Organization Publications, Melbourne, Australia/Academic Press, London, UK.

Tisdall, J.M. 1991. Fungal hyphae and structural stability of soil. *Aust. J. Soil Res.* 29:729-743.

Tisdall, J.M., and H.H. Adem. 1988. An example of custom prescribed tillage in south-eastern Australia. *J. Agric. Eng. Res.* 40:23-32.

Tisdall, J.M., and J.M. Oades. 1982. Organic matter and water-stable aggregates in soils. *J. Soil Sci.* 33:141-163.

Tisdall, J.M., and A.S. Hodgson. 1990. Ridge tillage in Australia. A review. *Soil Tillage Res.* 18:127-144.

Tiwari, K.N., B.S. Dwivedi, G.P. Upadhyay, and A.N. Pathak. 1984. Sedimentary iron pyrites as amendment for sodic soils and carrier of fertiliser sulphur and iron: A review. *Fert. News* 29:31-41.

Toogood, J.A. 1978. Fertility status of Solonetzic soils. In: J.A. Toogood, and R.R. Cairns (Eds.). *Solonetzic Soils Technology and Management in Alberta.* Bull. B-78-1, 2nd ed., pp. 32-50. University of Alberta, Edmonton, Alberta, Canada.

Torok, J., L.P. Buckley, and B.L. Woods. 1990. The separation of radionuclide migration by solution and particle transport in soil. *J. Contam. Hydrol.* 6:185-203.

Toth, S.J. 1964. The physical chemistry of soils. In: Bear, F.E. (Ed.). *Chemistry of Soils.* pp. 142-162. Van Nostrand Reinhold, New York, NY.

Tóth, T., F. Csillag, L.L. Biehl, and E. Michéli. 1991. Characterization of semivegetated salt-affected soils by means of field remote sensing. *Rem. Sens. Environ.* 37:167-180.

Tóth, T., S. Matsumoto, R. Mao, and Y. Yin. 1994. Plant cover as predictor variable of salinity and alkalinity of abandoned saline soils of the Huang-Huai-Hai Plain, China. *Agrokém. Talajtan* 43:175-195.

Tóth, T., and K. Rajkai. 1994. Soil and plant correlations in a solonetzic grassland. *Soil Sci.* 157:253-262.

Towler, R. 1994. Flocculation. *Proc. Soil and Water Manage. For Urban develop. Conf.,* Sydney, Australia.

Troedson, R. J., R.J. Lawn, D.E. Byth, and G.L. Wilson. 1989. Response of field-grown soybean to saturated soil culture. 1. Patterns of biomass and nitrogen accumulation. *Field Crops Res.* 21:171-187.

Treitz, P. 1924. *A sós és szikes talajok természetrajza.* Stádium. Budapest, Hungary.

Tsutsuki, K., and F.N. Ponnamperuma. 1987. Behaviour of anaerobic decomposition products in submerged soils: Effects of organic material amendment, soil properties, and temperature. *Soil Sci. Plant Nutr.* 33:13-33.

Tucker, B.M. 1983. Basic exchangeable cations. In: *Soils: An Australian Viewpoint.* pp. 401-416. Commonwealth Scientific and Industrial Research Organization Publications, Melbourne, Australia/Academic Press, London, UK.

Turner, F.T., and W.H. Patrick, Jr. 1968. Chemical changes in waterlogged soils as a result of oxygen depletion. *Trans. 9th. Int. Congr. Soc. Soil Sci.* 4:53-65.

Tyurin, I.V., I.N. Antipov-Karataev, and M.G. Chizhevskii. 1960. Reclamation of Solonetz soils in the USSR. Israel Program for Scientific Translations, Jerusalem, Israel.

Uehara, G. and G.P. Gillman. 1981. *The Mineralogy, Chemistry and Physics of Tropical Soils with Variable Charge Clays.* Westview Press, Boulder, CO.

Ulrich, B. 1980. Production and consumption of hydrogen ions in the ecosphere. In: Hutchinson, T.C., and M. Havas (Eds.). *Effects of Acid Precipitation on Terrestrial Ecosystems.* pp. 255-282. NATO Conf. 1 4. Plenum Press, New York, NY.

Uppal, H.L. 1955. Green manuring with special reference to *Sesbania aculeata* for the treatment of alkali soils. *J. Agric. Sci. Camb.* 25:211-235.

Uren, N.C., and H.M. Reisenauer. 1988. The role of root exudates in nutrient aquisition. *Adv. Plant Nutr.* 3:79-114.

US Bureau of Reclamation. 1977. *Design of Small Dams.* U.S. Department of the Interior, Washington, DC.

US EPA. 1990. *National Water Quality Inventory. 1988 Report to Congress.* Office of Water, U.S. Govt. Printing Office, Washington, DC.

USSL Staff. 1954. *Diagnosis and Improvement of Saline and Alkali Soils.* USDA, U.S. Govt. Printing Office, Washington, DC.

US Soil Survey Staff. 1975. *Soil Taxonomy: A Basic System of Soil Classification for Making and Interpreting Soil Surveys.* USDA, Soil Conservation Service, Washington, DC.

US Soil Survery Staff. 1992. *Keys to Soil Taxonomy.* Pocahontas Press, Blacksburg, VA.

van Beekom, C.W.C., C. van den Berg, T.A. De Boer, W.H. van der Molen, B. Verhoeven, J.J. Westerhoff, and A.J. Zuur. 1953. Reclaiming land flooded with salt water. II. Exchangeable cations. *Neth. J. Agric. Sci.* 1: 225-244.

van Breemen, N., J. Mulder, and C.T. Driscoll. 1983. Acidification and alkalinization of soils. *Plant Soil* 75:283-308.

van der Eyk, J.J., C.N. Macvicar, J.M. de Villiers. 1969. *Soils of the Tugela Basin.* Natal Town and Regional Planning Commission, Pietermaritzburg, South Africa.

van der Merwe, A.J., and R. Burger. 1969. The influence of exchangeable cations on certain physical properties of a saline-alkali soil. *Agrochemophysica* 1:63-66.

van der Merwe, C.R. 1956. *Soil Groups and Sub-groups of South Africa.* Department of Agriculture and Forestry Sci. Bul. 231, Government Printer, Pretoria, South Africa.

van der Watt, H.v.H., and C. Valentin. 1992. Soil crusting: The African view. In: Sumner, M.E., and B.A. Stewart (Eds.). *Soil Crusting: Chemical and Physical Processes.* pp. 301-338. Lewis Publishers, Boca Raton, FL.

van Olphen, H. 1977. *An Introduction to Clay Colloid Chemistry.* 2nd. John Wiley, New York, NY.

van Oss, C.J., M.K. Chaudhury, and R.J. Good. 1988. Interfacial Lifshitz-van der Waals and polar interactions in macroscopic systems. *Chem. Rev.* 88:927-941.

van Oss, C.J., R.F. Giese, and P.M. Costanzo. 1990. DLVO and non-DLVO interactions in hectorite. *Clays Clay Miner.* 38:151-157.

van Ouwerkerk, C., and P.A.C. Raats. 1986. Experience with deep tillage in the Netherlands. *Soil Tillage Res.* 7:273-283.

van Schilfgaarde, J. 1990. Irrigated agriculutre: Is it sustainable? In: Tanji, K.K. (Ed.). *Agricultural Salinity Assessment and Management.* pp. 584-594. ASCE Manuals and Reports on Engineering Practice No 71. American Society of Civil Engineers, New York, NY.

Verbeek, K. 1989. The Soils of Southeast Ngamiland. FAO/UNDP, AG: BOT/85/011, Field Document 14, Gaborone, Botswana.

Verburg, K., and P. Baveye. 1994. Hysteresis in the binary exchange of cations on 2:1 clay minerals: A critical review. *Clays Clay Miner.* 42:207-220.

Villafane, R. 1989. Evaluación de cuatro gramíneas forrajeras como recuperadoras de un suelo salino-sódico. *Rev. Fac. Agron. Univ. Cent. Venez.* 15:173-184.

Vinten, A.J.A., and P.H. Nye. 1985. Transport and deposition of dilute colloidal suspensions in soils. *J. Soil Sci.* 36:531-541.

Visser, S.A., and M. Caillier. 1988. Observations on the dispersion and aggregation of clays by humic substances. I. Dispersive effects of humic acids. *Geoderma* 42:331-337.

Waldron, L.J., and G.K. Constantin. 1970. Soil hydraulic conductivity and bulk volume changes during cyclic calcium-sodium exchange. *Soil Sci.* 110:81-85.

Walker, A., and D.V. Crawford. 1968. The role of organic matter in adsorption of the triazine herbicides by soil. In: *Isotopes and Radiation in Soil Organic Matter Studies.* pp. 91-108. International Atomic Energy Agency, Vienna, Austria.

Wallace, A., and R.T. Muller. 1978. Complete neutralization of a portion of calcareous soil as a means of preventing iron chlorosis. *Agron. J.* 70:888-890.

Wallace, A., G.A. Wallace, and A.M. Abouzamzam. 1986. Amelioration of sodic soils with polymers. *Soil Sci.* 141:359-362.

Warington, R. 1900. *Lectures on Some of the Physical Properties of Soil.* Clarendon Press, Oxford, UK.

Waters, A.G., and J.M. Oades. 1991. Organic matter in water-stable aggregates. In: Wilson, W.S. (Ed.). *Advances in Soil Organic Matter Research: The Impact on Agriculture and the Environment.* Royal Society of Chemistry, Cambridge, UK.

Weaver, R.J. 1969. Recharge Basins for Disposal of Highway Storm Drainage: Theory, Design Procedure, and Recommended Engineering Practices. NY State Dept. Transportation Research Rep. 69-2.

Weber, J.B. 1994. Properties and behavior of pesticides in soil. In: Honeycutt, R.C., and D.J. Schabacker (Eds.). *Mechanisms of Pesticide Movement into Ground Water.* pp. 15-41. Lewis Publishers, Boca Raton, FL.

Webster, G.R., and M. Nyborg. 1986. Effects of tillage and amendments on yields and selected soil properties. *Can. J. Soil Sci.* 66:455-470.

Wellings, F.M., A.L. Lewis, G.W. Mountain, and L.V. Pierce. 1975. Demonstration of virus in groundwater after effluent discharge onto soil. *Appl. Microbiol.* 29:751-757.

Wershaw, R.L., P.J. Burcar, and M.C. Goldberg. 1969. Interaction of pesticides with natural organic matter. *Env. Sci. Tech.* 3:271-273.

Westcot, D.W. 1988. Reuse and disposal of higher salinity subsurface drainage water: A review. *Agric. Water Manage.* 14:483-511.

Wetter, L.G., G.R. Webster, and J. Lickacz. 1987. Amelioration of a Solonetzic soil by subsoiling and liming. *Can. J. Soil Sci.* 67:919-930.

White, A.W., R.R. Bruce, A.W. Thomas, G.W. Langdale, and H.F. Perkins. 1984. Effects of soil erosion on soybean yields and characteristics of Cecil-Pacolet soils. Southern Piedmont Conservation Research Center Res. Rep. IRC 060184.

White, P.F., and A.D. Robson. 1989. Emergence of lupins from a hardsetting soil compared with peas, wheat and medic. *Aust. J. Agric. Res.* 40:529-537.

White, R.E. 1980. Retention and release of phosphate by soil and soil constituents. In: Tinker, P.B. (Ed.). *Soils and Agriculture.* pp. 71-114. Blackwell Scientific Publishers, Oxford, UK.

Whittig, L.D., and P. Janitsky. 1963. Mechanisms of formation of sodium carbonate in soils. I. Manifestations of biological conversions. *J. Soil Sci.* 14 322-333

Whittig, L.D., and P. Janitsky. 1964. Mechanisms of formation of sodium carbonate in soils. II. Laboratory study of biogenesis. *J. Soil Sci.* 15:146-156.

Wichelns, D., and J.D. Oster. 1990. Potential economic returns to improved irrigation infiltration uniformity. *Agric. Water Manage.* 18:253-256.

Wiklander, L. 1964. Cation and anion exchange phenomena. In: Bear, F.E. (Ed.). *Chemistry of Soils.* pp. 163-205. Van Nostrand Reinhold, New York, NY.

Wild, A. 1988. Plant nutrients in soil: Phosphate. In: Wild, A.(Ed.). *Russell's Soil Conditions and Plant Growth.* 11th ed. pp. 695-742. John Wiley, New York, NY.

Wilding, L.P., R.T. Odell, J.B. Fehrenbacher, and A.H. Beavers. 1963. Source and distribution of sodium in Solonetzic soils in Illinois. *Soil Sci. Soc. Am. Proc.* 27:432-438.

Wildman, W.E. 1981. Managing and modifying problem soils. University of California Coop. Ext. Leaflet 2791, Berkeley, CA.

Wildman, W.E. 1985. Site preparation and correction of soil problelms. In: D. E. Ramos (Ed.). *Walnut Orchard Management.* pp. 28-35. University of California Coop. Ext. Spec. Publ. 21410, Berkeley, CA.

Williams, C.H., and J.D. Colwell. 1977. Inorganic chemical properties. In: Russell, J.S., and E.L. Greacen (Eds.). *Soil Factors in Crop Production in a Semi-Arid Environment.* pp. 105-126. University of Queensland Press, St. Lucia, Australia.

Williams, C.H., and M. Raupach. 1983. Plant nutrients in Australian soils. In: *Soils: An Australian Viewpoint.* pp. 777-793. Commonwealth Scientific and Idustrial Research Organization Publications, Melbourne, Australia/Academic Press, London, UK.

Wilson, L.G. 1967. Sediment removal from flood waters by grass filtration. *Trans. ASAE* 10:35-37.

Wischmeier, W.H., and J.V. Mannering. 1969. Relation of soil properties to its erodibility. *Soil Sci. Soc. Am. Proc.* 33:131-137.

Wolman, M.G. 1977. Changing needs and opportunities in the sediment field. *Water Resour. Res.* 13:50-59.

Wright, G.C. 1985. Furrow irrigation of grain sorghum in a tropical environment. *Aust. J. Agric. Res.* 36:83-85.

Yaalon, D.H. 1987. Is gullying associated with highly sodic colluvium? Further comment to the environmental interpretation of Southern African dongas. *Paleogeogr. Palaeoclim. Palaeoecol.* 58:121-128.

Yadav, J.S.P. 1993. Problems and prospects of crop production and afforestation on salt-affected soils with special reference to India. *Agrokém. Talajtan* 42:157-172.

Yadav, J.S.P. and R.R. Agarwal. 1961. A comparative study on the effectiveness of gypsum and "Dhaincha" (*Sesbania aculeata*) in the reclamation of saline alkali soils. *J. Indian Soc. Soil Sci.* 2:151-156.

Yadav, B.R., and A.K. Singh. 1991. Amelioration of salt-affected soils through afforestation: An overview. In: Trivedi, R.N., P.K.S. Sarma, and M.P. Singh (Eds.). *Environmental Assessment and Management: Social Forestry in Tribal Regions.* Proceedings of 3rd Conf., Mendelian Society of India, Ranchi, April 1989. Scholarly Publications, Houston, TX/Today and Tomorrow's Printers and Publishers, New Delhi, India.

Yaron, B., and G.W. Thomas. 1968. Soil hydraulic conductivity as affected by sodic water. *Water Resour. Res.* 4:545-552.

Yeo, A.R., and T.J. Flowers. 1985. The absence of an effect of the Na/Ca ratio on sodium chloride uptake by rice (*Oryza sativa* L.). *New Phytol.* 99:81-90.

Young, R.A., and C.A. Onstad. 1978. Characterization of rill and interrill eroded soil. *Trans. ASAE* 21:1126-1130.

Yousaf, M., O.M. Ali, and J.D. Rhoades. 1987. Dispersion of clay from some salt-affected, arid land soil aggregates. *Soil Sci. Soc. Am. J.* 51:920-924.

Zahow, M.F., and C. Amrhein. 1992. Reclamation of a saline sodic soil using synthetic polymers and gypsum. *Soil Sci. Soc. Am. J.* 56:1257-1260.

Index